T0314068

PLASTICS AND ENVIRONMENTAL SUSTAINABILITY

PLASTICS AND ENVIRONMENTAL SUSTAINABILITY

ANTHONY L. ANDRADY, Ph.D
Adjunct Professor of Chemical and Biomolecular Engineering
North Carolina State University

Library of Congress Cataloging-in-Publication Data:

Andrady, A. L. (Anthony L.)
  Plastics and environmental sustainability / Anthony L. Andrady, PhD.
    pages   cm
  Includes bibliographical references and index.
  ISBN 978-1-118-31260-5 (cloth)
1. Plastics–Environmental aspects.   2. Plastics–Health aspects.   3. Plastics–Biodegradation.   I. Title.
  TD798.A53 2015
  628.4'4–dc23

                                                                        2014042233

Cover image courtesy of iStockphoto©Devonyu

1   2015

*This book is dedicated to my children
and my grandchildren.*

CONTENTS

PREFACE

How quickly a concept is grasped, adopted, and assimilated into the general culture is indicative of how germane a human need it addresses. If that is indeed the case, the notion of sustainable development seemed to have struck a vibrant sympathetic chord with the contemporary society. Since its emergence in the 1980s, the general tenet of sustainability has gained rapid worldwide salience and broad global appeal in some form or the other. Though it is easy to identify with and even subscribe to it in general terms, the goal of sustainability and how to achieve it remain unclear. In addition to being a dictionary term, "sustainability"[1] has also become a buzzword in the business world. Today's message of sustainability reaches way beyond that of the early environmental movements of the 1960s and 1970s in that it includes an ethical component based on social justice for future generations.

With the global carrying capacity already exceeded, energy/materials shortages looming in the medium term, and the climate already compromised by anthropogenic impacts, many believe that we have arrived at decisive crossroads with no time to spare. The only way out of the quagmire is a radical change in thinking that encompasses the core values of sustainable growth. The message of sustainable growth has also reached chemical industry at large including the plastics industry. In a recent global survey of consumer packaged goods companies by DuPont in 2011, a majority (40%) of the respondents identified attaining sustainability (not costs or profits) as the leading challenge facing their industry today. Environmental movements including the call for sustainability have hitherto evolved along strict conservationist pathways over the decades that saw economic development inextricably linked with polluting externalities and tragedy of the commons. This invariably pitted business

---

[1] The word is derived from the Latin root *sustinere*, which means to uphold.

enterprise against the health of global environment. Industry was still identified a significant polluter and generator of waste. This has lead to the plethora of environmental regulations promulgated in the United States during those decades aiming to "regulate" their operations. The knee-jerk response has been greenwashing, a mere defensive stance by industry, seeking to make small visible changes to nudge existing practices and products into a form that might be construed as being sustainable.

The entrenched belief that business and technological development must necessarily adversely impact the environment remained entrenched in the 1970s and 1980s. In 1992, at the UN Conference in Rio, this notion was finally challenged and the dictum that economic development (so badly needed to eradicate world poverty) can occur alongside environmental preservation was finally proposed. But preservation means maintaining the environmental quality and services at least in its current state for the future generations to enjoy. Without a clearly articulated mechanism of how to achieve this rather dubious goal or the metrics to monitor the progress along the path to sustainability, the notion blossomed out into a popular sociopolitical ideal. Consumers appear to have accepted the notion and are demanding sustainable goods and services from the marketplace.

The allure of sustainable development is that it promises to somehow disengage the market growth from environmental damage. It frees up businesses from having to continually defend and justify their manufacturing practices to the consumer and the environmentalists who continually criticize them. Industry and trade associations still continue under this old paradigm perhaps by the force of habit but the rhetoric and dialogue with environmentalists are slowly changing. Accepting in principal that the need for a certain metamorphosis in their operation that reshuffles their priorities is a prerequisite to fruitful collaboration with environmental interests. The effort toward sustainability is one where industry, the consumer, and the regulators work together, ideally in a nonadversarial relationship. In this awkward allegiance, the business will move beyond meeting the regulatory minima or "room to operate" in terms of environmental compliance and respond positively to burgeoning "green consciousness" in their marketplace. It frees up the environmental movements to do what it does best, and facilitates stewardship of the ecosystem in collaboration with business interest, rather than be a watchdog. This is not an easy transformation in attitudes to envision. Yet it is a change that needs to be achieved to ensure not only continued growth and profitability but the very survivability of the planet and life as we know it.

The Consumer

Primarily, it is the mindset of traditional consumption that determines the demand for market goods, that needs to change. Businesses do not exist to preserve the environment; they exist to make profit for their owners. But to do so, they must meet the demands in the marketplace. With the rich supply of easily-accessible (albeit sometimes erroneous) information via the internet, interested consumers are rapidly becoming knowledgeable. The consumer demand for sustainable goods will grow rapidly, automatically driving business into sustainable modes of operation. Consumers need to be well informed and educated so that they are aware of the need and know what exactly to change.

In such a future scenario, the industry will be called upon to justify not only their economic objectives but also explicitly consider environmental (and social) objectives. This shift from the solely fiscally-driven business plans to the triple bottom-line business plan will propel the marked shift in corporate function. To be successful, the change in corporate orientation must encompass the entire value chain with free flow of communication across the traditional boundaries and interphases with suppliers, customers, and waste managers. This cannot be achieved by a few analysts embedded within a single department but requires champions that represent all aspects of the value chain.

Plastics Industry and Change

Why would a growing, robust, and profitable industry providing a unique class of material that is of great societal value want to change? The plastic industry certainly is not an inordinate energy user (such as cement production or livestock management) and does not place a significant demand on nonrenewable resources. The benefits provided by plastics justify the 4% fossil fuel raw materials and another 3–4% energy resources devoted to manufacturing it. In building applications, plastics save more energy that they use. In packaging (where the energy/material cost can be high), plastics reduce wastage and afford protection from spoilage to the packaged material with savings in healthcare costs. Plastics are a very desirable invention in general. However, the customer base and operating environment are changing rapidly; responding to the challenge posed by these changes is a good business strategy.

The plastics industry has its share of environmental issues. It is based on a linear flow of nonrenewable fossil fuel resources via useful consumer goods into the landfills. Lack of cradle-to-cradle corporate responsibility and design innovations to allow conservation of resources is responsible for this deficiency. For instance, there is not enough emphasis on design options for recovery of post-use waste. The move toward bio-based plastics, an essential component of sustainability, is too slow with not enough incentive to fully implement even what little has been achieved. Though good progress has been made, over-packaging and over-gauging are still seen across the plastics product range. While the plastics litter problem is at its root a social-behavioral issue, the industry is still held at least partially accountable. The issue of endocrine disruptors and other chemicals in plastics potentially contaminating human food still remains a controversial issue. Complaints on plastics in litter, microplastics in the ocean, endocrine disruptors in plastic products, and emissions from unsafe combustion have been highlighted in popular press as well as in research literature. Proactive stance by industry to design the next generation production systems is clearly the need of the day.

Any effort toward sustainability must reach well beyond mere greening of processes and products. Not that greening is bad (unless it is "greenwashing" which is unethical) but because it alone will not be enough to save the day. Sustainability starts at the design stage. Visionaries in the industry need to reassess the supply of energy, materials, and operational demands of the products. Can the present products still remain competitive, profitable, and acceptable despite perhaps more stringent regulatory scrutiny in a future world? What are the ways to increase the efficiency of energy use, materials use, and processes for the leading products? What potential health hazards (perceived as well as real) can the product pose? What technologies are missing that need to be adapted to achieve sustainability? Sustainable growth is a process (not a goal) that has a high level of uncertainty as we are planning for the present as well as for a clouded undefined future. This uncertainty has forced it to be grounded on precautionary strategies.

This Volume

This work is an attempt to survey the issues typically raised in discussions of sustainability and plastics. The author has attempted to separate scientific fact from overstatement and bias in popular discussions on the topics, based on research literature. Strong minority claims have also been presented. Understandably, there are those where plastics have been unfairly portrayed in the media and those where sections of the industry in aggressively protecting their domain have understated the adverse environmental impacts of plastics. The author has attempted to remain neutral in this exercise and he was not funded either by the plastics industry or by any environmental organization in writing this volume.

A work of this nature can never expect to satisfy all stakeholders on all topics covered. Depending on his or her affiliation, the reader will either feel environmental impacts of plastics are exaggerated or that they are too conservatively portrayed and do not capture their full adverse impact. Despite this anticipated criticism, a discussion of the science behind personal judgments and public policy is critical to the cause of sustainability. If the work serves as a catalyst for engagement between industrial and environmental interests or at least generates enough interest in either party to dig deeper into the science behind the claims, the author's objective would have been served.

Raleigh, NC   ANTHONY L. ANDRADY
2014

We did not inherit the Earth from our fathers; we merely borrowed it from our children.

ACKNOWLEDGMENTS

I would like to acknowledge the help and support of many people in preparing this manuscript. Without their help and encouragement, the book would have not been possible. Special thanks to the many scientists who read through parts of the manuscript or offered helpful suggestions on the content. These include Professor Michelle A. Mendez (Department of Nutrition), University of North Carolina at Chapel Hill, Chapel Hill, NC; Professor Braden Allenby (School of Sustainable Engineering), Arizona State University, Tempe, AZ; Professor Linda Zettler (The Josephine Bay Paul Center for Comparative Molecular Biology and Evolution), Woods Hole, MA; Professor Kara Lavendar Law (Sea Education Association), Falthom, MA; Professor Halim Hamid (University of Petroleum and Minerals), Dhahran, Saudi Arabia; Professors Saad Khan (Chemical and Biomolecular Engineering); Richard A. Venditti (Pulp and Paper Science); Steven Sexton (Agricultural and Resource Economics) all of NCSU, Raleigh, NC; and Ted Siegler (DSM Environmental Services, Inc.), Windsor, VT.

I would especially like to thank my wife Lalitha Andrady who did not complain of my many solitary long hours at the laptop preparing this manuscript!

LIST OF PLASTIC MATERIALS

LDPE—Low density polyethylene
HDPE—High density polyethylene
PP—Polypropylene
PS—Polystyrene
PVC—Poly(vinyl chloride)
CPVC—Chlorinated poly(vinyl chloride)
PB—Polybutene
GPPS—General purpose polystyrene
HIPS—High impact polystyrene
EPS—Expendable polystyrene
PMMA—Poly(methyl methacrylate)
PET—Poly(ethylene terephthalate)
PBT—Poly(butylene terephthalate)
PC—Polycarbonate
PA—Polyamide
PA-6—Polyamide 6
PA-66—Nylon 66 or polyamide 66
CA—Cellulose acetate
EVA—(Ethylene-vinyl acetate) copolymer
SAN—(Styrene-acrylonitrile) copolymer
ABS—(Acrylonitrile-butadiene-styrene) copolymer
SBS—(Styrene-butadiene-styrene) copolymer

1

THE ANTHROPOCENE

We, the *Homo sapiens sapiens*, have enjoyed a relatively short but illustrious history of about 100,000 years on Earth, adapting remarkably well to its diverse range of geographical conditions and proliferating at an impressive pace across the globe. Easily displacing the competing relatives of the genus, we emerged the sole human species to claim the planet. It is a commendable feat indeed, considering the relatively low fertility and the high incidence of reproductive failures in humans compared to other mammals. A good metric of this success is the current world population that has increased exponentially over the decades and now standing at slightly over seven billion. It is estimated to grow to about 10 billion by 2100, given the increasing longevity worldwide. At this growth rate, the number of people added to the global community next year will now be equal to about the population of a small country (such as England or France) (Steck et al., 2013). The world population increased[1] by 26% just in the past two decades! The plethora of environmental issues we face today and the more severe ones yet to be encountered tomorrow are a direct consequence of this dominant human monoculture striving to survive on a limited base of resources on the planet. As we approach the carrying capacity[2] of the planet, competition for space and scarce resources, as well as rampant pollution, will increase to

---

[1]The increase was mostly in West Asia and in Africa according to UNEP estimates (United Nations Environment Programme, 2011a).
[2]Carrying capacity is the theoretical limit of population that the (Earth) system can sustain.

Plastics and Environmental Sustainability, First Edition. Anthony L. Andrady.
© 2015 John Wiley & Sons, Inc. Published 2015 by John Wiley & Sons, Inc.

unmanageable levels, unless the human race carefully plans for its future.[3] However, no global planning strategies have been agreed upon even at this late hour when irrefutable evidence of anthropogenic climate change, deforestation, and ocean pollution is steadily accumulating. Incredibly, no clear agreements are there on whether the looming major environmental problems are real or imaginary.

Though it did happen on Earth, the simultaneous occurrence of the conditions that support life as we know it is a very unlikely event, and even here, it is certainly a transient phenomenon. Life on Earth exists over the brief respite (in geological time-line) thanks to a cooling trend between the cauldron of molten metal the Earth was a few billion years back and the sun-scorched inhospitable terrain will turn into a few billion years from now. Even so, life spluttered on intermittently with a series of ice ages, geological upheavals, and mysterious mass extinctions regularly taking their toll on biodiversity. The last of these that occurred some 200 million years ago wiped out over 75% of the species! The resilient barren earth fought back for tens of millions of years to repopulate and reach the present level of biodiversity. Thankfully, the conditions are again just right to sustain life on Earth, with ample liquid water, enough solar energy to allow autotrophs to spin out a food web, a stratospheric ozone layer that shields life from harmful solar UV radiation, enough CO_2 to ensure a warm climate, and oxygen to keep the biota alive. We owe life on Earth to these natural cycles in complex equilibrium. However, the apparent resilience of the biosphere to human interference can often be misleading as the dire consequences of human abuse of the ecosystem might only be realized in the long term. Figure 1.1 shows the growth in world population along with 10-year population increments.

Clearly, human populations have already taken liberties with the ecosystem leaving deep footprints on the pristine fabric of nature. Biodiversity, a key metric of the health of the biosphere, is in serious decline; biodiversity fell by 30% globally within the last two decades alone (WWF, 2012). The current extinction rate is two to three orders of magnitude higher than the natural or background rate typical of Earth's history (Mace et al., 2005). Arable land for agriculture is shrinking (on a per capita basis) as more of the fertile land is urbanized.[4] Millions of hectares of land are lost to erosion and degradation; each year, a land area as large as Greece is estimated to be lost to desertification. Increasing global affluence also shifts food preferences into higher levels of the food pyramid. Though Earth is a watery planet, only 3% of the water on Earth is freshwater, most of that too remains frozen in icecaps and glaciers. Freshwater is a finite critical resource, and 70% of it is used globally for agriculture to produce food. Future possible shortage of freshwater is already speculated to spark off conflicts in arid regions of Africa. Evidence of global warming is mounting, there is growing urban air pollution where most live, and the oceans are clearly increasing

[3] There is a regional dimension for the argument as well. In the US, the birth rates are on the decrease, which will in the future result in lower productivity. Adding to the population in a resource-poor region (say, Sub-Saharan Africa) will result in lower standards of living as the available meager resources have to be now distributed over a larger population.
[4] In 2007, for the first time, global urban population outnumbers the rural populations. The figures for land area degradation are quoted from the World Business Council for Sustainable Development (2008).

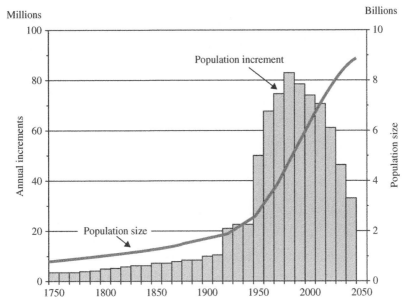

FIGURE 1.1 Projected world population and population increments. Source: Published with permission from UN Population Division. Reproduced with permission from World at Six Billion. UN Populations Division. ESA/P/WP.154 1999.

in acidity due to CO_2 absorption. Phytoplankton and marine biota are particularly sensitive to changes in the pH of seawater (Riebesell et al., 2000), and both the ocean productivity as well as its carbon-sink function might be seriously compromised by acidification. Some have suggested this is in fact the next mass extinction since the dinosaurs' die-off, poised to wipe out the species all over again.[5] Is it too late for the human organism to revert back to a sustainable mode of living to save itself from extinction in time before the geological life of the planet ends?

A driving force behind human success as a species is innovation. Starting with Bronze Age toolmaking, humans have steadily advanced their skills to achieve engineering in outer space, building supercomputers and now have arrived at the frontier of human cloning. Human innovative zest has grown exponentially and is now at an all-time high based on the number of patents filed worldwide. Recent inventions such as the incandescent light bulb, printing press, internal combustion engine, antibiotics, stem-cell manipulation, and the microchip have radically redefined human lifestyle.

[5]Major catastrophes that lead to mass extinction of species occurred five times in Earth's history during the last 540 million years; the last one was 65 million years back (end of the Mesozoic) when the dinosaurs disappeared. The high rate at which species are disappearing has led scientists to suggest that the sixth mass extinction is already under way; Barnosky estimates that in 330 years, 75% of mammalian species will be extinct! (Barnosky et al., 2011).

FIGURE 1.2   Rio Tinto (Red River) in Southwestern Spain devastated and tinted red from copper mining over several thousand years.

A singularly important development in recent years is the invention of the ubiquitous plastic material. It was about 60 years back when science yielded the first commodity thermoplastic material. It was an immediate and astounding success with increasing quantities of plastics manufactured each subsequent year to meet the demands of an expanding base of practical applications. There is no argument that plastics have made our lives interesting, convenient, and safe. But like any other material or technology, the use of plastics comes with a very definite price tag.

Mining anything out of the earth creates enormous amounts of waste; about 30% of waste produced globally is in fact attributed to mining for materials. In 2008, 43% of the toxic material released to the environment was due to mining (US Environmental Protection Agency, 2009). For instance, the mining waste generated in producing a ton of aluminum metal is about 10 metric tons (MT) of rock and about 3 MT of highly polluted mud. The gold in a single wedding band generates about 18 MT of such waste ore left over after cyanide leaching (Earthworks, 2004)! The complex global engine of human social and economic progress relies on a continuing supply of engineering materials that are mined out of the earth and fabricated into diverse market products. At the end of the product "life cycle" (often defined merely in terms of its unacceptable esthetics rather than its functionality), it is reclassified as waste that has to be disposed of to make room for the next batch of improved replacements. The mining of raw materials and their preprocessing, whether it be oil, metal ore, or a fuel gas, are also as a rule energy intensive operations. Air and water resources used are "commons resources" available at no cost to the miners (Fig. 1.2). With no legal ownership, the users tend to overexploit these resources (or pollute it) to maximize

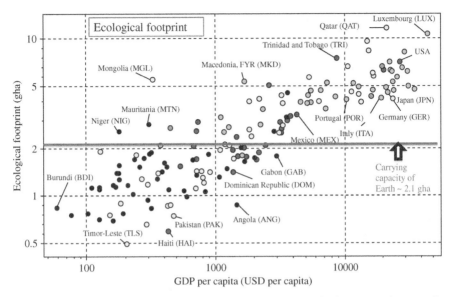

FIGURE 1.3 The ecological footprint of nations (hectares required per person) versus the per capita GDP of the nation. Source: Reproduced with permission from Granta Design, Cambridge, UK. www.grantadesign.com

individual gain. Naturally, in time, the resource will be compromised.[6] Externalities[7] associated with mining or other industrial processes, however, are not fully reflected in what the users pay for in a given product. Often, a community, a region, or even the entire global population is left to deal with the environmental effects of the disposal of waste generated during manufacture. The use of these ever-expanding lines of products, made available in increasing quantities each year to serve a growing population, presents an enormous demand on the Earth's resource base.

The notion of "ecological footprint (EF)" (Reese, 1996, 1997) illustrates the problem faced by the world at large. EF is defined as the hectares of productive land and water theoretically required to produce on a continuing basis all the resources consumed and to assimilate all the wastes produced by a person living at a given geographic location. For instance, it is around 0.8 global hectares (gha) in India and greater than 10 gha in the United States. By most estimates, the footprint of the population has already exceeded the capacity of the planet to support it. In 2008, the EF of the 6 billion people was estimated at 2.7 gha/person, already well over the global biocapacity of approximately 1.8 gha/person in the same year (Grooten, 2013)! In North America, Scandinavia, and Australia, the footprint is already much larger (5–8 gha/capita) (Fig. 1.3). The largest

---

[6] A good example of this "tragedy of the commons" is the state of the global fishing industry. The deep-sea fishery is a common property available to all nation players. Rampant overfishing by different nations without regard to agreed-upon quotas and ecologically safe practices has seriously depleted the fishery.

[7] An externality is a cost or benefit resulting from a transaction that is experienced by a party who did not choose to incur that cost/benefit. Air pollution from burning fossil fuel, for instance, is a negative externality. Selecting renewable materials in building can in some instances be cheaper and delivers the positive externality of conserving fossil fuel reserves.

component of the footprint is availability of sufficient vegetation to sequester carbon emissions from burning fossil fuels.

Plastics, being a material largely derived from nonrenewable resources such as oil, are not immune from these same considerations. Their production, use, and disposal involve both energy costs and material costs. The process also invariably yields emissions and waste into the environment that can have local or global consequences. Plastics industry is intricately connected and embedded in the various sectors that comprise the global economy. Its growth, sustainability, and impact on the environment ultimately depend on what the future world will look like. Therefore, to better understand the impacts of the use of plastic on the environment, it is first necessary to appreciate the anthropogenic constraints that will craft and restrict the future world. The following sections will discuss these in terms of the future energy demand, the material availability, and the pollution load spawned by increasing global population and industrial productivity.

1.1 ENERGY FUTURES

Rapid growth in population accompanies an inevitable corresponding increase in the demand for food, freshwater, shelter, and energy. Supporting rapid growth of a single dominant species occupying the highest level of the food chain must invariably compromise global biodiversity. Humans naturally appropriate most of the Earth's resources, and to exacerbate the situation, the notion of what constitutes "comfortable living" is also continually upgraded in terms of increasingly energy- and material-intensive lifestyles. Invariably, this will mean an even higher per capita demand on materials and energy, disproportionate to the anticipated increase in population. An increasing population demanding the same set of resources at progressively higher per capita levels cannot continue to survive for too long on a pool of limited resources.

Energy for the world in 2012 was mainly derived from fossil fuels: 36.1% from oil, 25.7% from natural gas, and 19.5% from coal, with 9.7% from nuclear power and about 9% from renewable resources (Fig. 1.4). The global demand is projected by the Energy Information Administration (EIA) to rise from the present 525 quads/year[8] in 2010 to 820 quads/year by 2040; over half of this energy will continue to be used for transportation[9] (Chow et al., 2003). Even this estimate is likely an underestimate given the rate of growth in China and the developing world. In the developing countries, residential heating/cooling demands most of the energy followed by industrial uses. The pattern is different in the developed world where transportation is often the leading sector for energy use. How will this large annual energy deficit of about over 295 quads of energy be covered in the near future? Given our singular penchant for

---

[8] A "quad" is a quadrillion (10^{15}) BTUs of energy and is the energy in 172 million barrels of oil, 51 million tons of coal or in 1 trillion cubic feet of dry natural gas.
[9] Internal combustion engine is a particularly inefficient converter of fuel into useful energy. About 75% of energy input into an automobile is lost as heat. Only about 12% is translated to energy at the wheels!

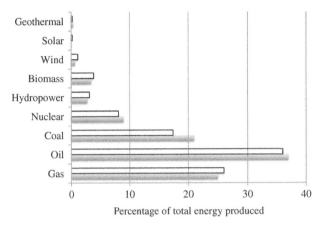

FIGURE 1.4   Global energy use (open bars) and US energy use (filled bars) by source. Source: US 2011 data based on US Energy Information Administration. Web: www.eia.gov. World 2011 data based on International Energy Agency 2012 Report, www.iea.org.

energy, this presents a particularly vexing problem. The most pressing problem will be the huge demand for electricity, the world's fastest growing form of high-grade energy. About 40% of our primary energy (more than half of it from fossil fuel) is spent on generating electricity in an inefficient process that captures only about half their energy content as useful electrical energy. Satisfying electricity demand in next 20 yrs will use as much energy as from bringing on line a 1000 MW power station every 3.5 days during that period (Lior, 2010).

The United States was the leading consumer of energy in the world (~95 quads in 2012) until recently. Since 2008, however, China has emerged in that role with the United States in the second place. Naturally, the same ranking also holds for national carbon emissions into the atmosphere. By 2035, China alone is expected to account for 31% of the world consumption of energy (US EIA, 2010). Around 2020, India will replace China as the main driver of the global energy demand. On a per capita basis, however, the United States leads the world in energy use; 4.6% of the world population in United States consume approximately 19% of the energy, while 7% in the European Union consume 15%. While most of this (~78%) is from fossil fuels approximately 9% of the energy is from renewable sources. But in the medium term, the United States is forecasted to have ample energy and will in fact be an exporter of energy, thanks to the exploitation of natural gas reserves.

Increased reliance on conventional fossil fuel reserves appears to be the most likely medium-term strategy to address the energy deficit, assuming no dramatic technology breakthrough (such as low-temperature fusion or splitting water with solar energy) is made. But it is becoming increasingly apparent that any form of future energy needs to be far less polluting and carbon intensive relative to fossil fuel burning. If not, there is a real possibility that humankind will "run out of livable environment" long before they run out of energy sources! About 26% of the global greenhouse gas (GHG) emissions (mostly CO_2) is already from energy production.

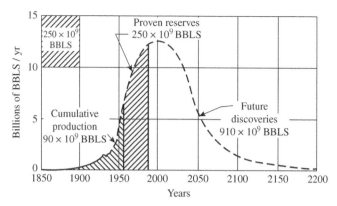

FIGURE 1.5 Hubbert's original sketch of his curve on world oil production. Source: Reprinted with permission from Smith (2012).

1.1.1 Fossil Fuel Energy

Fossil fuels, such as coal, oil, and natural gas, were created millions of years ago by natural geothermal processing of primitive biomass that flourished at the time. Thus, fossil fuel reserves are in essence a huge savings account of sequestered solar energy. Since the industrial revolution, we have steadily depleted this resource to support human activity, relying on it heavily for heating and generating power. About 88% of the global energy used today is still derived from fossil fuels,[10] and that translates primarily into burning 87 million barrels of oil a day (bbl/d) in 2010 (estimated to rise to nearly 90 bbl/d in 2012).[11]

1.1.1.1 Oil Since Edwin Drake drilled the first oil well at the Allegheny River (PA) in 1859, we have in the United States ravenously consumed the resource also importing half of our oil needs. Global reserves of oil presently stand only at about 1.3 trillion barrels, over half of it in the Middle East and Venezuela. The US oil reserves that stand only at 25 billion barrels (2010) are continuing to be very aggressively extracted at the rate of 5.5 million (bbl/d) and can therefore only last for less than a decade. Hubbert (1956)[12] proposed a bell-shaped Gaussian curve (see Fig. 1.5) to model US oil production and predicted it to peak in 1970 (and ~2005 for the world). Estimating the future oil supplies is complicated as new reserves are discovered all the time, improvements are made to extraction technologies, more oils being classified as proven resources, and due to fluctuating demands for oil in the future.

---

[10]The data are from the Statistical Review of World Energy (British Petroleum, 2007). The remaining 12% is from nuclear and hydroelectric power plants. The estimates of global oil reserves are also from the same source.

[11]Based on figures published in 2011 by the US Energy Information Administration.

[12]As with Malthus's famous predictions, Hubbert's timing was off by decades, but his arguments were sound. Estimates of new reserves are upgraded each year, but we may have finally reached peak production.

In the United States, we likely have already reached the peak production rate for oil [or the Hubbert's peak] and are fast reaching the same for world oil production (see Fig. 1.5); thereafter, we can expect escalating prices. As prices rise, the recovery of heavy crudes in unconventional oil sand reserves will become increasingly economical. As expected, price increases driven by scarcity may result in both the lowering of the minimum acceptable quality of product and exploitation of poor reservoirs hitherto considered unprofitable to work on. Burning lower quality oil will result in emissions with an adverse effect on the environment. Our addiction to oil in the United States is such that we have seriously considered drilling for oil in the Arctic National Wildlife Refuge off the Northern Alaskan coast, the largest protected wilderness in the United States.

However, fossil fuel will likely be in good supply in the United States in the immediate future because of aggressive policies in place to exploit shale oil and gas reserves. These include particularly the shale gas (within layers of rock) and "tight oil and gas" trapped in low-permeability rock formations. Hydraulic fracturing or "fracking" might be the only way to exploit these presently inaccessible resources. The potential for "tight oil" and "tight gas" is so high locally that within the next few years, the United States could well be the leading oil producer in the world (replacing Saudi Arabia) and soon thereafter a net energy exporter. In spite of its attractiveness, however, hydrofracking is associated with serious environmental risks. Some of these are its link to earthquakes, the relatively high water demand for the process, limitations of environmentally acceptable disposal choices for spent process wastewater, risk of groundwater contamination, and high potential for GHG release. Despite the opposition from environmental groups, fracking is gaining pace in the United States.

Thankfully, the world still has considerable coal and shale gas reserves; the United States is believed to have 261 billion tons of coal and around 827 trillion cubic feet of shale gas. The United States is presently the second largest producer of coal, and at the present rate of consumption, reserves of coal should last the United States about another 500 years. Not only can coal be burnt to derive power but can also be converted to oil via the Fischer–Tropsch chemistry. Developed in the 1920s, the Fischer–Tropsch process converts CO and H_2 (called syngas) into liquid paraffin hydrocarbons using transition metal catalysts. Syngas is obtained from coal:

$$C + H_2O \longrightarrow CO + H_2$$
$$nCO + (2n+1)H_2 \longrightarrow C_nH_{2n+1} + H_2O$$

The paraffin produced is upgraded into fuel by hydrocracking into smaller molecules.

1.1.1.2 Coal Already, by the mid-decade, 43% of the world's electricity supply was derived from burning coal.[13] In the United States, 21% (and globally close to 30%) of the energy consumed in 2010 was derived from coal. At some future higher

---

[13] UNEP/GRID Arendal.

level of oil prices, the use of coal to produce synthetic oil may become cost-effective, and the relevant Hubbert's curve would have been pushed back a few years or decades into the future. Coal is a cheap direct energy source for the United States, but this reassurance of a few more centuries of fossil fuel comes with a forbidding environmental price tag. Coal plants are more polluting and less costeffective compared to state-of-the-art natural gas plants. At least 49 GW of existing coal power plants in the United States can be retired and replaced with natural gas plants or even with wind-power plants with significant cost savings as well as improved environmental emissions (UCS, 2012).

There is a good justification for closely examining large-scale coal burning, especially without capture or sequestration of CO_2 as a future strategy for generating energy. Coal-fired power plant emissions include particulate matter, sulfur dioxide (SO_2), and nitrogen oxides (NOx) as well as mercury. They are already the largest source of mercury (Hg) pollution in the United States; the emissions from US plants in 2009 exceeded 134,000 lbs. Organic mercury is already present in human blood though at levels below those associated with health effects. Ingestion of mercury-contaminated food can result in serious neurological damage in humans especially children (Counter and Buchanan, 2009). A national standard on limiting mercury emissions from power plants is presently being drawn up by the USEPA; it is already being challenged by the power industry.[14]

1.1.1.3   Gas   About a quarter of the domestic as well as global energy consumed is derived from natural gas where the proven global reserves have been estimated to meet about 64 years of production. In recent years, the domestic production of natural gas has increased, and in 2011, the United States was the leading producer of natural gas in the world. The dramatic growth of natural gas industry has been even more apparent in China with investment in upscale technologies for its cost-effective exploitation. Reserves are the largest in these two countries. Interestingly, in the first half of the twentieth century, natural gas was thought of as a virtually useless by-product of oil production! Natural gas is often hailed as an example of cheap "green energy," but methane (the primary component of natural gas) that escapes during the drilling process is a potent global warming gas.

Extraction of shale gas by fracturing the porous rock (fracking) involves pumping a slurry of water, sand, and chemicals into the rock to crack them to release the trapped gas. The process requires drilling vertically, often through aquifers, and horizontally below them. The slurry pumped into the ground has additive chemicals (such as acids, surfactants, and methanol) that can leak into aquifers creating a very serious, regrettably underestimated water pollution problem (Cooley and Donnelly, 2012). The water demand for fracking is high (2–5 million gallons/well),

---

[14]In October of 2011, 25 states urged a federal court to require that the USEPA delay implementing the rule on emission of Mercury and other pollutants from (coal-fired) power plants by at least a year, as the changes will be too costly. Gardnery, T. Reuters. Monday, October 10, 2011. Available at http://www.reuters.com/article/2011/10/11/us-25-states-urge-court-to-make-us-epa-d-idUSTRE79A0E520111011. Accessed July 1, 2013.

and wastewater management can also be an issue. The slickwater is collected in lined ponds and disposed of away from aquifers of potable water. Scientific data on the costs of fracking to the environment are sparse as large-scale fracking is just starting. But the potential damage is serious enough (Boudet et al., 2014; Mackie et al., 2013) to adopt a precautionary attitude and closely observe the development of this technology.

Perhaps a bright spot in the energy future is the huge untapped fossil fuel resource of methane hydrate (or clathrate) trapped in icy marine sediments in places such as the outer continental shelf of the United States (Chatti et al., 2005). These reserves are larger than all fossil fuel reserves combined (Collett, 2002). The US reserves alone are estimated to be sufficient to replace the current global natural gas demand for a century. Global warming is slowly disrupting under sea clathrate supplies and releasing methane, a GHG, into the atmosphere. Harvesting the methane therefore serves two purposes: producing energy and avoiding global warming. The technology to use the methane hydrate as an energy source is being aggressively developed; for the first time in the world, Japan successfully extracted methane from clathrate fields off the central coast (Nankai Trough) in mid-2013. However, methane too is a fossil fuel that when burnt will add to the carbon load in the atmosphere of an uncontrollably warming earth.

1.1.1.4 Nuclear Energy
Increased use of nuclear energy (a nonrenewable source) might be a potential short-term solution especially if the penalty cost of carbon emission (under Kyoto Protocol) increases. Decommissioning of nuclear weapons can of course be a low-cost short-term source for enriched uranium that can be diluted and used as reactor fuel. It is an option being aggressively pursued in China, India, and Russia. In the United States, 104 nuclear plants are presently operational. However, like with oil reserves, the known U_{235} reserves are not adequate to meet the projected global energy demand (world uranium resources are estimated to be only ~5 million tons). As already discussed, mining is particularly damaging to the earth and results in the release of particulates carrying heavy metal residues into air and acid mine drainage into groundwater. The overwhelming negative effects of these on native wildlife and plant populations cannot be overstated.

A majority of the plants in operation today (over 200 worldwide) are 20–30 years old and have a residual lifetime of only 10–20 more years. Uranium ore is presently used to generate nearly 15% of the world's electricity (and ~20% of US electricity) and will likely last only a few more decades. Even if more ore becomes available, nuclear energy can be an environmentally high-risk technology as illustrated by the nuclear accidents in Russia's Chernobyl plant in 1986 and Japan's Fukushima Daiichi power plant in 2011. The ecological devastation and the cost of cleanup of inevitable spills of radioactive material are far greater and more complicated compared to managing oil spills. Alternative reactors (such as those based on thorium[15]) might be used in the future despite the safety risks they pose. Nuclear waste disposal is another

---

[15]Thorium-232 can be used in specially designed nuclear reactors that use U_{235} or Pu_{239}. But the World Thorium reserves also stand around 6 million tons.

daunting problem. Nuclear plants are shut down every 12–24 months to replace "spent" fuel with fresh uranium. The still radioactive spent fuel has to be stored safely for thousands of years in robust underground storage facilities (1000 ft deep site within in Yucca Mountain, NV, is being considered for the purpose).[16] The plutonium waste, for instance, has a half-life of 24,000 years!

1.1.2 Renewable Energy

The only route to sustainable development is via renewable energy technolgies that are generally carbon-neutral. In the United States, hydroelectric power is used to generate approximately 7.9% of the electricity. A significant amount of our energy already comes from renewable sources (19% of the energy used globally in 2011) primarily as hydroelectric power. The new 22.5 GW plant at the Three Gorges Dam in China is a remarkable example of the technology, which also illustrates the socioeconomic costs of displacement of people and loss of land use associated with such projects. Worldwide, however, the best hydroresources are already exploited, and a natural limit to growth in hydroelectric power generation might be anticipated. Yet, at this time, the highest growth rate in electricity generation worldwide is with hydropower. In the United States and in the West, the water resources are nearly fully tapped already and the growth will be much slower.

1.1.2.1 Wind Energy Wind energy that accounts for a respectable 2% of the worldwide electricity generation (and ~1.2% of US energy) has the potential to grow and be deployed rapidly. It is an economical option; cost/MWh compares well with that of conventional coal or hydropower installations. Several small countries generate 10–20% of their power needs from wind energy. The present technology can be relied upon to deliver about 2 W/m^2 of wind farm. It is a fundamentally attractive option with the potential power proportional to the third power of wind speed. A recent report (Hansen et al., 2013) finds wind energy to be the leading or renewable energy source for electricity production until 2035. The main constraint will be availability of land in windy areas to locate such farms. In regions with adequate wind resources, the technology holds promise as a supplementary power source. Offshore wind farms, especially in deep-sea areas, might be more efficient than land-based facilities.

A potential environmental problem with wind farms is their negative impact on migrating bird populations. A recent estimate suggests the mortality to be about 0.27 deaths/GWh generated (Marris and Fairless, 2007). Given that the bird deaths by fossil fuel plants are greater than 5 per GWh and that for nuclear power plants are 0.42 per GWh (Sovacool, 2012), the cost is modest compared to energy derived.

---

[16]This one facility designed for 77,000 tons of waste will not be enough even for our present needs. Furthermore, the area is seismically active, and one needs to worry about the buried canisters of waste being compromised in an earthquake.

1.1.2.2 Solar Energy All of the Earth's processes are ultimately energized by solar energy (excluding chemosynthesis in the seabed). Solar energy reaching the Earth's surface in a single hour is estimated to be more than the annual global energy demand (US Department of Energy, 2005). Presently, only a paltry 0.01% of the global energy demand (0.1% of the US demand in 2011) is met by solar energy. The sunlit half of the globe receives a solar flux of $680 \, W/m^2$ of radiation and is for the most part captured by plants that very inefficiently (typically <2%) convert it into biomass. (Interestingly, the incandescent lamp also converts only 5–10% of the input energy into light!) Biomass generated in turn serves as food to herbivores with the food energy transferred up the complex food chain to the human consumer. The energy transfer across trophic levels (say, herbivores to predator carnivore) is particularly inefficient, approximately only 10%. The rest (~90%) of solar energy captured by plants is dissipated as low-value heat. The efficiency of using installed[17] solar cells converting sunlight into useful energy as electricity is at least an order of magnitude higher in efficiency compared to photosynthesis.

Solar energy can be harvested either using photovoltaic (PV) cells that convert the light directly into electricity or using solar thermal collectors. The latter is much lower in capital cost and is far more efficient in that they convert nearly half the impinging solar radiation into heat. Heat, however, is a low-grade energy not as convenient to store or use as electricity. Commercial efficiency of PV modules based on polycrystalline silicon (over 70% of modules produced in 2010) is approximately 14% (and 7–11% for the newer thin-film modules) (U.S. Department of Energy (US DOE, 2011)). Research cells under development show higher efficiencies, as much as 42% in a multijunction concentrator. The 3D solar PV panels still under development can generate 20 times more energy compared to conventional flat panels and are claimed to push the efficiencies close to the theoretical maximum for silicon. Emerging printed electronic technologies are also likely to soon deliver roll-to-roll production of flexible, fully printed solar cells on plastics substrate.

The global installed PV capacity in 2010 stood at 40 GW with Europe, the market leader, and the United States, a minor producer, with a capacity of only 2.5 GW. The world's largest facility in Bavaria, Germany, produces 10 MW of electricity from its 3 acre solar farm. The capital cost of installed PV cannot as yet effectively compete with fossil fuel energy; a robust PV farm installed over less than 0.05% of Earth's surface should be able to generate the annual fossil fuel energy budget of the world. Despite the low cost of energy in the near future, solar energy technology is likely to grow into a very significant player in the future energy markets around the world (Fig. 1.6).

1.1.2.3 Solar Biomass Energy Indirect harvesting of solar energy via biomass (ineffective as the conversion might be as pointed out already) is a growing renewable energy strategy. The best-known example is the use of corn-based alcohol as fuel. In the United States, up to 10% ethanol is typically blended into gasoline, and in 2010,

---

[17]The capital cost of growing corn is very different from that of installing a field of solar cells. The comparison is therefore based on installed solar cells versus growing corn.

FIGURE 1.6   Sprawling solar energy complex in San Luis Valley, CO.

40% of the US corn production was diverted from feed/food uses to make about 13 billion gallons of fuel-grade ethanol.[18] While biomass technologies can be scaled up and implemented quickly, an approach that converts agricultural food-producing land into fuel-producing acreage has obvious drawbacks. In 2007, 47% of vegetable oil in the EU was used for biodiesel production but still contributed only to 0.36% of the global energy supply (UNESCO, 2009)! Biomass such as marine algae, cellulosic waste, or rapidly growing nonfood land species (switchgrass) can also be converted into fuel. Using these in place of corn biomass will make far better economic sense in the future. Available crop varieties presently yield less than $1 \, W/m^2$ of land used.

1.2   MATERIALS DEMAND IN THE FUTURE

Thomas Malthus (Cambridge University) in 1798 predicted a catastrophic fate for the human race due to a growing population outrunning its subsistence. More recently, Paul Ehrlich (1974) predicted materials shortages worldwide that will result in sky-rocketing prices even for basic commodities. However, his timing has been proven to be inaccurate; population has grown, but the austere times of severe shortages they envisioned had thankfully not materialized. Except in some remote regions of Asia and Africa, adequate food supplies are still available. Then was Malthus wrong? What Malthus did not fully take into account and what effectively countered the predicted shortages thus far is technology or human ingenuity. The same winning trait that out-witted the *Neanderthals*, tamed fire, and developed tools to conquer nature in the dawn of human civilization has continued on, unabated, in modern times.

[18] US total installed capacity of fuel-grade ethanol to 15.0 billion US gallons.

Great strides have been made in high-yield agricultural technology, in post-harvest management of produce, as well as in packaging and distribution of food. This has allowed the inconceivable achievement of producing increasingly more food despite the depleting acreage of arable agricultural land. Science and technology has thus far allowed humankind to be a step ahead of Malthus's ominous prediction. Material consumption too has become increasingly efficient across the board, with substitution for scarce materials, design improvements to use less of the more expensive materials, and learning to better locate new ore reserves. But can we always count on technology to keep us a step ahead of a Malthusian catastrophe? In the short term, it is probably so. But it cannot be assured over long periods of time as there are inherent limits to improving a system using progressively better technologies (Brown, 2009; Evans, 2009). The limit will probably be due to either the shortage of nonrenewable resources needed to fuel the industrial machinery or the pollution load associated with more sophisticated technologies that have to be practiced at high intensity.

1.2.1 Materials of Construction

The dominant materials in demand worldwide are materials of construction including wood, gravel, clay, and aggregate, followed by of course the fossil fuel materials. Several metals are in high-volume use; iron is the most important of these followed by aluminum, copper, zinc, manganese, chromium, nickel, titanium, and lead. As opposed to these, there are metals that are used in very small quantities but are nevertheless indispensable in high-technology applications, especially in the energy industry.

Being a renewable material, the availability of hardwood for construction can be relied on as long as enough acreage is available for forestry development. Other materials such as sand, steel, cement, aluminum, and plastics that rely on mineral supplies need to be mined out of the ground. As long as extractible resources are available and the energy expenditure in tapping/processing of these is affordable, these too are likely to be in good supply. Based on today's conditions and technology, the embodied energy and CO_2 footprint for representative materials are given in Figure 1.7. Plastics are only moderately energy intensive to use, though not as economical as materials such as concrete or wood.

Industry requires a continuing supply of raw material to produce goods and services. In effect, the purpose of the complex industrial machinery is to use materials (some renewable and other not) to continuously produce goods and services for consumers. With renewables, managed harvesting or use should not present a special problem. But with scarce materials such as rare earth oxides, a problem that parallels that of fossil fuels exists. Metals occur as concentrated ores; their use entails extracting them and using them often in minute quantities in various products and then dispersing them as waste into the environment. Post-use metals are very expensive and tedious to recover. The global and US consumption of selected materials of construction in 2011 is given in Table 1.1.

It is interesting to calculate the total embodied energy (GJ) in different types of building materials based on the data in Table 1.1. The bar diagrams in Figure 1.8 compares the global materials-use energy and carbon emissions for selected materials based on 2011 data. In numerous applications, the functionality demands of the

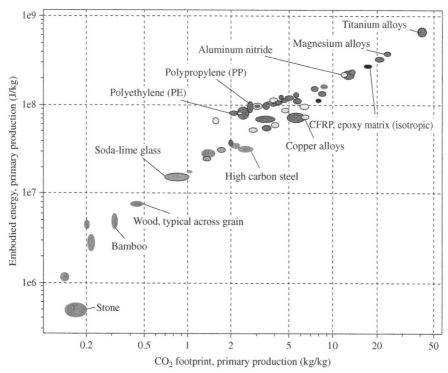

FIGURE 1.7 Comparison of the embodied energy (J/kg) and CO_2 footprint for different materials. Source: Reprinted with permission from Ghenai (2012).

TABLE 1.1   Approximate Global Use of Selected Building Materials (2011 Data)

Region	Cement	Roundwood	Steel	Plastic	Aluminum
World (BMT)	3.6	1.74	1.52	0.28	0.04
United States (MMT)	72	145	90	47.5	3.6

The table is based on data from "Materials and the Environment" (www.forestinfo.org)

product can be delivered using less-scarce and lower energy-intensive substitutes such as plastics. In Chapter 3, the advantage of using plastics as a substitute material will be discussed in greater detail.

1.2.2   Metal Resources

Metal resources are nonrenewable and their long-term availability depends on the known reserves and the cost of extraction. With some metals such as uranium, the fraction of the oxide present in earth is approximately 0.1–0.2%. This means that a large area of earth has to be processed to extract the metal. This would result in relatively larger environmental impact compared to producing a metal such as aluminum

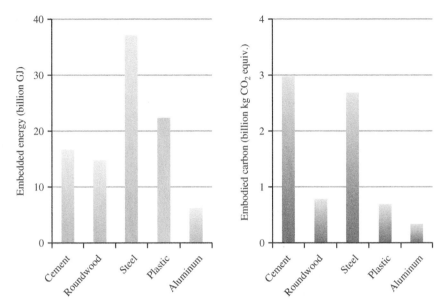

FIGURE 1.8 Estimated embodied energy (left) and carbon emissions (right) of classes of building materials globally consumed in 2011. Source: Calculations based on data in Hammond and Jones (2011). See http://www.circularecology.com/ice-database.html.

TABLE 1.2 Estimated Future Global Supply of Some Common Metals

Metal	Estimated supply (years)	Metal	Estimated supply (years)
Iron ore	178	Lead	19
Aluminum	219	Copper	35
Zinc	19	Nickel	51
Manganese	43	Uranium	65

Source: Data from Richards (2009).

where the ore is 30–50% oxide. Richards (2009) reported the world reserves of selected industrially important metals. These numbers (Table 1.2) are based on the current rate of use and can change in the future because of accelerated use and the discovery of new reserves.

Hubbert's theory of diminishing reserves on exploiting a finite resource beyond a certain level might also be used to estimate how long rare metal resources might last. Unlike base metals such as iron and copper, the rare metals show a peak in the production versus time curves. Some, such as mercury, zircon, selenium, and gallium, have already peaked by year 2000. Metals such as indium, hafnium, gallium, germanium, and arsenic are estimated to deplete within two decades! Indium is used in solar panels as well as in liquid crystal displays, while hafnium is used in computer chips and in nuclear engineering. A majority of the metals in short supply are expected to deplete within the next 100 years (Rhodes, 2008).

However, such estimates will be approximations for a variety of reasons. First, the resource base is not constant as new supplies are being added to the reserve base each year and the annual demand for metals also changes with time. Secondly, the rate of exploitation being demand-sensitive is highly variable. Physically running out of these materials is unlikely in the medium term; more intensive and increasingly polluting new technologies will ensure their supply. The United States is heavily dependent on foreign sources[19] of these materials; this suggests that, as with oil political realities may also play a role in their future supply. Despite these observations, the next century is unlikely to be an austere metal resource-strapped world as more of the Earth's crustal reserves will probably be exploited more intensively and with even better technology. Learning to use less of the scarce materials and substituting for them will be a slower process, and changes in technology may ultimately shift the demand away from the scarce metals.

Generally, the base metals used in high volume are not "consumed" in the sense that they are at the end of use converted into an irrecoverable state. Thus, metals such as iron or aluminum used in construction or packaging can be recovered and recycled extending the lifetime of the resource. While there are inevitable losses in reuse or recycling operations, it can save energy and reduce the pollution load on the environment (Gordon et al., 2006). Energy savings in recycling of steel, aluminum, copper, and lead are estimated to be 74, 95, 85, and 65%, respectively (Steinbach and Wellmer, 2010).

1.2.3   Critical Materials

A class of materials increasingly used in a variety of high-technology and emerging energy applications is the rare earth and platinum group metals.[20] However, future technologies such as electric vehicles, displays, next-generation solar panels, and advances in wind power rely on the availability of these metals. For instance, the world demand for neodymium used in magnets and laser applications will be 40,000 tons/year by 2030; this compares with the demand of only 7000 tons in 2006 (European Environment Agency, 2010). The US DOE has identified several critical materials in this category that will be in short supply in the United States within the next couple of decades. A strong growth in short-term demand is expected at the very least for Te, In, and Ge. The periodic table in Figure 1.9 highlights these and also indicates those regarded as being critical materials in the European Union.

Applications of critical metals are dissipative, and post-use recovery is either impractical or impossible. In theory, the low-volume, high-value critical metals can also be recycled effectively. Often, these are used in complex constructs such as thin layers used in solar panels. The processes to separate out the components in recycling

---

[19]Over 95% of mineral commodities used in the US is imported from China.
[20]Communication services, such as the operation of satellites, GPS, computers, and even cellphones, all depend on the availability of specialized materials such as semiconductor materials, phosphorus, and battery technology.

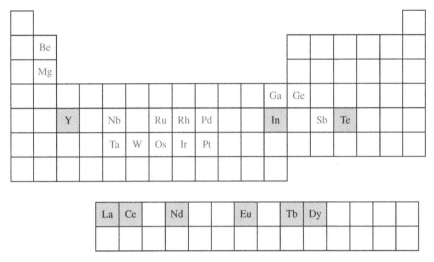

FIGURE 1.9 Critical elements likely to be in short supply in the near future. The shaded boxes are those identified by the US DOE study (2010). The others are additional critical elements identified by a European Commission (2010).

are complicated and costly. A recent UNEP report (UNEP, 2011b) suggests global rates of post-use recycling of these to be less than 1% (1.35% for tantalum). Table 1.3 summarizes some typical exemplary uses of critical metals.

With future shortages anticipated, a strategy for using these scarce resources prudently and an aggressive plan to recover/recycle these are needed. Inevitably, recovery and reuse of the higher-value materials will be an attractive propositions. With those materials beyond Hubbert's type peak recycling will be the only way to ensure a continuing supply in the near future. For instance, rhenium (Hubbert's peak in the late 1990s) is already being extracted and recycled. In the West with an ingrained "disposable" consumer culture, reuse–recycle will be a difficult task. But at some cost point, it will be cost-effective and even be very lucrative to recycle these than to process fresh ore.[21] Figure 1.9 shows the critical elements that are likely to be in short supply in the near future.

1.2.4 Plastic Materials

Though commercially introduced into the market as a commodity material some 60 years ago, the design versatility, low cost, formability, and bio-inertness of plastics have made it the material of choice in a broad range of applications. The demand for plastics is linked to economic development of nations. With some conventional materials such as cement or steel, the demand on materials is decoupled with economic development and is decreasing on a per capita basis in developed countries. This can

---

[21] Already, the Pt content derived from automobile catalytic converters in road dust has reached the levels it is present in South African Platinum ore! (Cohen, 2007).

TABLE 1.3   The Use Sectors, Global Reserves, and Production of Selected Critical Materials

Electrical and electronic	E	P	Ba	Cat	Reserves 1000 MT	Annual production 1000 MT	Main uses
Tantalum	X				130	1.4	Capacitors, carbide tools, and alloys
Indium	X	X			11	0.6	LCD and OLEDs, alloys, and solder
Ruthenium	X				5	0.04	Magnetic (hard drive) media
Gallium	X	X			6.5 (in Zn ore)	0.08	LEDs, cell phone displays
Germanium	X	X			0.45	0.11	
Palladium	X						
Tellurium		X			21	0.45	Alloys, solar energy
Cobalt			X		7,000	55.5	Superalloys and cemented carbide
Lithium			X		4,100	25	Batteries, ceramics, and lubricants
Platinum				X	27	0.20	Cat converters, jewelry
Palladium				X	26	0.37	Cat converters, jewelry
Rare earth oxides			X	X	88,000	124	Cat converters, refineries, and alloys

Source: Data compiled from UNEP (2009).
Primary sectors of application.
Ba, batteries; Cat, catalysts; E, electrical and electronic uses; P, photovoltaic.

occur when the gross domestic product (GDP) of the country becomes dominated by the service sector (vonWeizsäcker et al., 1997). However, this is not the case with plastic materials that continue to grow coupled and in tandem with regional or national economic growth. The future supply of commodity plastics is generally tied closely to that of fossil fuels. This need not necessarily be the case as plastics can also be manufactured from renewable resources, but the processes are as of yet not cost competitive. As petroleum resources continue to dwindle, the cost of plastics will undoubtedly increase, and recycling of some of these may become cost-effective.

Historically, the rate of growth of material consumption has outpaced that of population. During the period (1961–2012) that saw a population increase of about 230%, that of wood, steel, and cement consumption grew by 160, 426, and 1100%, respectively. Plastics consumption in the same period grew by over 4800%.[22] Plastics are so common a material that today it is difficult to imagine living in a world with no plastics. If all plastics were instantaneously removed from modern lifestyle, we would certainly miss the material. Most of our clothing including footwear, consumer goods and building products (plumbing, siding, some glazing, and

[22] Data quoted from www.forestinfo.org by Dovetail Partners Inc., Minneapolis, MN.

electrical components), parts of vehicles (some bodywork, all seat covers, lamps), critical residential services (electricity, water/sewer, gas, telephone), most packaging and healthcare products will fade away. Of course, some of these can be substituted with other materials such as glass, metal, wood, or paper but generally at a higher materials and life cycle energy costs.

There are several key characteristics of plastics that make them highly competitive as a material in the marketplace and will guarantee its continued growth:

1. Strength and low density

   Plastics, though lightweight, are exceptionally strong materials. Some specialty plastics such as Kevlar are stronger than steel and are used in bulletproof vests. Carbon-fiber and other composite materials (including nanocomposites) are used in transportation applications that require lightweight and high strength. As will be elaborated in Chapter 5, reducing weight, especially in vehicles, is particularly profitable because of savings in fossil fuel use and related GHG emissions.

2. Moldability into complex shapes

   Advanced molding techniques allow both thermoplastic and thermoset materials to be fabricated into complex 3D objects. This has the engineering advantage where, unlike with materials such as steel (used in aircraft construction, for instance), individual pieces need not be fastened together to create such shapes. Large objects like hot tubs, shower stalls, or marine vessels can be fabricated easily as a single piece with no joints that can fail during use. These can be colored in any hue desired and in some instances even coated with a protective surface layer.

3. Durability can be designed into the material

   Depending on the specific application, the plastic material can be compounded with antioxidants, light stabilizers, and flame retardants to ensure that the service requirements of products are met. For instance, the same plastic can be compounded for durability or long service lifetimes as well as degradability or controlled loss of properties (as with enhanced photodegradable plastics) using appropriate additives.

4. Biological and chemical inertness

   Common plastics are not affected by aggressive chemicals, and some (such as bleach, lubricant oils, solvents, and acids) are even safely packaged in plastic bottles. It is their bio-inertness that allows them to be extensively used in food packaging and medical devices. Common plastics do not support the growth of microorganisms. In cases where this sometimes is seen to occur (such as in PVC shower curtains), it is the additive plasticizer that supports the growth, rather than the PVC plastic material.

5. Electrical and thermal insulators

   Plastics do not conduct electricity[23] and are used to make electrical hardware such as switches and household power outlets. However, where needed,

---

[23] A special group of plastics that include examples such as polyaniline and polythiophene are inherently electrically conducting and are used in electronic applications.

conductive fillers such as metal whiskers or carbon nanotubes can be used as fillers in the plastic to impart electrical conductivity at the desired level. As they are good thermal insulators as well, plastics are used for retort packaging and microwavable cookware. The inherent low thermal conductivity can be improved further by making the plastic into a closed-cell foam that traps air or other gas to make it an excellent insulation material. Again, where needed, a desired level of thermal conductivity can be imparted into the plastics using a conductive filler.

From sustainability considerations, all materials are not created equal. A distinction needs to be made between materials derived from:

1. fossil fuel;
2. other limited resources such as critical metal and oxide ores;
3. raw materials available in abundant or renewable supply (e.g., wood, sand, sodium chloride); and
4. waste or residues from the use or consumption of the three other categories earlier.

Wagner and Wellmer (2009) have formalized this classification into a hierarchy, suggesting that substituting for materials placed higher in the hierarchy with those at a lower level constitutes a sustainable change. This included recycling a material or reusing within its own class within the hierarchy. The hierarchy is based on the valuation of the materials alone and lacks the dimension of environmental and human safety that should also be important considerations in material selection and use. The use of a material or a product that contains a potentially toxic residue/leachate or does not have a means of recovery is certainly not a sustainable practice.

1.3   ENVIRONMENTAL POLLUTION

Emissions that pollute the environment, particularly the air and water resources, are inextricably linked with the life cycle of a generic product especially where the raw material has to be extracted from the earth's crust as shown in the generalized diagram in Figure 1.10. Each step involves the use of energy and some emissions into the environment and generates a residue that invariably needs to be disposed of. These costs are neither immediately evident or are fully accounted for in the cost of the product or reflected in assessing the GDP of the producer nations. It is this shortcoming of GDP as a measure of development that helped the popularity of the better[24] index, Genuine Progress Indicator, used in full cost accounting.

---

[24] It is indeed a better index compared to GDP. For instance, GDP counts pollution as income from the abatement and cleanup of pollution is a business activity in the economy. GPI, however, counts it correctly as a cost. Cut down a thriving forest into lumber; the GDP counts this as income, while GPI counts it as cost.

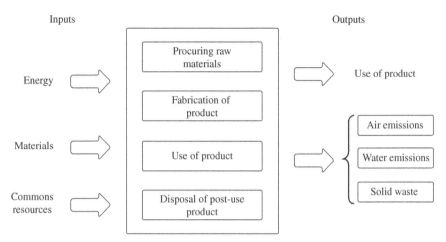

Inputs   Outputs

Energy   Procuring raw
                          materials   Use of product

                          Fabrication of
                          product
                                                                Air emissions

Materials   Use of product   Water emissions

                                                                Solid waste
Commons
resources   Disposal of post-use
                          product

FIGURE 1.10 Illustration of the life cycle of a product showing different steps. Residues are the externalities associated with each phase. Each phase also requires the input of energy.

1.3.1 Classifying Pollution Impacts

In assessing the significance of different types of pollution, it is useful to classify them in terms of their spatio-temporal impacts. The spatial dimension defines the extent of the ecosystem impacted (and therefore also the population affected) by the pollutant in question, while the temporal aspect takes into account the kinetics of the impact. Spatial effects are easier to estimate and where needed, weighted to take into account any effects on human health. Temporal impacts are far more difficult to quantify as the impacts are felt by different populations or even different generations. The fact that the polluter and the affected can be so markedly separated in space and time introduces an ethical dimension to environmental issues. Generally, consumers tend to pay more attention to the needs of the present generation compared to future generations.

An attempt is made to capture this distinction in the following four-group classification of environmental impacts (Table 1.4) of (i) short-term local impacts, (ii) short-term global impacts, (iii) long-term local impacts, and (iv) long-term global impacts.

Local effects as well as the effectiveness of remedial strategies implemented are fairly easy to monitor and validate. For instance, when 700 oil wells in Kuwait were set ablaze by the retreating Iraqi forces in 1991, it spawned a severe local short-term environmental catastrophe. But the extent of damage and the success of control measure used by the fire control teams could be easily monitored. The same was not true of long-term effects of mercury waste being dumped into Minamata Bay in Japan; the more serious impacts occurred in the future, while there was no immediate recognition of a threat to local community.[25]

---

[25]But the pollutant may affect other localities in the general region. For instance, the release of Mercury from coal-fired power plants in the Upper Ohio River basin resulted in high blood Mercury levels in Eagles (especially eaglets) in Catskill area of New York. The pollutant was transferred, biomagnified, via contaminated fish that the eaglets consumed. The report is reminiscent of the Minamata Bay incident in Japan in 1960s (Nearing, 2008).

TABLE 1.4   Classification of Environmental Pollution Events

	Local impact	Global impact
Short term	Eutrophication of lakes due to fertilizer pollution	Oil or chemical spills during ocean transport of materials
	Deterioration of indoor air quality by VOC[a]	Nuclear fallout or the release of active or waste nuclear material into air or sea
	Strip mining for metal ore releasing aerosols	Accidental release of genetically modified cultivars or animal species into the environment
	Destruction of coral reefs by fishing	
Long term	Discharge of organic Hg into the Minamata Bay (Japan) leading to neurological disease	Stratospheric ozone depletion resulting in increased solar UV-B radiation at the ground level
	Overfishing resulting in the depletion of preferred fish stocks	Global warming and climate change
	Deforestation and loss of biodiversity	Mercury pollution of air and water from coal-powered plants

[a]Volatile organic compounds.

Despite any classification adopted for convenience of the discussion, it is still a single interconnected global environment that needs to be protected. Pollution reduction that entails the removal of a pollutant from one part of the environment (say, air) only to increase its concentration in another (say, the ocean) is of no practical benefit. For instance, remediation of groundwater contaminated with fuel (perhaps from a leaking fuel tank) by air stripping to volatilize the hydrocarbons merely shifts the problem from one medium to another. The argument that in an alternative medium the pollutant will have a reduced risk is not a robust one. Emphasis needs to be on pollution prevention in industrial processes to make sure that less waste is generated, emitted, and waste does poses the minimum damage to the environment.

1.3.2   Climate Change and Global Warming

The temperature at the Earth's surface depends on how much solar radiation reaches the surface and how much of it is reemitted back into space. Only about half the incoming solar radiation reaches Earth's surface due to scattering by clouds and absorption by atmosphere. A fraction of the incoming radiation absorbed by the Earth is emitted back as longer-wavelength heat into the atmosphere. Some of this is reflected back into space. It is the delicate balance between the incoming radiation absorbed by Earth and that emitted back into space that maintains the average temperature at Earth's surface within a hospitable range. How well the emitted heat can traverse the atmosphere and escape into space depends on the composition of the

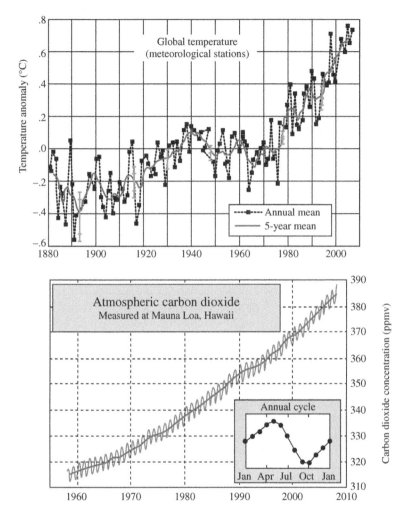

FIGURE 1.11 Global average temperature variation and global CO_2 emissions over time. Source: Reproduced with permission from Akorede et al. (2012).

atmosphere. Molecules such as water vapor, CO_2, and CH_4, the GHGs, are in the upper atmosphere, impair this process, and the heat is reflected back toward the Earth's surface. It is this natural "greenhouse effect" that maintains the temperatures at Earth's surface within a range that supports life.[26]

However, the levels of these contaminants in the environment, especially that of CO_2 and methane emissions from human activity, have steadily increased by about 40%, from 280 ppm, at the time of industrial revolution, to about 337 ppm today (see Fig. 1.11). Higher levels of CO_2 have not been seen in Earth's atmosphere for nearly

[26] Without the greenhouse effect, the temperature on Earth's surface will plummet to 0°F (−18°C), and all water on Earth, including the oceans, will freeze! Life as we know it will not be possible.

the last one million years! In response, the average global temperature rose by 0.8°C over the last century. In the twenty-first century, it can increase by a further 1.5–6.1°C, depending on how well we control the emission of these gases (Solomon et al., 2007). The potential of a gas to cause global warming (GWP) depends on its lifetime in the atmosphere and how well it absorbs the infrared radiation (especially in a wavelength window where the atmosphere itself does not absorb such radiation). For instance, CO_2, CH_4, NO_2, and the Freon HFC-23 have 100-year GWP values of 125, 298, and 14,800, respectively (IPCC, 2010). The same mass of these different gases released into the atmosphere can contribute to dramatically different extents of warming.

Despite the best concerted effort the world can muster, the average global temperatures will still likely be at least 2°C higher than the pre-industrial level by 2100 (this is an agreed-upon climate goal[27]). There is no assurance that being under this limit of warming (<2°C) will avoid serious deleterious impacts (Hansen et al., 2013; Richardson et al., 2009). Severe and sustained climate change will probably occur with that much, or even smaller, increases in the average temperature. These include changes in weather patterns across the globe, melting glaciers (Arctic sea ice dropped to its lowest levels in 2011 (Kinnard et al., 2011)), sea level rise (now at ~2.1 mm/ year), changes in precipitation patterns, increase in average ocean temperatures, and increases in UV flux reaching earth. The 10 warmest years on record have all been within the last two decades. Effects of warming are readily apparent in the melting of snow and ice masses (especially in Greenland and the Arctic), higher incidence of heat waves as well as droughts, tropical storms, changes in sea level, and flooding. These changes are expected to reduce the hydroelectric power output, decrease agricultural productivity, increase incidence of disease, and disrupt freight transportation in the coming decades. But international commitment to hold the goal of warming lesser to less than 2°C is essential for the well-being of the planet and, very likely, the survival of the human race.

Both globally and in the United States, man-made GHGs are dominated by CO_2 from combustion of fossil fuels (reaching a record of 31.6 Gt in 2012),[28] mainly for production of energy. However, much larger loads of carbon dioxide are emitted by natural processes that have been around for millions of years. The respiratory emissions of biota alone amount to over 750 Gt of CO_2 a year, but at least until the industrial revolution, the carbon cycle efficiently removed the gas, allowing only about 290 ppm of it to remain in the atmosphere. The issue is with the *additional* 34 Gt CO_2 equiv./year[29] of the gas from anthropogenic activity. The carbon cycle cannot easily accommodate this added load placed on it in the short period of less than a quarter century. With the cycle overwhelmed, the concentration of CO_2 in the atmosphere is on the rise. Globally, about a fifth of the CO_2 emitted is from

[27]Governments agreed at the United Nations Framework Convention on Climate Change (UNFCCC) Conference of the Parties in Cancun, Mexico in 2010 (COP-16) on this level.

[28]A ton of CO_2 gas has the volume inside a typical two-story US home.

[29]Carbon dioxide is the dominant greenhouse gas, but others such as methane and oxides of nitrogen also contribute to the greenhouse effect. For convenience, the total effect is quantified in terms of CO_2 equivalents.

industrial and manufacturing activity. Most of it is derived from burning oil, while the contributions from coal and natural gas fuels are also significant. The annual US emissions of about 30 MMT of CO_2 (2007 data) presently amount to about 20% of the global emissions. In order to be under the <2°C limit, by 2050, the global GHG emissions must be reduced to 80% of that in 1990. This can only be achieved if conventional energy used today is replaced substantially by renewable energy. If we continue business as usual, the limit will be surpassed within the next 50 years (Joshi et al., 2011).

REFERENCES

Akorede MF, Hizam H, AbKadir MZA, Aris I, Buba SD. Mitigating the anthropogenic global warming in the electric power industry. Renew Sust Energ Rev 2012;16 (5):2747–2761.

Barnosky AD, Matzke N, Tomiya S, Wogan GO, Swartz B, Quental TB, Marshall C, McGuire JL, Lindsey EL, Maguire KC, Mersey B, Ferrer EA. Has the earth's sixth mass extinction already arrived? Nature 2011;471:51–57.

Boudet H, Clarke C, Bugden D, Maibach E, Roser-Renouf C, Leiserowit A. "Fracking" controversy and communication: using national survey data to understand public perceptions of hydraulic fracturing. Energy Policy 2014;65:57–67.

British Petroleum. BP Statistical Review of World Energy. London: British Petroleum; 2007.

Brown LR. Could food shortages bring down civilization? Sci Am 2009;300(5):50–57. Available at www.scientificamerican.com/article.cfm?id=civilization-food-shortages. Accessed October 6, 2014.

Chatti I, Delahaye A, Fournaison L, Petitet J. Benefits and drawbacks of clathrate hydrates: a review of their areas of interest. Energy Convers Manage 2005;46 (9–10):1333–1343.

Chow J, Kopp RJ, Portney PR. Energy resources and global development. Science 2003;302:1528–1531.

Cohen D. Earth audit. New Sci 2007;194:34–41.

Collett TS. Energy resource potential of natural gas hydrates. AAPG Bull 2002;86: 1971–1992.

Cooley, H, Donnelly K. Hydraulic Fracturing and Water Resources: Separating the Frack from the Fiction. Oakland: Pacific Institute; 2012. Available at www.pacinst.org. Accessed October 6, 2014.

Counter SA, Buchanan LH. Mercury exposure in children: a review. Toxicol Appl Pharmacol 2009;198:209–230.

Earthworks. Dirty Metals. Mining, Communities and the Environment: A Report by Earthworks. Boston: Oxfam America; 2004.

Ehrlich P. The End of Affluence. Cutchogue: Buccaneer Books; 1974.

European Commission (EC). Critical Raw Materials for the EU. Report of the Ad-Hoc Working Group on Defining Critical Raw Materials. Geneva: EC; 2010.

European Environment Agency (EEA). Selected Raw Materials: World Use and Rare Earth Elements, Germanium and Tantalum. Copenhagen: EEA; 2010.

Evans A. The Feeding of the Nine Billion: Global Food Security for the 21st Century. London: Chatham House; 2009.

Gardnery TR. 25 States urge court to make EPA delay power plant rule. 2011. Available at http://www.reuters.com/article/2011/10/11/us-25-states-urge-court-to-make-us-epa-d-idUSTRE79A0E520111011. Accessed July 1, 2013.

Ghenai C. Life cycle assessment of packaging materials for milk and dairy products. Int J Therm Environ Eng 2012;4 (2):117–128.

Gordon RB, Bertram M, Graedel TE. Metal stocks and sustainability. Proc Natl Acad Sci U S A 2006;103 (5):1209–1212.

Grooten M. Living Planet Report 2012. Biodiversity, Biocapacity and Better Choices. Gland: Report by World Wildlife Fund in collaboration with Global Footprint Network; 2013.

Hammond GP, Jones CI. 2011. Inventory of carbon & energy. Version 2.0. Institution of Civil Engineers. Available at http://www.circularecology.com/ice-database.html. Accessed October 8, 2014.

Hansen J, Kharecha P, Sato M, Masson-Delmotte V, Ackerman F, Beerling D, Hearty PJ, Hoegh-Guldberg O, Hsu S-L, Parmesan C, Rockstrom J, Rohling EJ, Sachs J, Smith P, Steffen K, Van Susteren L, von Schuckmann K, Zachos JC. Assessing "dangerous climate change": required reduction of carbon emissions to protect young people, future generations and nature. PLoS One 2013;8 (12):e81648.

Hubbert MK. Nuclear energy and the fossil fuels. American Petroleum Institute, Drilling and Production Practices, Proceedings of the Spring Meeting, San Antonio; 1956. p 7–25.

IPCC. Intergovernmental panel on climate change working group I contribution to the IPCC fifth assessment report climate change 2013: The physical science basis, Report no, University Press, Cambridge, UK and New York, NY; 2010.

Joshi M, Hawkins E, Sutton R, Lowe J, Frame D. Projections of when temperature change will exceed 2°C above pre-industrial levels. Nat Clim Chang 2011;1:407–412.

Kinnard C, Zdanowicz CM, Fisher DA, Isaksson E, de Vernal A, Thompson LG. Reconstructed changes in Arctic sea ice over the past 1,450 years. Nature 2011;479 (7374):509–512.

Lior N. Sustainable energy development: the present (2009) situation and possible paths to the future. Energy 2010;35:3976–3994.

Mace, G., H. Masundire, J. Baillie, (2005). Biodiversity. Pages 79–115 in H. Hassan, R. Scholes, and N. J. Ash, editors. Ecosystems and Human Wellbeing: Current State and Trends. Island Press, Washington, DC.

Mackie P, Johnman C, Sim F. Hydraulic fracturing: a new public health problem 138 years in the making? Public Health 2013;127 (10):887–888.

Marris E, Fairless D. Wind farms' deadly reputation hard to shift. Nature 2007;447:126.

Nearing B. 2008. Mercury a concern in eagles, Times Union. Available at http://earthhopenetwork.net/forum/showthread.php?tid=1586. Accessed October 6, 2014.

Rees WE. Revisiting carrying capacity: area based indicators of sustainability. Popul Environ 1996;17:195–215.

Rees WE. Urban ecosystems: the human dimension. Urban Ecosystems 1997;1:63–75.

Rhodes CJ. Short on reserves. Chemistry and Industry 2008, August 25, p 21–23.

Richards, J. (Ed.) (2009) Mining, Society, and a Sustainable World., XXV, 506 pages Heidelberg/New York Springer.

Richardson K, Steffen W, Schellnhuber H-J, Alcamo J, Barker T, Kammen DM, Leemans R, Liverman D, Munasinghe M, Osman-Elasha B, Stern N, Waever O. Synthesis report. Climate change: global risks, challenges & decisions. Summary of the Copenhagen Climate Change Congress; 2009 Mar 10–12; Copenhagen: University of Copenhagen; 2009.

Riebesell U, Zondervan I, Rost B, Tortell PD, Zeebe RE, Morel FMM. Reduced calcification of marine plankton in response to increased atmospheric CO2. Nature 2000;407:364–367.

Smith JL. On the portents of peak oil (and other indicators of resource scarcity). Energy Policy 2012;44:68–78.

Solomon S, Qin D, Manning M, Chen Z, Marquis M, Averyt KB, Tignor M, Miller HL. Climate Change 2007: The Physical Science Basis. Contribution of Working Group I to the Fourth Assessment Report of the Intergovernmental Panel on Climate Change. Cambridge: Cambridge University Press; 2007.

Sovacool BK. The avian and wildlife costs of fossil fuels and nuclear power (June 30, 2012). J Integr Environ Sci 2012;9(4):255–278. Vermont Law School Research Paper No. 04–13. Available at SSRN: http://ssrn.com/abstract=2198024. Accessed October 29, 2014.

Steck T, Bartelmus P, Sharma A. 2013. Human population explosion. In: Cleveland Cutler J, editor Encyclopedia of Earth. Washington, DC: Environmental Information Coalition, National Council for Science and the Environment. Available at http://www.eoearth.org/article/Human_population_explosion?topic=54245. Accessed October 6, 2014.

Steinbach V, Wellmer FW. Consumption and use of non-renewable mineral and energy raw materials from an economic geology point of view. Sustainability 2010;2 (5): 1408–1430.

Union of Concerned Scientists (UCS). 2012. Ripe for retirement. The case for closing America's costliest power plants. Cambridge, MA. Available at http://www.ucsusa.org/assets/documents/clean_energy/Ripe-for-Retirement-Executive-Summary.pdf. Accessed October 6, 2014.

United Nations Environment Programme (UNEP). Critical Metals for Future Sustainable Technologies and their Recycling Potential. Freiburg: Öko-Institut e.V.; 2009.

UNESCO UN World Water Report3: Water in a Changing World. Paris: 2009.

UNEP. Keeping Track of our Changing Environment. Nairobi: UNEP; 2011a.

UNEP. Recycling Rates of Metals: A Status Report. Nairobi: UNEP; 2011b.

US Department of Energy (US DOE). Report on the Basic Energy Sciences Workshop on Solar Energy Utilization. Washington, DC: US DOE; 2005.

US DOE. Critical Materials Strategy. Washington, DC: US DOE; 2010.

US DOE (2011) The 2010 Solar Market Trends Report. Washington, DC: US DOE, Office of Solar Electric Technology. Available at http://www.nrel.gov/docs/fy12osti/51847.pdf. Accessed October 29, 2014.

US Energy Information Administration (US EIA). Energy in Sweden 2010: Facts and Figures. Washington, DC: US EIA; 2010. Available at http://www.eia.gov. Accessed October 17, 2014.

US EIA. International Total Energy Outlook in 2011. Washington, DC: US EIA; 2011. Report Nr: DOE/EIA-0484.

US Environmental Protection Agency (USEPA). 2007 Toxic Release Inventory Data. Washington, DC: USEPA; 2009.

Von Weizsäcker EU, Lovins AB, Lovins LH. Factor Four: Decoupling Wealth, Halving Resource Use. London: Earthscan; 1997.

Wagner, M. and Wellmer, F.W. (2009) A hierarchy of natural resources with respect to sustainable development: a basis for a natural resources efficiency indicator. In Mining, Society and a Sustainable World; Richards, J.P., editor.; Heidelberg: Springer; pp 91–121

World Business Council for Sustainable Development (WBCSD). Agricultural Ecosystems: Facts and Trends. Geneva: WBCSD; 2008.

World Wildlife Fund (WWF). 2012 Living Planet Report 2012. Gland: World Wide Fund for Nature (formerly World Wildlife Fund). Available at www.panda.org, http://wwf.panda. org/about_our_earth/all_publications/living_planet_report/2012_lpr/. Accessed October 6, 2014.

2

A SUSTAINABILITY PRIMER

Rampant population growth coupled with a concurrent increase in per capita resource consumption must be invariably unsustainable (Bartlett, 2006), as the nonrenewable resource base that supports growth is finite and will eventually be depleted. Even the might of human innovation and technology cannot push back this natural limit indefinitely. There is good evidence that the global carrying capacity has already been compromised, signaling an urgent need to change our patterns of consumption to ensure sustainability (Huang and Rust, 2011). If the present levels of per capita consumption in the United States or Western Europe are extrapolated to the entire global population, it will even at present, demand the resources of 1.5 Earths.[1] This suggests that affluent lifestyles in some parts of the world are only possible at the expense of less than modest lifestyles at other locations. An implication even more serious than the geographic inequity in resource consumption is the anticipated intergenerational equity. The present rate of depletion of natural and environmental resources seriously threatens the well-being of our future generations.

This is where the somewhat elusive notion (Hannon and Callaghan, 2011) of sustainable growth or sustainable development[2] comes in. It is often presented as a general prescription that somehow allows continuing development and expansion while

---

[1] Presently, some of the world's citizenry make do with less than their fair share of resources, allowing the society at large to make do with a single Earth. Pollution is also rising in almost all environmental compartments.

[2] The term "growth" means a quantitative increase in the size of an entity, while "development" means more than mere growth and suggests the growth to help it reach its full potential.

Plastics and Environmental Sustainability, First Edition. Anthony L. Andrady.
© 2015 John Wiley & Sons, Inc. Published 2015 by John Wiley & Sons, Inc.

ensuring inter-generational equity in terms of the availability of resources and eco-system services. An endeavor can only be sustainable in the long term only if it consumes resources needed by the process at a rate that is equal to or slower than that for the regeneration of the resources (Daly, 1990). (Today's global industry uses fossil fuel energy and material resources at unprecedented rates dramatically faster than their rates of regeneration.) The somewhat vague definition of "sustainability" first proposed in 1987 (Brundtland) does not explicitly articulate this. Despite its present-day political salience of the notion of sustainability, this definition remains unclear, ambiguous, and qualitative (Kajikawa, 2008; Taylor, 2002). It essentially requires growth or development to be structured to meet the needs of the present generation while not compromising the ability of future generation to meet their own needs. This open-ended definition raises several questions:

1. Does the definition really mean *all* future generations? If so, the goal is clearly unachievable because population growth must eventually exceed the global carrying capacity at some point in time (Bartlett, 2006; Odum, 1983).

2. Resource needs of future generations could be quite different from those we anticipate. What specific resources that we use today need to be conserved for their future?

3. What metrics of sustainability do we use? Are global or even regional metrics appropriate? National players operate in a global anarchy and report progress in environmental sustainability strictly in local terms, sometimes achieved by exporting the problems to a different locale.[3]

The numerous alternative definitions of sustainability in the literature (over 300 had emerged by 2007; Johnston et al., 2007) reflect this inadequacy of the original. Proliferation of various, often inconsistent, definitions is, however, not productive in that it questions the credibility of the concept itself (Bolis et al., 2014) and dilutes its interpretation and meaning (Hopwood et al., 2005). Definitions that emphasize particular aspects of sustainable growth have been proposed but are not completely satisfactory because these do not capture the entire spectrum of impacts (Dale, 2001; Mebratu, 1998).

At the very heart of sustainable growth lies the demand for both intra- and inter-generational equity. It seeks stability, development, and social justice in the present generation, but the main focus is on future generations. Even at present, global resources such as energy (and the power that their ownership bestows) and critical materials are inequitably distributed between the nations. Poverty stemming from this imbalance leads to environmental exploitation in the poorer nations and is already a threat to the survival of this generation.

The conundrum of the concept of sustainable growth is that it plans for an essentially unknown future. The underlying assumption that the contentment of future

---

[3] Rambo (1997) cites an example where air and water pollution is achieved by exporting polluting industries into other countries. The benefits of vigorous reforestation in Japan are countered by deforestation elsewhere in Asia to keep Japanese economy supplied with timber.

generations must necessarily require the same set of indispensable goods and services we rely on today, is untested. This might be true in a broad sense such as the need for primary production or breathable air, but there is no crystal ball. For instance, even the energy futures in the next century are not that easy to foresee.[4] Cheap and abundant solar energy, hydrogen from ambient-temperature splitting of water, or abundant methane from clathrates can markedly affect tomorrow's energy supply. Critical but rare materials identified today may become irrelevant with advances in technology with better substitutes taking their place.

2.1   THE PRECAUTIONARY PRINCIPLE

Sustainable growth is a conservative risk-averse strategy that is best understood in terms of the "precautionary principle" implicit in the process.[5] This principle was first formerly affirmed by the European Union back in 1990 in its Bergen Declaration on Sustainable Development. It requires steps be taken to avoid or diminish serious or irreversible damage to the environment (or to human health) from human activities, if such damage is scientifically plausible but remains uncertain.

In simpler terms, the lack of a preponderance of evidence should not preclude regulation of suspect practices or phasing out the use of certain chemicals in products, if enough credible scientific evidence suggests potentially serious threats to the environment (Raffensperger and Tickner, 1999). This is often the case with emerging environmental problems where complete data on their full potential impacts are simply not available. There are two possible responses one can adopt in such situations: assume no adverse outcome and therefore not respond, or assume adverse outcomes and take mitigating action. Precautionary principle prescribes the latter conservative position, in effect minimizing statistical type II errors.[6] Any policy action pursuant to the principle must of course be evidence-based and the precautionary principle does not propose to do away with that requirement. It merely suggests that policy action even during the emergence of that evidence, well ahead of its conclusion, can be justified in some instances. The principle proposes a rational exercise grounded in scientific plausibility of damage and urges action even when the probability of damage remains unknown because of incomplete knowledge.

We have entered a unique epoch in history where human activity is the key driver of change to the Earth system. The sheer magnitude of anthropogenic perturbations to the system obligates the adoption of a precautionary stance with respect to consumption of resources and in dealing with externalities associated with it. Sustainable

---

[4] As recently as in 2003–2004, imports of energy kept the US economy going. Today, the United States is a net exporter of energy with plenty of shale gas and oil available!

[5] Precautionary stance is essentially being "better be safe than be sorry later" and is a risk-averse, conservative position based on intrinsic, potential risks rather than proven perceived damage.

[6] Statistically, the null hypothesis is Ho = the phenomenon (say, global warming) has no adverse impact. If we reject Ho and adopt a precautionary posture but in reality there was no adverse impact, we would have made a type I error. Not taking any corrective action and later finding out that there is an adverse impact would amount to a costly type II error.

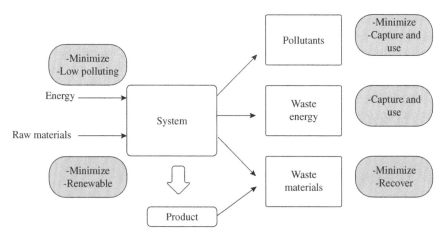

FIGURE 2.1   Linear flow of materials supporting an expanding consumer base.

development emerged as the collective normative response to this obligation. In the present model of consumerism, resources are used to make products for short-term use to be often disposed of well ahead their useful life into landfills. This linear model of "extract–use–dispose" cannot be indefinitely supported by a fixed asset base of energy and materials (see Fig. 2.1). Weaning the consumer societies off this legacy pattern of consumption will indeed be difficult. Sustainable consumerism seeks to replace this with a circular model where the product space is a series of interrelated overlapping cycles like those found in nature. Therefore, material resources as well as most of the embodied energy in post-use products need to be captured and reused to make other useful products, ensuring maximum use of the material before it is perhaps reprocessed into new feedstock or energy. Tenets of "industrial ecology" are based on the circular model. A circular model of consumption too will be invariably unsustainable, as the cycles themselves never produce zero waste, but the conservation it fosters will extend the useful life of the key ecological services.

   A valid counterargument against tinkering with social systems or the free markets is that it can seriously stifle development. Clearly, the world needs rapid economic growth in most regions to eliminate global poverty and inequity. If not, the social system itself may not survive into future generations, and the well-being of the ecosystem will be irrelevant. This social justice argument may have substance in a regional sense, and economic disparity indeed has to be addressed. Growth, however, can still be accommodated within environmental accords that make large concessions to the developing world to allow them to "catch up" in industrial development. This has been demonstrated in the Montreal and Kyoto Protocols, crafted with major concessions that differentiate the commitments of the developing world and the developed countries.

   It is the notion that economic development is necessarily pitted against environmental health (Caccia, 2001) that has resulted in the exploitation of the commons and spawned outrageous incidents of social injustice across the globe. If growth and environmental well-being were really antipodes, a path toward sustainable growth cannot exist. It was the Rio Conference (UNCED, 1992) that recast this debate in a different light, proposing that successful business expansion and healthy environmental policies can coexist. But this can only be in the context of a more liberal definition of sustainable growth (Crittenden et al., 2011).

The best "sustainable growth" can hope to achieve in practice is a strategy that allows the environmental goods and services to survive in a form somewhat close to their present state, for as long as possible into the future. Arbitrary durations of 500 years (Hansen et al., 2007) or 1000 years (MacKay, 2009) have been suggested in lieu of the "as long as possible" above. Will the future generations beyond this duration then be necessarily destined to live in a "dead" world? The expectation is that this "time window" bought at the cost of our environmental prudence can then be used to come up with lasting solutions to the impending situation, so the future generations can still be served. Given the human innovative acumen, this is not an unreasonable expectation.

2.1.1 Objectives in Sustainability

Environmental sustainability is a component embedded within the overarching objective of the sustainability of civilizations. The larger goal calls for global management of material flows and energy expenditure to achieve equitable distribution and efficient allocation. Among its goals are not only the preservation of the ecosystem but also the growth of the global economy and achievement of social equity. In fact, early discussions of overall sustainability in the post-Rio years focused more on the latter goals. Environmental sustainability, however, is very much a key aspect of recent discussions on the subject, and the emergence of ecological economics (Costanza et al., 2004) provides a strong foundation for managing sustainable growth.

Sustainable growth in the context of a business entity, such as a plastic processing operation, has a more limited scope and ideally seeks to decouple the profitability and financial growth of the entity from adverse environmental and social impacts of its operations. This is not a trivial undertaking as the two are intricately connected. In practical terms, it involves a sharp transition in managerial thinking and business planning. This change in thinking has to do with assessing business success; instead of the simple metric of profitability, the businesses now needs to also take into account environmental and social merits of the venture.

In planning, this translates into optimization of three sets of objectives simultaneously, as often depicted in Figure 2.2 (*left*) and summarized in terms of the interrelated dimensions of people, planet, and profit, sometimes referred to as the "triple bottom line" (Elkington, 1998).

Some examples of typical goals within each of these "bottom lines" are as follows:

1. People (social equity): Community development, education, career training, human rights, fair wages, worker rights, employment security, employer advancement, ethical practices, and lack of discrimination

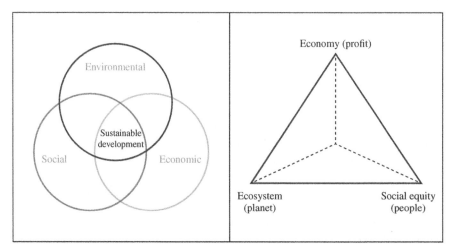

FIGURE 2.2   Sustainable development depicted in simple diagrams.

2. Planet (environmental well-being): Efficient use of energy and material in the production of goods and services. Replenishing resources used, end-of-life waste treatment, designing for reuse, and sustaining biodiversity. Minimizing pollution and toxicity associated with the products and process

3. Profit (fiscal well-being): Long-term profitability, competitive advantage, efficient processes, creativity and innovation, global expansion of business, multinational collaborations, and attracting capital investments

The "triple bottom line" requires the performance of a business be assessed on all three sets of objective and not in terms of profit alone. Though attractive in principle, there is no accepted methodology as yet that allows this metric to be widely used in practice. Present accounting practices are adequate only for calculating economic indicators of a business. A main difficulty is that the yield in each bottom line is expressed in a different currency. Monetizing all impacts is an attempt in finding a common metric, but not a satisfactory one because of the many intangible impacts involved. Though attempts at developing sustainability indices (Cabezas-Basurko, 2010; Missimer et al., 2010) are emerging, these are not as yet developed enough to be used in corporate planning.

2.2   MICROECONOMICS OF SUSTAINABILITY: THE BUSINESS ENTERPRISE

Businesses do not exist to ensure environmental well-being; they exist to create financial capital for their owners. In the process, they may perhaps enrich the community through job creation or human development and in some instances, incidentally, have a positive impact on the environment. However, the consumers are

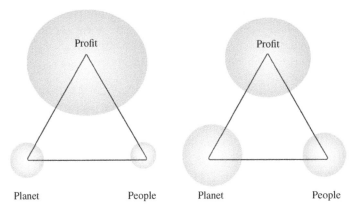

FIGURE 2.3 Schematic illustration of the emphasis in business planning and implementation.

no longer passive receivers of goods and services. Increasingly, they engage with the business to actively demand and shape the products they consume. Ideally, guiding a business toward sustainability is therefore best achieved via market forces, which in turn can only be realized with a change in consumer thinking. Product design and operational aspects of a business, presently focused exclusively on profit, just one of the Ps in the pyramidal model, will no longer be sustainable. This present corporate practice is illustrated in the qualitative emphasis map in Figure 2.3 (*left*) where the focus is almost exclusively on profit. Corporate environmental and social goals are included but generally only as specific responses to regulatory pressure or to be in step with sustainability expectations of trading partners (who are often well ahead of the United States in implementing sustainability practices).

Present corporate thinking that "people" and "planet" are important only to the extent that they define the "freedom to operate" or the regulatory constraints, has to change in moving toward sustainable growth. In the new paradigm, resources are assigned to all three areas as shown qualitatively in Figure 2.2 (*right*), with the objective of creating a healthy triple bottom line. The relative areas of circles qualitatively indicate the change in focus with more equitable distribution of attention and resources to all three Ps. Business plans of the future will explicitly account for the cost of natural capital used—that of all externalities generated in the value chain and have good labor relations (Beard et al., 2011) and engage constructively the community (Sen et al., 2006).

Schrettle et al. (2014) recognize a set of endogenous factors in addition to the obvious exogenous factors such as regulation and market drivers that may promote corporate interest in sustainable development. A robust corporate culture with strong champions across the management team is essential to ensure a planned, focused transition into sustainable growth (avoiding ad hoc "greenwashing" projects).

Accepting the new paradigm with changed emphasis is easier when market success of the company is strongly linked to the triple bottom line. This might be through branding advantages (where consumers will prefer the "sustainable" products over an unsustainable one) or shareholder preference in response to or ahead of possible changes in the regulatory environment. With a highly wired set of consumers and stakeholders, who are becoming increasingly environmentally conscious, the future success of a business can be influenced by the triple bottom line.[7]

2.3   MODELS ON IMPLEMENTING SUSTAINABILITY

Corporate repositioning that can facilitate this change can be appreciated using a production possibilities frontier (PPF). PPF is a conceptual graphic used to illustrate the trade-offs between the production of a pair of alternative products (e.g., guns vs. bread) by a notional business entity. Assuming the available resources and the manufacturing technology to be invariant, making more of one will necessarily reduce the amount of the other. The business must decide the mix of these products it must produce. In this case, one of the products is "environmental quality" (EQ), and the other is "market profitability" used as a proxy to profitability. The former is notional product that is essentially a measure of the avoided adverse impacts on the environment. Thus, minimal use of energy and increased use of renewable resources, recyclability at end of life, lack of toxic components, and low levels of externalities are all assumed to contribute to a hypothetical product reflected by this single measure of EQ. As it is inversely related to environmental deterioration, high numerical values of EQ mean low adverse impacts of production on the environment.

The plot (Fig. 2.4) is a qualitative representation of the relationship between EQ (the avoided adverse impact on environment) and the market goods produced. The assumption here is that increasing the production of the latter will lead to an increase in adverse impacts on the environment. Multiple choices or trade-offs are possible within this simple model. The area under the curve indicates all the production possibilities available to the firm. Points on the curve (such as C) represent a set of trade-offs with the highest efficiency, using the best available technology, and are therefore Pareto efficient. The point O lying beyond the curve, however, is technology constrained and not feasible, while all points below the curve (such as X) are possible but relatively inefficient trade-offs. At X, the firm could choose to produce more market goods without sacrificing EQ by moving to a point on the frontier. Other combinations such as P_1 and P_2 are theoretically possible but are constrained by practical considerations. Environmental regulations preclude production at P_2, and the cost of production at P_1 will price the product too high to be competitive. These are not choices a rational entity will select. Convexity of the curve to the origin shows that sacrificing one product does not result in the same magnitude of increase of the other (the phenomenon of diminishing returns). Higher investment will expand the business with the convex PPF curve moving outward, making higher volume of production possible without altering its shape.

---

[7] A good illustration is how the attempted sale of Hershey Foods Corporation (Hershey, PA) in 2002 was foiled by social and community pressure.

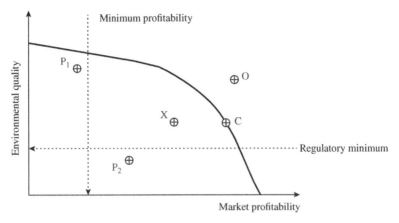

FIGURE 2.4 Production possibilities frontier with illustrative placement of business entities.

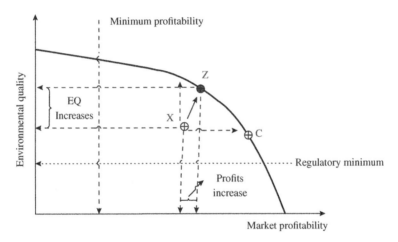

FIGURE 2.5 Improving the environment quality of product also increases profit.

The PPF is particularly valuable in illustrating the different phases of adoption of the sustainable developmental paradigm by businesses. Three such phases might be anticipated.

In the first phase, an entity such as *X* placed below the curve will strive to move on to the curve, the efficient frontier, to operate at full efficiency and maximum profitability. Often, this is by reassessment leading to changes in the wasteful legacy systems. This can theoretically occur along two extreme paths: along *X–A* where the market product is increased at a constant EQ or along *X–B* where EQ is increased at constant market product value. In practice, a course *X–Z*, where some improvement is obtained for both goods, is the likely choice (Fig. 2.5). Even where the move is motivated by improving EQ, it can still increase profitability (US DOE and SPI, 2003). This is a common observation in the introductory phase of corporate

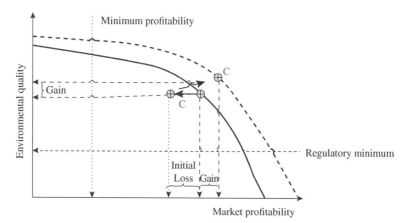

FIGURE 2.6 Investment in better technology allows the choice of simultaneous gains in both goods to be secured but at a short-term cost.

sustainability awareness for most business entities. Not much momentum within the corporate structure is needed to justify and implement strategies that increase both EQ and profits simultaneously. While desirable, this move toward PPF is merely good business practice (with incidental EQ benefits) and should have been achieved as a part of operational excellence rather than as an exercise in strategic planning for increased sustainability.

In the second phase of sustainable development, a business entity already operating somewhere along the PPF (perhaps as point *C*) considers improving its environmental performance or increasing EQ. Repositioning at a different point on the curve toward higher values of EQ, though theoretically possible, is unacceptable; it can only be at the expense of market goods and profitability. The only way to increase production of both goods will be to invest in new technology or through management innovations that would change the shape of the PPF curve (shown by the broken line). The functional dependence of variables in the new curve allows higher EQ at a higher level of market goods as well. The change or the migration to the new curve (*C–Z*) generally involves a short-term setback because during the migration new capital assets are acquired and depreciated. However, once the migration onto the new PPF curve is complete, its competitiveness and long-term market share will increase (Fig. 2.6). The difficulty at this stage is in convincing management and shareholders of the trade-off involving short-term costs but no tangible reward in terms of corporate profits. This phase therefore requires the full commitment of sustainability champions across the corporate team.

The third phase of the paradigm change involves the design of products and planning of production to ensure that sustainability principles will be adhered to throughout the value chain. The same has to be demanded of the suppliers as well as the fabricators or distributors who interface with the corporate value chain. The entire life cycle of products from cradle to cradle will then be re-examined not only in terms of profitability but also from environmental and social perspectives. Realistic end-of-life options must be built into the design of products to ensure that consumers can use the product in a sustainable manner. The product and service concepts must be developed explicitly using the triple bottom line to estimate market yield.

2.4   LIFE CYCLE ANALYSIS

Life cycle analysis (LCA) is the primary tool used to develop that guide sustainable growth (Klöpffer, 2003). It is a recent (dates back to the 1970s), still evolving, and flexible methodology with a general framework standardized in ISO 14040 (and ASTM E1991-05 in the United States) series of standards. A holistic comprehensive analysis has value as a management tool. Unfortunately, it often can be complex and an expensive undertaking and is therefore limited to larger organizations. LCA can be effectively used within an organization or an industry to recognize areas of improvements from a sustainability standpoint.

Perhaps, the most important caution in using LCA is to recognize that its findings are particularly sensitive to the scope and objectives adopted in the assessments. The results depend heavily on the scope, specific system boundaries adopted, simplifying assumptions made, and the quality of data used (Hottle et al., 2013). Metrics for an identical product (e.g., a plastic bag) manufactured by different fabricators or even by a single manufacturer but studied by different analysts can yield very different LCA outcomes.

At the outset, the "life cycle" itself needs to be defined, usually selecting one of the three popular formats illustrated in Figure 2.7. All three start with the extraction

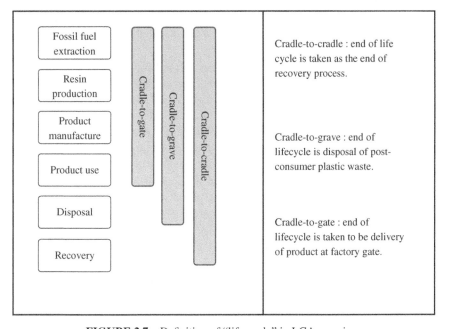

FIGURE 2.7   Definition of "life cycle" in LCA exercises.

of fossil fuel raw materials (*cradle*) but adopt different endpoints in the life of a product. Cradle-to-gate considers the lifetime up to the factory gate; cradle-to-grave includes the use and disposal (*grave*); and cradle-to-cradle includes recovery of the postuse waste and recycling. It is these boundaries that indicate the scope of analysis adopted in these LCA studies. The longer the life cycle to be covered, the more information is required, meaning a more complex and expensive study.

For example, LCA studies on identical poly(ethylene terephthalate) (PET) soda bottles produced at two plants can have very different embodied energies because their process electricity is derived from different sources (coal vs. hydropower) or because modes of transportation or the distances in the distribution networks are different or even because they use different disposal methods. Mechanical recycling, where available, can make a very significant differences to both the energy and carbon footprint of the product. A technically "recyclable" product should not be credited with saved energy in the LCA unless the infrastructure is actually present to recycle it.

A detailed discussion of how to carry out an LCA for a product is outside the scope of this volume. Several excellent sources of such information are available (Curran, 2012; Horne et al., 2009). In general, the process is made up of four steps:

1. Definition of goals and the scope of analysis. Goals clearly formulate the reasons to carry out the study and how the findings will be used. Often, the goals also specify the audience expecting to use the findings (Winkler and Bilitewski, 2007). Scope identifies the functional unit of product being studied and establishes the boundaries (Eriksson et al., 2002) of the study.

2. Building the life cycle inventory (LCI). This includes the collection of data, assigning data to different processes, and calculating, converting, and aggregating the data in relation to functional units of the product (Consonni et al., 2005). A complete set of data covering all phases of the value chain is essential for a good LCA (Hassan, 2003).

3. Impact assessment. The collected data analyzed for physical, chemical, and biological impacts on the environment. These are generally quantified in different units and therefore need to be normalized (Tugnoli et al., 2008) into dimensionless values. Only the comparable impacts are aggregated together to reduce the number of parameters involved in the analysis. For instance, data relating to resource depletion, ecosystem degradation from externalities, and human health impacts might be aggregated and normalized, separately.

4. Interpretation of findings. Conclusions are drawn from the LCI analysis and impact analysis. Where products are compared, this provides a comparative ecoprofile. The attributes of a product that needs to be changed to make it more sustainable are identified in the analysis. The limitations and weaknesses of the LCA are addressed usually via a series of iterative executions of these four steps.

LCA models may synthesize environmental indicators using data from LCI or by using midpoint or endpoint indicators. LCI results can compare different products on the basis of energy costs, resource demand, and the externalities generated. But they do not describe impacts on the environment completely or even substantially. Midpoint indicators use models to determine the impact of flows on a selected environmental predictor such as the global warming potential (GWP). Endpoint indicators show the final impact of the flows from LCI and not only are more complex and more difficult to model but also have a high level of uncertainty. Tool for the Reduction and Assessment of Chemical and Other Environmental Impacts is a midpoint indicator model by the USEPA and recognizes 12 impact areas including ozone depletion, global warming, acidification, eutrophication, tropospheric ozone (smog) formation, ecotoxicity, human particulate effects, human carcinogenic effects, human noncarcinogenic effects, fossil fuel depletion, and land use effects. The choice of which impacts to use in a given study, however, is in the hands of the analyst. Where only short-term product differentiation or "ecolabeling" is needed, only a few selected impacts might suffice. Alternatively, a large number of impacts for which the data are available might be used in a more comprehensive analysis. Including additional impacts is only useful if sufficient data is available to support them. Often, a sensitivity analysis is undertaken to assess the magnitude of changes in the findings when selected parameters in the LCA model are varied. It is an important addition to the analysis; in fact, LCAs lacking explicit interpretation of the degree of uncertainty are of limited value (Guo and Murphy, 2012).

The findings from LCA can be conveniently displayed as a simple radar or polygon plot (see Fig. 2.8) where comparable products plotted on the same set of scales allow a sustainability comparison to be made. Alternatively, principal component analysis (Bersimis and Georgakellos, 2013) might be used for a more elegant analysis.

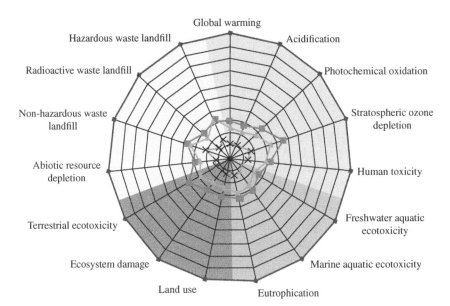

FIGURE 2.8 An example of a polygon plot summarizing LCA results on three products, based on 15 attributes. Source: Courtesy of the US Department of Energy, Building Technologies Office. http://apps1.eere.energy.gov/buildings/publications/pdfs/ssl/lca_factsheet_apr2013.pdf

A potential problem of LCA is in the value judgments involved in the last two steps above, where weighting functions might be used to sum up the total impact from different categories (Graedel and Allenby, 2010). There is no sound basis to select these weights, but results of the analysis often depend on the choice of weights (Carlsson Reich, 2005). Also, there is no standard methodology to aggregate inventory data into cumulative indicators (step 2). Different impacts, such as acidification of oceans or human carcinogenicity or smog formation, have very different consequences on the ecosystem. LCA study on polystyrene (PS) and recycled-paper egg packages found the PS to contribute more to acidification potential and winter and summer smog, while paper packages contributed more to heavy metal and carcinogenicity impacts (Zabaniotou and Kassidi, 2003). In such situations, it is the weighting factors of attributes that determine the overall environmental merit of the products (Goedkoop and Spriensma, 2001; Graedel and Allenby, 2010). In the "paper versus plastic bag" debate, individual impacts such as "litter reduction," "no threat to animal life via entanglement," or "use of renewable resources" clearly favor single-use degradable paper bags over the plastic bags. But in terms of other attributes such as "energy use," "water use during manufacture," or "greenhouse gas emission (GHG) emissions," plastic bags are markedly superior.[8] It is the relative importance of the different impacts in the analyst that determine how these multiple impacts might be synthesized into a meaningful metric.

Carrying out LCA assessment in practice is considerably simplified by available software. The most used LCA software packages globally (Cooper and Fava, 2006) at the present time are GaBi and SimaPro. Others such as Sustainable Minds (www.sustainableminds.com) or Compass (http://www.greenerpackage.com) support product design exercises. Using different LCA tools (software) can also yield inconsistent findings. Ranking of the environmental merits of four types of diapers using three different LCA tools yielded very different rankings (the lowest ranked diaper in one assessment was in fact found to be the best in another!) (Simon et al., 2012). LCA cannot be used to find out which product is "greener" in absolute terms; it can at best only compare individual impacts related to two comparable products.

Even with the best available inventory data, very good software, and conscientious analysis, LCA can still yield inconsistent conclusions on environmental impact (Lenzen, 2000), and the best quantitative estimates have an error of at least ±10%. Despite these limitations, LCA can be a valuable tool to identify and compare products and processes that have critical environmental impacts and guide the inevitable changes to make these sustainable.

2.5   THE EMERGING PARADIGM AND THE PLASTICS INDUSTRY

While the general directions of the needed changes are fairly easy to envision, articulating these in concrete practical terms to plan for their implementation is more difficult. What is particularly useful is a simple means of determining

[8]Producing a plastic bag results in only 39% of GHG emissions, consumes <6% of the water, generates almost five times less solid waste, and uses 71% less energy compared to manufacturing a paper bag (ULS Report, 2008).

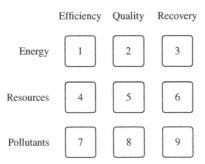

FIGURE 2.9 Sustainability matrix for assessing environmental sustainability.

if a proposed change in business practice has value in terms of environmental sustainability. If a methodology that classifies or groups the different options based on sustainability considerations is available, it might be used as an aid in planning or decision making.

The generalized matrix (Fig. 2.9) strives to serve this purpose. The row labels in the matrix refer to the three broad impact areas of concern to sustainability: energy, resources (materials), and pollutants (externalities) associated with a process or product. Responsible use of energy and material resources and pollution prevention are the cornerstones of sustainable growth (Geiser, 2001).

(a) **Energy**: Global energy demand will rise from the present 524 quads (2010) to 630 quads by 2020 and 820 quads in 2040.[9] Given the growth rates of GDP in populous nations such as China and India, these are probably under-estimates. Even by 2040, about 80% of this energy will still be derived from fossil fuel sources.

(b) **Resources**: Both energy production itself and a myriad of other industries rely on critical metals and oxides in their processes. For instance, wind turbines, solar energy, and battery technology depend on the use of rare earth metals. Conservation, especially of non-renewable resources, through economy, reuse, and substitution with less rare alternatives, is therefore a key component of sustainable growth.

(c) **Pollution**: Externalities associated with manufacturing including gaseous emissions (CO_2, NO_x, SO_x, Hg, etc.) as well as water emissions of raw materials, unused products, and process waste. Also included in this category are leachable constituents in food packaging, residential coatings, and toys that result in adverse human health problems. With new information available, the role of low levels of endocrine disruptor chemicals (EDCs) used in plastic products (see Chapter 7) and their potential contamination of human food supplies is of particular concern.

---

[9] International Energy Outlook 2013 (IEO, 2013).

The three-column labels refer to the types of changes needed under each row category. These are

(a) Efficiency: Increasing the efficiency of use and minimizing wastage,
(b) Quality: Considering the quality (of energy, materials, or pollution) in terms of minimum environmental footprint, and
(c) Recovery: Reusing resources as well as waste. This in essence equates waste to a raw material removing disposal as an end-of-life option. This column incorporates the principles of circular economy.

These nine elements are generally stated as follows:

[Element 1]: Use the minimum energy needed to manufacture,[10] use, or dispose of products.

[Element 2]: Select the least polluting, preferably renewable, forms of energy.

[Element 3]: Capture and reuse waste energy in the same process or elsewhere.

[Element 4]: Use the minimum amount of material (especially nonrenewable material) to achieve the required functionality.

[Element 5]: Select the material with minimum environmental footprint that can deliver the required functionality.

[Element 6]: Reuse and recycle postconsumer products into material, feedstock, or energy.

[Element 7]: Minimize undesirable externalities associated with processes. Reduce air and water emissions as well as solid waste from manufacturing.

[Element 8]: Avoid or at least minimize the release of chemicals from processes or products, especially those toxic to humans or disruptive to the ecosystem. Avoid hazardous chemicals in products.

[Element 9]: Convert emissions otherwise released into the environment (such as CO_2) into useful raw materials or products.

Avoiding hazardous materials or chemicals in product design is a particularly important element (8) in the matrix. Hazardous in this context means one or more of the following: acute or chronic health hazard, environmental hazard, explosion risk, exposure potential, fire hazard and potential for accidents, using the material. Redesigning products to ensure judicious use of limited material resources[11] as well as to encourage transition into renewable energy and monomer feedstock is an overarching aspect of sustainable growth.

---

[10] Using the state-of-the-art equipment in plastic processing operations can trim energy use by 30% (www.omnexus.com).

[11] Over the decades, the cost of labor involved in making plastic products has consistently decreased and now stands below 10% of the cost of product. It is the material and overhead costs that dominate production costs. Resource efficiency is therefore critical to sustainability of the business as well and provides an effective means of competing with production in regions where labor costs are lower.

A change or strategy that fits any of the nine spaces will be a sustainable move. But even best practice does not of course address all nine elements. In fact, it is unlikely that sustainable strategies that address a majority of the elements are easy to come up with or to implement. Those that address at least a single element effectively contribute to environmental sustainability.

The nine strategies taken together make the "sustainability matrix" that qualitatively describes the different aspects of sustainable growth. It is a convenient way to summarize sustainable growth and perhaps has use as a tool to assess if a given change or strategy (such as improved product design, changed processes, or a different packaging of a product) has merit in terms of sustainable growth. Yet, the metrics needed to make decisions will still have to be derived from LCA or LCI exercises. It does not replace the value of a comprehensive LCA in comparing present and envisioned practice but provides a qualitative, inexpensive, initial means of evaluating changes to a product or process.

2.5.1 Examples from Plastics Industry

In this section, each of the nine elements in the sustainability matrix will be defined and illustrated using practical feasible examples from the plastics industry. In most instances, the examples given have actually been implemented, sometimes even at an economic gain to the business involved.

2.5.1.1 Using the Minimum Energy Needed to Manufacture Products A company[12] that initiated a disciplined approach to improve the efficiency in its operations reduced its energy demand for lighting, water heaters, and air conditioning in its plants. It also increased energy efficiency in its molding operation and expanded its recycling efforts. These energy savings were estimated to reduce their carbon emissions by over 40,000 tons annually.

In injection molding of plastics, switching over to all-electric molding machines results in improved energy efficiency, saving as much as 60% of the energy demand by existing machinery.[13] Substantial energy savings result from not having to cool the hydraulics in all-electric machines. Reduced water consumption in electric machines is an added bonus. Similar energy savings can also be achieved with other processing equipment such as extrusion or blow molding.

2.5.1.2 Using the Energy Mix with a Minimal Environmental Footprint The mix of energy used in a manufacturing operation is dictated by prevailing costs and will therefore change with the location of the plant. Using mixes of energy that are less polluting from a life cycle perspective is a valuable sustainable strategy. Switching to renewable energy whenever it is cost competitive is clearly the best choice.

---

[12]Techtronic Industries Company (TTI). The company won the Think Green Initiative award in 2010 for their greening efforts (www.ttigroup.com).
[13]Energy efficiency in plastic processing: practical worksheets for industry. Tangram Technology Ltd. (www.tangram.co.uk).

Back in 2002, a plastic molding company[14] in Ontario, New York, used wind energy to supplement its process energy needs. The 250 kW wind turbine is driven by the consistent winds off the Lake Ontario. More recently, the same company set up a larger 850 kW wind turbine, increasing its commitment to renewable energy.

2.5.1.3 Recovering Waste Process Energy for Reuse Processing operations typically produce waste heat that is allowed to escape contributing to thermal pollution. Where it is practical, the waste low-grade heat energy might be captured for reuse.

A company (Naitove, 2012) in Southern California uses gas turbines to generate in-house some of the electrical energy used to run its plastic processing operation. The waste heat from boilers was recovered and used in the crystallizer/dryer unit to condition the PET resin prior to sheet extrusion. No additional electric heating was therefore needed for the purpose, resulting in energy as well as cost savings.

2.5.1.4 Using Only as Much Material as Is Needed to Ensure Functionality With plastic packaging, this translates into downgauging and avoiding overpackaging of products. Downgauging has been a consistent trend in the industry over the decades. The average weight of a single-use two liter PET soda bottle, for instance, dropped from 68 gauge in the 1970s to the present 49 gauge, saving 200 million pounds of virgin resin annually.[15] Leading soft drink manufacturers are following suit and coming up with innovative lighter bottles (Bauerlein, 2009).

Another example is downgauging LDPE shrink wrap material from 80 to 63 gauge, resulting in substantial material saving (Connolly, 2009). Downgauging is possible because of improvements in materials technology that allow less plastic to be used without sacrificing functionality of the product (Fig. 2.10). Impressive downgauging is also common in nonplastic packaging industry. The largest manufacturer[16] of aluminum cans, for instance, reduced the weight of its 12 oz can from 0.048 lbs in 1970 to 0.029 lbs in 2012 without any loss in functionality.

2.5.1.5 Using More of Renewable and Recycled Raw Materials Replacing fossil fuel-based plastics with identical resins derived from renewable sources is receiving the serious attention it deserves from industry. Revolutionary package design by Coca-Cola (IPF Inc, 2009) to introduce a PET beverage bottle where one of the two monomers used in the resin is derived from renewable plant material as opposed to from fossil fuel, is a particularly good example. These bottles are closed-loop recyclable[17] or downcycled into other products. A premier apparel

[14]The company is Harbec Inc., and the turbine is located at the Wayne Industrial Sustainability Park, built in 2008 in rural New York. In 2013, it installed an 850 kW wind turbine with significant savings in their energy costs.

[15]Data from ExxonMobil Corporation. Available at http://www.exxonmobilchemical.com/Chem-English/sustainability/sustainability-plastics-downgauging.aspx

[16]Ball Corporation also reduced the weight of their steel cans by 33% over a 25-year period. Its sustainability goals are set in terms of explicit energy efficiency and GHG goals. However, they have not as yet switched to BPA-free liners.

[17]Closed-loop recycling here means bottle-to-bottle recycling. One hundred percent post-consumer recycled PET was cleared by the FDA for direct food contact in March of 2009. See, for instance, IPF Inc. (2009).

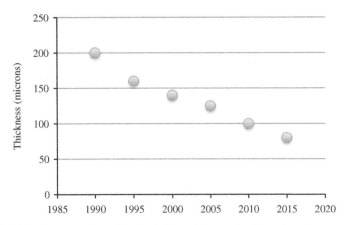

FIGURE 2.10 Downgauging of polyethylene film in plastic garbage bag applications. Source: Exxon Mobil Corporation. *Life Cycle Assessment of Metallocene Polyethylene in Heavy Duty Sacks.* ExxonMobil Chemical Company; May 2011.

manufacturer[18] also produced a pair of jeans that contains at least 20% recycled polyester fiber derived from recycled PET (i.e., eight bottles recycled into each pair).

2.5.1.6 Reusing and Recycling Postuse Products

Designing plastic products that allow convenient reuse as well as material recycling is an indispensable aspect of sustainable growth and underscores the "cyclic" consumption model already alluded to. Ensuring that products are recyclable or amenable to feedstock chemical recovery is a significant step toward sustainability. Chemical recycling of waste into valuable building blocks or the recapture of embodied energy (via incineration), also falls under this topic. What is to be avoided is landfilling and indiscriminate burning of waste.

Cosmetic packaging waste (tubes, bottles, lids) is generally of little interest to recyclers who tend to focus on high-volume plastic waste such as soda/water bottles and milk jugs. A leading cosmetics company[19] started accepting all cosmetic empties, regardless of brand, in their stores, for sorting and recycling, preventing them from ending up in the landfill. This allowed the recovery of single-use plastic cosmetic bottles from the thousands of hotel rooms across the United States.

2.5.1.7 Minimizing Externalities at Source: Green Chemistry

Designing products where the materials and processes contribute to minimal externalities is the best practice in this strategy. The challenge is to use green chemistry, bio-based resources, and clean energy to minimize the environmental footprint while still maintaining market competitiveness for the product.

A flooring manufacturer (Tarkett North America) announced a switch from conventional phthalates to alternative plasticizers in their vinyl flooring product

---

[18]Levi Strauss and Co. markets Waste-Less jeans. Others such as Patagonia, a manufacturer of outdoor clothing, and Puma have also incorporated recycled PET into clothing.

[19]Estée Lauder Company through its Origins brand encourages consumers to return postconsumer packages to the store and compensates them for the effort.

worldwide in 2014. This reduces human exposure to unsafe phthalates in residential and commercial buildings. Some European manufacturers (Upofloor OY) were promoting phthalate-free flooring as early as 2010.

2.5.1.8 Avoiding Toxic Components and Potential Hazards Associated with Products and Processes Chemical species used in plastics and rubber industry (including the resin itself) can be released inadvertently, polluting the environment. If some of these in water, air, or food can potentially reach the human consumer, the risk is clearly unacceptable. Where conventional designs of products include known hazardous chemicals, it is incumbent upon the industry to voluntarily phase out these and redesign products that are innocuous but functionally equivalent:

1. *The epoxy liners in tin cans leach bisphenol A (BPA), an endocrine disruptor, into the food contents. Ingesting even low levels of BPA is a health concern. While BPA-free cans with no epoxy liners are available, they are about 15% more expensive compared to conventional cans. Despite this disadvantage, some canned-food manufacturers[20] have switched to the BPA-free cans.*

2. *Some skin care preparations use polyethylene microbeads that serve as exfoliants. The particles in municipal waste reach the oceans via wastewater and are potent marine pollutants as well as a pathway that introduce concentrated persistent chemicals (POPs) in seawater to the marine food web (microparticles; see Chapter 9). On recognizing the environmental damage posed by these, several leading manufacturers[21] have pledged to phase out their use of these and to undertake research to identify nonpolluting replacement.*

2.5.1.9 Converting the Pollutants into Resources Carbon dioxide is a pollutant of immediate interest, and most operations in the plastics supply chain emit CO_2 into the atmosphere. Emerging research suggests that emissions from industry might be used to produce polymer raw materials:

1. *Waste CO_2 can be captured and converted into poly(propylene carbonate) [PPC] plastic. Large-scale conversion of the waste gas into PPC has been demonstrated.[22]*

[20] In January 2011, Eden Foods switched from metal cans to amber glass jars for high-acid organic crushed tomatoes and tomato sauces to avoid BPA (McTigue Pierce and Caliendo, 2012).
[21] Procter & Gamble, Johnson & Johnson, and Unilever.
[22] The first successful large-scale production of a polypropylene carbonate (PPC) polymer using waste carbon dioxide (CO_2) was conducted by Novomer in collaboration with specialty chemical manufacturer Albemarle Corporation (Orangeburg, SC) in May 2013 (http://energy.gov/fe/articles/recycling-carbon-dioxide-make-plastics).

The critical component is a proprietary catalyst developed by the innovators. As more uses for this plastic and its copolymers and blends are found, this can be a valuable means of converting waste CO_2 into a resource.

2. *Using a different technology (Joule Unlimited, MA), waste CO_2 is mixed with algae or cultures of microorganism exposed to sunlight to directly generate biofuel. This novel approach bypasses the costly intermediate steps (via lipid hydrolysis or carbohydrate fermentation) to obtain fuel from waste.*

REFERENCES

Bartlett AA. Reflections on sustainability, population growth and the environment. In: Keiner M, editor. *The Future of Sustainability*. the Netherlands: Springer; 2006.

Bauerlein V. Pepsi to pare plastic for bottled water. Wall Street Journal Online 2009. Available at online.wsj.com/article/SB123791618253927263.html. Accessed February 8, 2014.

Beard A, Hornik R, Wang H, Ennes M, Rush E, Presnal S. It's hard to be good. Harvard Bus Rev 2011;89 (11):88–96.

Bersimis S, Georgakellos D. A probabilistic framework for the evaluation of products' environmental performance using life cycle approach and principal component analysis. J Clean Prod 2013;42:103–115.

Bolis I, Morioka SN, Sznelwar LI. When sustainable development risks losing its meaning. Delimiting the concept with a comprehensive literature review and a conceptual model. J Cleaner Prod 2014;83 (15):7–20.

Brundtland GH Sustainable development is development that meets the needs of the present without compromising the ability of future generations to meet their own needs. The World Commission on Environment and Development. Oxford: Oxford University Press; 1987. The report was published as Our Common Future.

Cabezas-Basurko O. Methodologies for sustainability assessment of marine technologies [PhD thesis]. Newcastle upon Tyne: School of Marine Science and Technology, Newcastle University; 2010. p 220.

Caccia C. *Hammond Lecture Series: The Politics of Sustainable Development*. Guelph: Guelph University; 2001.

Carlsson Reich M. Economic assessment of municipal waste management systems—case studies using a combination of life cycle assessment (LCA) and life cycle costing (LCC). J Clean Prod 2005;13:253–263.

Connolly KB. Stretching to be green. Food and Beverage Packaging; 2009. Available at www.foodandbeveragepackaging.com/Articles/Feature_Articles/BNP_GUID_9-5-2006_A_10000000000000673376. Accessed February 8, 2010.

Consonni S, Giugliano M, Grosso M. Alternative strategies for energy recovery from municipal solid waste. Part B: Emission and cost estimates. Waste Manag 2005;25 (2):137–148.

Cooper JS, Fava JA. Life-cycle assessment practitioner survey: summary of results. J Ind Ecol 2006;10 (4):12–14.

Costanza R, Stern D, Fisher B, He L, Ma C. Influential publications in ecological economics: a citation analysis. Ecol Econ 2004;50 (3–4):261–292.

Crittenden VL, Crittenden WF, Ferrell LK, Ferrell OC, Pinney CC. Market-oriented sustainability: a conceptual framework and propositions. J Acad Mark Sci 2011;39 (1):71–85.

Curran MA. *Life Cycle Assessment Handbook: A Guide for Environmentally Sustainable Products*. New York: Wiley-Scrivener; 2012.

Dale A. *At the Edge: Sustainable Development in the 21st Century*. Vancouver, BC: UBC Press; 2001.

Daly HE. Commentary: toward some operational principles of sustainable development. Ecol Econ 1990;2:1–6.

Elkington J. Partnerships from cannibals with forks: the triple bottom line of 21st century business. Environ Qual Manag 1998:37–51.

Eriksson O, Frostell B, Björklund A, Assefa G, Sundqvist J-O, Granath J, Carlsson M, Baky A, Thyselius L. ORWARE—a simulation tool for waste management. Resour Conserv Recycl 2002;36 (4):287–307.

Geiser K. *Materials Matter: Toward a Sustainable Materials Policy First*. Cambridge, MA: MIT Press; 2001. p 479.

Goedkoop M, Spriensma R. *Eco-Indicator 99-A Damage-Oriented Method for Life Cycle Impact Assessment: Methodology Report*. Amersfoort: Pré Consultants; 2001. p. 132. Available at www.pre-sustainability.com/download/misc/EI99_annexe_v3.pdf. Accessed October 7, 2014.

Graedel TE, Allenby BR. *Industrial Ecology and Sustainable Engineering*. Upper Saddle River: Pearson; 2010.

Guo M, Murphy RJ. LCA data quality: sensitivity and uncertainty analysis. Sci Total Environ 2012;435–436:230–243.

Hannon A, Callaghan EG. Definitions and organizational practice of sustainability in the for-profit sector of Nova Scotia. J Clean Prod 2011;19:877–884.

Hansen J, Sato M, Kharecha P, Russell G, Lea D, Siddall M. Climate change and trace gases. Phil Trans R Soc A 2007;365:1925–1954.

Hassan OAB. A value-focused thinking approach for environmental management of buildings construction. J Environ Assess Policy Manage 2003;5 (2):247–261.

Hopwood B, Mellor M, O'Brien G. Sustainable development: mapping different approaches. Sustain Dev 2005;13:38–52.

Horne R, Grant T, Varghese K. *Life Cycle Assessment: Principles, Practice and Prospects*. Collingwood: CSIRO Publishing; 2009.

Hottle TA, Bilec MM, Landis AE. Sustainability assessments of bio-based polymers. Polym Degrad Stab 2013;98 (9):1898–1907.

Huang M-H, Rust RT. Sustainability and consumption. J Acad Mark Sci 2011;39 (1):40–54.

IPF Inc. Recycled PET rollstock is cleared for food-contact use. Food and Beverage Packaging; 2009. Available at www.foodandbeveragepackaging.com. Accessed February 8, 2010.

Johnston P, Everard M, Santillo D, Robert K-H. Reclaiming the definition of sustainability. Environ Sci Pollut Res Int 2007;14 (1):60–66.

Kajikawa Y. Research core and framework of sustainability science. Sustain Sci 2008;3:215–239.

Klöpffer W. Life-cycle based methods for sustainable product development. Int J Life Cycle Assess 2003;8 (3):157–159.

Lenzen M. Errors in conventional and input-output-based life-cycle inventories. J Ind Ecol 2000;4 (4):127–148.

McTigue Pierce L, Caliendo H. 2012. Packaging Digest on July 1, 2012. http://www.packagingdigest.com/shipping-containers/bpa-packaging-defying-pressure. Accessed October 27, 2014.

MacKay D. *Sustainable Energy without the Hot Air*. Cambridge, UK: UIT Cambridge; 2009.

Mebratu D. Sustainability and sustainable development: historical and conceptual review. Environ Impact Assess Rev 1998;18:493–520.

Missimer M, Robèrt K-H, Broman G, Sverdrup H. Exploring the possibility of a systematic and generic approach to social sustainability. J Clean Prod 2010;18 (10–11):1107–1112.

Naitove M. Thrifty California processor dries pet with 'waste' heat. Plast Technol 2012. Available at www.ptonline.com/articles/thrifty-california-processor-dries-pet-with-waste-heat. Accessed October 7, 2014.

Odum EP. *Basic Ecology*. New York: Saunders College Publishing; 1983.

Raffensperger C, Tickner W, editors. *Protecting Public Health and the Environment: Implementing the Precautionary Principle*. Washington, DC: Island Press; 1999.

Rambo AT. *The Fallacy of Global Sustainable Development*. Asia Pacific: Analysis from the East-West Center Honolulu. East-West Center; (1997, March 30) 1–8. www.ewc.hawaii.edu

Schrettle S, Hinz A, Scherrer-Rathje M, Friedli T. Turning sustainability into action: explaining firms' sustainability efforts and their impact on firm performance. Int J Prod Econ 2014;147 (Part A):73–84.

Sen S, Bhattacharya CB, Korschun D. The role of corporate social responsibility in strengthening multiple stakeholder relationships: a field experiments. J Acad Mark Sci 2006;34 (2):158–166.

Simon R, Rice E, Kingsbury T, Dornfeld D. *A Comparison of Life Cycle Assessment Software in Packaging Applications*. Berkeley: Laboratory for Manufacturing and Sustainability. College of Engineering, University California; 2012.

Taylor J. *Sustainable Development: A Dubious Solution in Search of a Problem*. Policy Analysis 449. Washington, DC: Cato Institute; 2002, p 1–49.

Tugnoli A, Santarelli F, Cozzani V. An approach to quantitative sustainability assessment in the early stages of process design. Environ Sci Technol 2008;42 (12):4555–4562.

ULS Report. 2008. Available at http://www.use-less-stuff.com/Paper-and-Plastic-Grocery-Bag-LCA-Summary-3-28-08.pdf. Accessed October 27, 2014.

UNCED. United Nations Conference on Environment and Development (UNCED), Rio de Janeiro, Brazil, June 3–14, 1992.

US DOE and SPI. *Industrial Technologies Program: Improving Energy Efficiency at U.S. Plastics Manufacturing Plants*. Washington, DC: US DOE; 2003.

Winkler J, Bilitewski B. Comparative evaluation of life cycle assessment models for solid waste management. Waste Manag 2007;27:1021–1031.

Zabaniotou A, Kassidi E. Life cycle assessment applied to egg packaging made from polystyrene and recycled paper. J Clean Prod 2003;11 (5):549–559.

3

AN INTRODUCTION TO PLASTICS

Back in 1922, Hermann Staudinger, a German chemist, first realized that polymers were in fact made up of long covalently bonded molecules. Until then, polymers were generally assumed to be colloids. Not only are polymer molecules very large with average molecular weights reaching millions of grams per mole (and hence called macromolecules), but they also have long-chain-like molecular architecture. The recognition of this class of unique giant linear molecules marked the birth of a new science and technology that has yielded, among other materials, the ubiquitous plastics widely used in numerous applications. The unique, impressive mechanical properties of plastics (or polymers in general), such as high strength and modulus, are the result of their distinctive chain-like molecular architecture.

The following discussion on polymers is not intended to be a comprehensive introduction but provides only basic information on the subject. Many important aspects of polymer structure, properties and characterization have been left out. Aspects discussed here are primarily those likely to assist the reader in understanding the content in subsequent chapters. The reader is, however, directed to excellent monographs on polymer science for more detailed and complete information (Coleman and Painter, 1998; Rudin and Choi, 2012; Young and Lovell, 2011).

Plastics and Environmental Sustainability, First Edition. Anthony L. Andrady.
© 2015 John Wiley & Sons, Inc. Published 2015 by John Wiley & Sons, Inc.

3.1 POLYMER MOLECULES

Structurally, polymers are long-chain-like molecules made up of repeating structural segments (called "repeat units") linked end to end. The desirable properties of plastics (a subcategory of polymers)[1] are primarily a consequence of this unique molecular geometry. With large molecules where many atoms are in such close proximity, the van der Waals forces of attraction are quite large; this is what is behind the exceptional mechanical properties of plastics. Most common plastics, for instance, have the same structural unit repeated along the macromolecular chain, and are called homopolymers. In copolymers, two or more such repeat units can be mixed within a single polymer chain. Common plastics polyethylene (PE) and polypropylene (PP) are, for instance, homopolymers, while plastics such as acrylonitrile–butadiene–styrene (ABS) resin and styrene–butadiene–styrene (SBS) are copolymers with three and two types of repeat units, respectively.

PE is the most-used plastic in the world and is made by polymerizing ethylene gas. Ethylene, used as the raw material, is called the monomer, and n molecules of it join together to form a PE chain of n repeat units as shown in Figure 3.1. As n is typically a very large number, writing out the full structural formula for the molecule is uninformative and impractical. Also, n is not even the same for all PE molecules produced in a single polymerization reaction (as we will see later). Therefore, only a single repeat unit of PE is typically used to denote its structure along with a subscript n to indicate that it is a polymer (see Fig. 3.1). Typically, the term "poly" is used in front of the parenthesis carrying the name of repeat unit (e.g., poly(vinyl chloride) (PVC) or poly(ethylene terephthalate) (PET) to denote it is a polymer). The parenthesis is often dropped in case there is no structural ambiguity as with PE or polystyrene (PS).[2]

The chemistry of the repeat unit is the key determinant of the physical and chemical characteristics of the polymer.

Structural formulae of polymers as in Figure 3.1 are incomplete in that they do not provide any geometric information on the chain molecule. Macromolecular geometry is better visualized using a simple three-dimensional (3D) model (or a

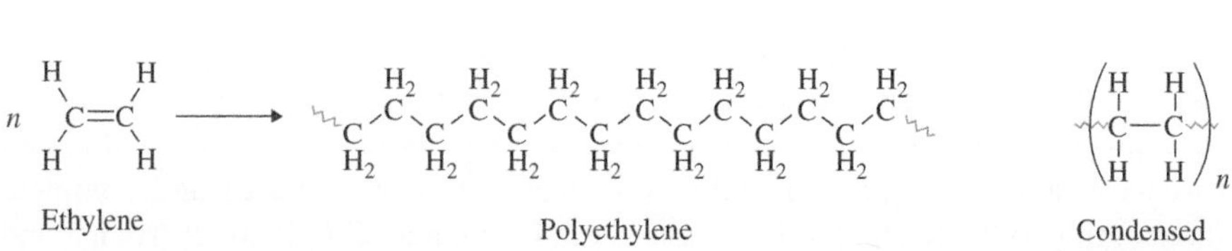

Ethylene   Polyethylene   Condensed
                                                                               formula for PE

FIGURE 3.1   The polymerization reaction of ethylene yielding polyethylene.

[1] The group "polymers" include plastics, rubbers, and fibers. But, we will use the terms polymer and plastic interchangeably in this work.
[2] With poly(vinyl chloride) or poly(methyl methacrylate), however, the parentheses should always be used because the term without them is ambiguous. Writing all polymer names with parenthesis (e.g., poly(ethylene)) would still be correct.

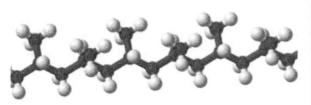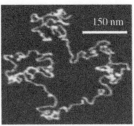

FIGURE 3.2 *Left*: A ball and stick model of a section of a PP chain. *Right*: An AFM image of a single polymer chain suggesting flexibility. Reprinted with permission from Kiriy et al., (2002). Copyright (2002) American Chemical Society.

space-filling model) that takes bond angles and atomic sizes into account. In such models, it is easy to see that covalent bonds can rotate, allowing the polymer chain to be very flexible (Fig. 3.2). The molecules can even bend and fold over themselves behaving more like a chain of beads rather than a linear rigid rod. This flexibility allows for profuse entanglement of polymer chains leading to the excellent mechanical properties of polymers. Images of single molecules of polymers by atomic force microscopy (AFM) illustrate this flexibility. Figure 3.2 shows an AFM image of a polyelectrolyte polymer molecule deposited from a dilute solution onto a smooth surface (Kiriy et al., 2002).

3.1.1 Size of Polymer Molecules

Given such flexibility, the contour length of the polymer chain is not a good measure of the size of a polymer chain made up of n segments each of length ℓ linked end to end. The common metric of molecular size is based on the distance between ends of the chain R.[3] But polymer molecules can rearrange in different geometries, each with a different end-to-end distance, R, and its mean value $\langle R \rangle$ is used to quantify macromolecular size. The freely jointed model of polymers therefore yields only a mean value for square[4] of the end-to-end distance as $\langle R^2 \rangle = n\ell^2$. But in real chains, the bonds are also able to rotate in any axis, allowing the chain to take diverse conformations in 3D space. The above expression, when modified to take free rotation about the covalent bonds into account, yields a slightly more complicated expression 3.1 for the $\langle R^2 \rangle$ of a very long chain (Kawakatsu, 2004; Rubinstein and Colby, 2003; Strobl, 2007):

$$\langle R^2 \rangle = n\ell^2 \left(\frac{1 + \cos\Theta}{1 - \cos\Theta} \right) = C_\infty n\ell^2 \tag{3.1}$$

---

[3]However, the contour length L of the chain is not as useful a measure of the "size" of polymer molecules as $\langle R^2 \rangle$. An alternate measure of the size of a polymer molecule is the root mean square of its "radius of gyration," Rg. It is defined as the average distance of a chain element from the center of gravity of the chain.
[4]The mean end-to-end distance $\langle R \rangle$ for such a chain is zero. Therefore, the second moment $\langle R^2 \rangle$ is used instead.

where Θ is the valance angle defined by the bond vectors and C_∞ is called the characteristic ratio. The value of C_∞ is essentially a measure of the stiffness of polymer chains; for PE, it is 5.0, whereas for PS, it is greater than 10 and for DNA, it is approximately 600! The expression is further refined by also including perturbations and limitations imposed by torsional angles on the different chain conformations. All these torsional angles are not equally likely due to steric effects of substituents on neighboring repeat units. Taking these perturbations into account yields the following refinement:

$$\langle R^2 \rangle = n\ell^2 \left(\frac{1-\mathrm{Cos}\,\Theta}{1+\mathrm{Cos}\,\Theta} \right) \left(\frac{1+\langle \mathrm{Cos}\,\phi \rangle}{1-\langle \mathrm{Cos}\,\phi \rangle} \right) \tag{3.2}$$

where ϕ is the torsional angle. However, a detailed discussion of these models and the statistical treatment of polymer chains are beyond the scope of the present discussion. It is sufficient to appreciate that all representations are crude models of polymer chains and that $\langle R^2 \rangle$ invariably depends on the bond length, the bond angles, as well as the substituents on adjacent repeat units. In 2D space, freely-jointed chain of a high polymer looks like the "random-walk" path or the tracing of the movement of a particle in Brownian motion (see Fig. 3.3). This picture is very different from the simple structural models used to draw polymer structures but closer to the 3D image obtained using profiling microscopy (Fig. 3.2).

In any event one does not encounter single polymer chains (except in theoretical discussions); what is more relevant to the present discussion is the solid state of a

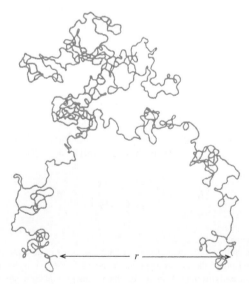

FIGURE 3.3   Approximate simulation of a polymer chain with freely jointed chain model. The value of r is the end-to-end distance.

large ensemble of polymer molecules. When considered as a collection, these long, flexible molecules are highly entangled and strongly attracted to each other to form a unique solid.

3.2 CONSEQUENCES OF LONG-CHAIN MOLECULAR ARCHITECTURE

This long-chain molecular geometry has interesting consequences in terms of properties and is invariably the reason behind the success of plastics as a material. Some important practical consequences of linear architecture of molecules will be considered in this section.

3.2.1 Molecular Weight of Chain Molecules

When a collection of monomer molecules such as ethylene is converted into polymer, the result is a mix of long-chain molecules. Different macromolecules in the mix will have a different chain length; their individual values of n will be different. But since each PE molecule can be represented by the same chemical structure, $-(CH_2-CH_2-)_n$, they are all PE molecules but with different molecular weights. Unlike organic compounds such as ethylene (mol. wt. = 28), table salt (mol. wt. = 58.54), or ammonia (mol. wt. = 17), PE (and polymers in general) does not have a unique characteristic molecular weight. One consequence of this is that we have to always refer to an average molecular weight of the ensemble of polymer molecules. Two such averages are commonly used. These are **Mn**(g/mol),[5] the number average molecular weight, and **Mw**(g/mol), the weight average molecular weight. We use bold letters here to indicate the average values. The definitions of these measures are elaborated below. Typically, the average molecular weight of polymers increases with **n** (which is the average degree of polymerization (DP)). For addition polymers such as polyolefins, the **Mw**(g/mol) increases with DP:

Number average molecular weight $\mathbf{Mn} = \dfrac{\text{Total weight of all polymer molecules}}{\text{Total number of polymer molecules}}$

$$\mathbf{Mn} = \frac{\sum_{i=1}^{\infty} N_i M_i}{\sum_{i=1}^{\infty} N_i} = \sum_{i=1}^{\infty} X_i M_i \tag{3.3}$$

where N_i chains have the molecular weight of M_i and $\{N_i / \sum N_i\}$ is the molar fraction (a number fraction) X_i. However, some properties of polymers are governed by the fraction of the largest chain molecules in the sample. In such situations, the polymer is better characterized by a weight average molecular weight **Mw**.

---

[5]A mole (abbreviated mol.) is a number like a dozen or a gross. It is a large number: 6.022×10^{23}. The molecular weight of any compound is the mass in grams of the above large number of molecules of that compound. This number is called Avogadro's number.

The weight average molecular weight can be similarly defined as follows:

$$\mathbf{Mw} = \frac{\sum_{i=1}^{\infty} w_i M_i}{\sum_{i=1}^{\infty} w_i} = \frac{\sum_{i=1}^{\infty} N_i M_i^2}{\sum_{i=1}^{\infty} N_i M_i}$$

where w_i is the weight of chains of molecular weight w_i.

Figure 3.4 shows a smooth curve of the frequency distribution of molecular weights of individual chains in a polymer sample. In practice, such a distribution can be determined for polymers using gel permeation chromatography (GPC). This is essentially a modified column chromatography experiment where a dilute polymer solution is injected into the top of a column packed with porous (cross-linked PS) beads. Molecules elute out from the bottom of the column at different times; larger molecules do not penetrate the "holes" in porous beads and elute relatively faster. The shorter chains tend to get trapped in the "holes" and are relatively slower to elute. A detector based on refractive index changes or other properties (e.g., changes in UV absorption) monitors the concentration of polymer in solution exiting the column. Multiple detectors are used in the newer instruments. Polymer fractions with the same molecular size (hydrodynamic volume) elute at the same time. The column, however, has to be calibrated using polymers of known average molecular weight.

The readout obtained is a distribution of weight fractions (Fig. 3.4, *left*). The right-hand end of the distribution has short-chain molecules that can hardly be called polymer, and the left-hand end of it has the longer-chain high polymer molecules in the mix. The two averages, **Mn** and **Mw**, are indicated with lines; note that always **Mw** > **Mn** and (**Mw/Mn**) is the polydispersity index (PDI) of the chains. The smaller this value, the narrower will be the distribution of chain lengths. If all the molecules in the polymer had exactly the same chain length or molecular weight, then, **Mw** = **Mn** and PDI = 1. Polymers made by chain reaction generally have values of PDI ranging from 2 to 20, while those made by step reaction have narrower distributions with P.I. ~ 2.0.

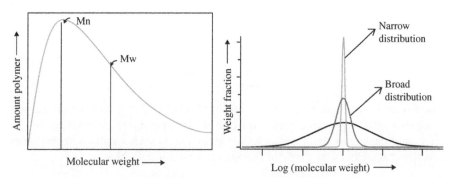

FIGURE 3.4 *Left*: Schematic drawing of the molecular weight distribution of a polymer indicating the two averages **Mn** and **Mw**. *Right*: Schematic diagram of the molecular weight distribution for polymer samples with low and high PDI.

Average molecular weight and PDI are both important as they determine the properties of the polymer. The mechanical properties of plastics that make them so useful depend on the material having a high molecular weight. As the average molecular weight is increased, mechanical properties like tensile strength, stiffness, and hardness will also increase. As **Mn** increases, the melting range as well as the viscosity of the melt also similarly increases. The higher the **Mn**, the better will be the mechanical properties, but more difficult will be the processing (which involves molten plastic) due to the high melt viscosity. Depending on the intended application, usually, a compromise has to be made between high **Mn** of the plastic and the ease of processing. At very high or infinite molecular weight, a given property, P, will have a limiting value of P_0. Then at a given average molecular weight **Mn**, the property P will have a value:

$$P = P_0 - \frac{k}{\mathbf{Mn}}$$

where k is a constant.

3.2.2   Tacticity

Chain-like geometry also allows for different stereoregular arrangement of the repeat units. The regularity in structure or the "tacticity" of the polymer therefore strongly influences the degree to which it has rigid, crystalline long-range order or flexible, amorphous disorder and therefore the bulk properties of the polymer (Stevens, 1999; Chang, 2011). New catalysts are continually being developed to increase the stereoregularity of plastics such as PP with better mechanical properties.

With vinyl polymers such as PP or PVC, the substituent group Z (Z=$-CH_3$ and $-Cl$, in PP and PVC, respectively) might be expected to be randomly oriented along the polymer chain. This yields a polymer with no stereoregularity in its structure or an *atactic* polymer. Using special catalysts (Ziegler–Natta catalysts or the new metallocenes) to facilitate the polymerization reaction, it is also possible to synthesize the very same polymer with the same molecular weight but as a stereoregular or *isotactic* polymer, where all the substituents are located on the same side[6] of the carbon chain backbone. Where the substituent groups have alternative positions on the chain, a *syndiotactic* polymer is obtained. These structures are illustrated in Figure 3.5.

For instance, the Tg of PP depends strongly on its tacticity. *Atactic* PP has a Tg ~ −19°C and is used in bitumen and hot-melt adhesive formulations, while for *isotactic* PP, it is approximately −8°C. The latter is a tough, semicrystalline plastic used in packaging films. The Tg of s*yndiotactic* form of PP can be as high as 130°C! Sophisticated catalyst systems even allow synthesis of polymers that have ordered blocks of *atactic* materials and *isotactic* materials on the polymer chains.

---

[6]In a polymer chain where all backbone carbon atoms have a tetrahedral molecular geometry, the zigzag backbone is in the plane of the paper with the substituents either sticking out of the paper or retreating into the paper.

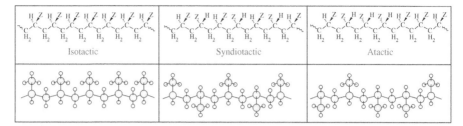

FIGURE 3.5   Illustration of the stereochemistry in a vinyl polymer. Below each structural formula is an illustration of the stereochemistry with a "ball and stick model" for polypropylene.

3.2.3   Partially Crystalline Plastics

Polymer chains are too long to be arranged into discrete single crystals. Single crystals of polymers can only be formed in very dilute solutions in the laboratory, but these have no particular practical significance in plastics technology. But when bundles of chains align closely with each other, short segments of polymer chains can form small highly-ordered regions given their strong interchain attraction. This yields small crystalline phases embedded in the surrounding amorphous polymer matrix. These form into crystalline domains with short-range 3D order and yields a *partially crystalline* morphology. Polymers are therefore uniquely capable of "fractional crystallinity" that can vary from a few percent to as high as 90%. The crystallites have a size distribution but are always embedded in and covalently linked to the amorphous matrix around them (Flory, 1962; Painter and Coleman, 1997; Strobl et al., 1980).

When these crystallites grow out symmetrically from a nucleating point, they assume a spherical shape and are called "spherulites." Still, even within a spherulite, an amorphous disordered phase can be trapped in, particularly as intercalated material between the lamellar crystallites radiating from its center. Crystalline fractions act as reinforcing fillers contributing to the high modulus and strength of such materials. Fully amorphous polymers can be identified experimentally as they do not show an X-ray diffraction pattern nor a first-order (melting) transition on heating in differential scanning calorimetry (DSC), both characteristic of crystallinity.

The densities of the crystalline region ρ_c and amorphous regions ρ_a are very different. The density of the polymer, P, will therefore depend on its percent crystallinity x:

$$P = x\rho_c + (1-x)\rho_a$$

Polymers such as low-density polyethylene (LDPE) where the molecules are branched are difficult to crystallize compared to high-density polyethylene (HDPE) and typically have only 35–55% crystallinity. Virtually unbranched longer molecular form of PE, HDPE, typically has a percent crystallinity as high as 85% and is used in applications that demand high strength. Percent crystallinity can be further increased by uniaxial drawing of polymer or by slow heating below the glass transition (annealing).

The percent crystallinity of commodity plastics generally increases as follows:

{HDPE; PP; PMP} > {Chlorinated PE; VLDPE} > {LDPE; LLDPE; EVA copolymer}

For engineering thermoplastics, the order is as follows:

$${PET; PBT} > {Nylon} > {PB; UHMWPE}$$

where UHMWPE is ultrahigh-molecular-weight polyethylene; PB, polybutene; PBT, poly(butylene terephthalate); PMP, poly(4-methylpentene); VLDPE, very-low-density polyethylene; LLDPE, linear low-density polyethylene; and EVA copolymer, (ethylene vinyl acetate) copolymer.

As a general rule, *atactic* polymers are of too irregular geometry to crystallize. Plastics such as PS, PVC, and nylon are amorphous; some amorphous polymers such as polycarbonate (PC), however, can be made partially crystalline by slow annealing. PE, PP, and PET are partially crystalline and are generally difficult to get in 100% amorphous form. Their degree of crystallinity depends on their thermal history as well. A polymer that is structurally amenable to crystallization, when heated into a melt and quenched quickly in a cold fluid, will still yield an amorphous glass. Slow cooling of the same polymer will often yield a semicrystalline material.

Figure 3.6 illustrates the embedded crystallites in a plastic matrix and a schematic of a semi-crystalline polymer.

In common with those of simple organic compounds, polymer crystallites also have melting points defined by characteristic temperatures. When melting occurs in the crystalline phase of the polymer, it is converted into amorphous polymer. Under suitable conditions such as annealing, it can slowly recrystallize. The crystalline melting point is characterized by its thermal signature but generally does not result in a visible phase change of the entire solid resin into a fluid as with organic compounds. Above the melting point,[7] the modulus and the strength of the plastic drop very significantly. Figure 3.7 illustrates these changes schematically. E is the elastic modulus, a measure of stiffness of the material.

Extending a plastic material in one direction (anisotropically) results in forcing the chains into aligning themselves in that direction. As the chains get closer to each other, further crystallization is encouraged. At very high extensions, the plastic becomes a highly crystalline, anisotropic fiber such as those used in textiles.

3.2.4 Chain Branching and Cross-Linking

In conventional organic compounds, any branching is characteristic of the molecular structure and is reflected in its IUPAC name. For instance, the alkane molecule 3-ethyl hexane has a branch in the structure, an ethyl group on the third carbon. With a polymer such as PE, short, medium length, long, or even complex branches can be

---

[7]Melting point of semicrystalline plastics should not be confused with the term "melt" as used in plastic processing. The latter is the viscous liquid obtained when plastics are heated to high processing temperatures. A great majority of plastics, even amorphous ones, heated to high enough temperatures will form viscous "melts."

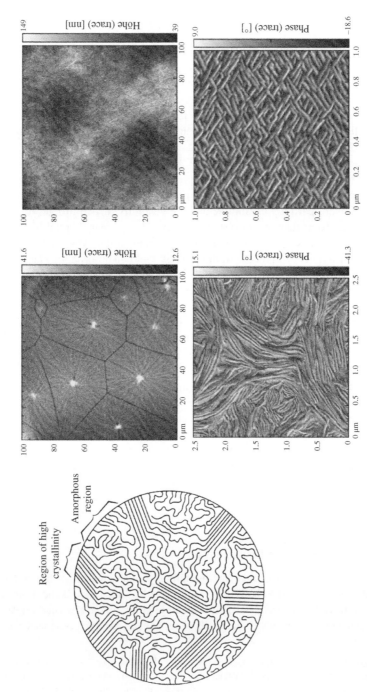

FIGURE 3.6 *Left:* An illustration of crystallites embedded in an amorphous polymer matrix. *Right:* Crystallites in plastic crystals imaged by AFM. Source: Reproduced with permission from Mouras et al. (2011).

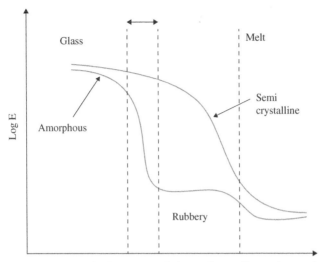

FIGURE 3.7   The change in elastic modulus E of a semicrystalline and amorphous polymers with the temperature.

present on a small fraction repeat units on the chain, but still, *all* these molecules are identified as "polyethylene."

Branch points get incorporated into the long polymer chains in the free radical polymerization process itself. But it can be controlled by using branched monomers in the reaction mixture. Since the extent and nature of chain branching influences the properties of plastic materials, it is closely controlled during manufacture to yield different grades of the same polymer. PE, for instance, is available in different basic grades, such as LDPE, HDPE, medium-density polyethylene (MDPE), and linear low-density polyethylene (LLDPE) with very different properties and applications. The differences in chain branching and the consequent difference in density characterize different grades of the same polymer. As branching interferes with close packing of polymer chains, higher levels of branching reduce the percent crystallinity and decrease the density. Figure 3.8d shows an illustration of a branched block copolymer chain.

Individual polymer chains in an ensemble can also be covalently joined to other chains around it at discrete points along it. This yields a 3D network of chains (or open-tree structures of chains or a mix of both). Cross-linking is desirable where insolubility and high mechanical strength are demanded of a plastic. Ideally, each and every chain will be linked to each other so that the entire ensemble of chains is a single giant molecule (this actually does occur in natural rubber when vulcanized or cross-linked.) An automobile or aircraft tire[8] is an example of a fully cross-linked polymer. On heating, cross-linked polymers do not convert into a viscous liquid "melt" as the molecules are chemically linked to one another and cannot flow independently.

---

[8] Aircraft tires as well as those for off-road vehicles are still made of 100% natural rubber because of the high heat and abrasion they need to endure. Automotive tires are made of synthetic as well as natural rubber.

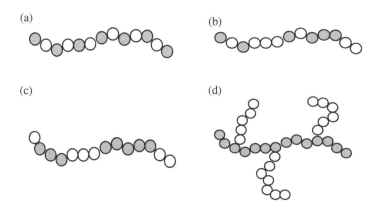

FIGURE 3.8 Illustration of different types of copolymers. Sections of polymer chains are shown and each circle represents a repeat unit. (a) Alternating copolymer, (b) random copolymer, (c) block copolymer, and (d) branched block copolymer.

From an engineering perspective, plastics fall into two broad categories: *thermoplastics* and the *thermosets*. Thermoplastics refer to those plastics that soften and flow on heating, allowing them to be molded or formed into different shapes. The change on heating is a physical change (as with heating candle wax), and the chemical nature of the polymers remains unaltered. Thermoplastics can therefore be recycled as they can be melted and reformed into different products (Subramanian, 2000). Thermoset plastics on the other hand are cross-linked polymers such as vulcanized rubber, polyurethanes, glass-reinforced polyester, or epoxy resins that do not melt or flow on heating and cannot therefore be remolded into a different shape. When heated to high temperatures, the material simply degrades chemically into small molecular products.

3.2.5 Glass Transition Temperature

Another interesting phenomenon that is unique to polymer materials is the phenomenon of glass transition. In amorphous glassy plastics, the chains are frozen in place and unable to move around (translational movement) within the bulk of the material at ambient temperatures. On gradual heating, the thermal or vibrational energy absorbed by the molecules results in increased motion of sections (short segments that are only 6–10 carbons long) of the polymer chains. With segments of this length moving or flexing as a single unit, additional free volume is created in the bulk, and the polymer characteristics abruptly change. The temperature at which this change occurs is the glass temperature[9] Tg of the polymer. As the name suggests, the hard glass-like microstructure of the plastic is converted into a softer rubbery material

---

[9] Though Tg is usually referred to as the "glass transition temperature," no thermodynamic transition such as melting occurs at this temperature.

(it is also called glass–rubber transition) at this temperature. Other modes of mobility of even shorter segments of the chains or just the side chains can occur at specific lower (sub-Tg) temperatures in the amorphous glassy state.

The Tg of the polymer is not a characteristic property of the material but depends on the rate of heating during its measurement. With polymers that are partially crystalline, this transition can only occur in the amorphous fraction, leaving the crystallites virtually untouched. As partial crystallinity, cross-linking of the polymer chains and tacticity of polymer all affect the freedom of the chain segments to move in response to heating, these also increase the value of Tg. Cross-linking, for instance, restricts such motion increasing the Tg of the polymer, while the higher concentration of chain ends (shorter average chain lengths) with decreasing average molecular weight or the presence of certain low-molecular-weight compounds (plasticizers) increases free volume, plasticizing the polymer and reducing the Tg. Below its Tg, the polymer is either a glass or a glass embedded with crystallites; these are hard and brittle (low impact strength) solids with high strength and modulus. While above Tg, the same polymer will turn into rubbery and tough materials with good elasticity.

We generally observe and use polymers at ambient temperatures; our perception of their being glassy or rubbery will depend on if they happen to be above or below their Tg at the ambient temperature of observation. For instance, PS has a Tg of approximately 100°C and is a hard "plastic" material at ambient temperature T °C ($T < Tg$). The Tg of natural rubber cis-polyisoprene is −70°C (and $T > Tg$), which makes the polymer have a soft rubbery feel at ambient temperature, which is ~90°C above its Tg. If the temperature of a rubber is decreased below its Tg (e.g., immerse a rubber ball in liquid nitrogen), it will behave like a glass and even shatter on being dropped on the floor. Similarly, PS, a hard plastic, heated above its Tg of 100°C will yield a soft rubbery material. Values of Tg for common plastics given in Table 3.1 illustrate that these polymers have values well above the ambient temperatures. These values depend on average molecular weight, polydispersity, and the heating rate used to determine it.

Processing generally involves heating the polymer well above the Tg, and on heating, a plastic material will transition through three phases distinguished by very different moduli (or stiffness; see Fig. 3.7). The high modulus of the polymer in the glassy state changes somewhat abruptly when it reaches the Tg, and a low-modulus "rubbery" plateau is reached. At even higher temperatures, the amorphous polymer becomes a "melt" or a viscous liquid that can be easily extruded or molded. The modulus of the glassy polymer decreases by several orders of magnitude during the heating process.

3.3 SYNTHESIS OF POLYMERS

Two broad classes of polymers are based on the two types of polymerization reactions: addition polymerization and condensation polymerization of monomers. Most high-volume thermoplastics such as PP, PE, PS, and PVC are addition polymers. Those such as nylons or PC are examples of condensation polymers.

TABLE 3.1   Glass Transition Temperature of Common Plastics

Polymer	Tg (°C)	Applications
Low-density polyethylene (LDPE)	−100	Plastic film, bags, and containers
High-density polyethylene (HDPE)	−100	Bottles, milk jugs, toys
Polypropylene (PP)	−20	Bottle caps, tape, film, fiber
Poly(ethylene terephthalate) (PET)	70	Soda bottles, plastic film
Polycarbonate (bisphenol A polycarbonate) (PC)	144	Bottles, containers, and glazing
Poly(vinyl chloride) (PVC)	80	Siding, pipes, flooring
Polystyrene (PS)	95	Foam cups and food service items (EPS), cutlery, containers
Poly(tetrafluoroethylene) (PTFE)	115	Nonstick surfaces
Poly(vinylidene chloride) (PVDC)	−35	Saran wrap
Polyacrylonitrile (PAN)		Textile fibers
Poly(methylmethacrylate) (PMMA)	102–117	Glazing in skylights, windows, and picture frames. Used also in lighting fixtures
Poly(vinylacetate) (PVAc)	30	Chewing gum and bubble gum base, adhesives
Polyamide (nylon 6) (PA)	40	Fishing gear and fiber

3.3.1   Addition or Chain Growth Reaction

Most monomers are either stable gases or liquids; but when mixed with an initiator compound, they yield active radicals or an ion. Organic peroxides [R–O–O–R′], hydroperoxides [R–O–O–H], and azo compounds [R–N=N–R′] are generally efficient initiators. The radicals react readily with unsaturated monomers to yield a monomer radical. Addition or chain growth polymerization of unsaturated monomers (with a double or a triple bond in the structure) progresses by adding on repeat units sequentially at the end of the growing radical. The initiation reaction is the formation of free radical species **R•** and it propagates via sequential addition of more monomer to create a growing chain (hence the name). The DP of the chain increases with each addition of an unsaturated monomer M:

$$(M\bullet) + M \rightarrow M - M\bullet$$

There are no by-products of this reaction as addition is a mere rearrangement of atoms. Eventually, the macroradials terminate by a combination of a pair of these long-chain radicals. The free radicals in reaction mixture can react not only with a monomer but also with the polymer chain. In this case, it abstracts a hydrogen from the chain creating a chain radical. Polymerization initiates at this point too and gives rise to branches, either short or long branches depending on how far removed the branch point is from the growing chain end (Mishra and Yagc, 2009). Common addition polymers and their structures are listed in Table 3.2.

TABLE 3.2   Structure of Common Addition Polymers and their Applications

Polymer	Repeat unit	Applications
Low-density polyethylene (LDPE)	$-CH_2-CH_2-$	Plastic film and bags
High-density polyethylene (HDPE)	$-CH_2-CH_2-$	Bottles, milk jugs, toys
Polypropylene (PP)	$-CH(CH_3)-CH_2-$	Caps, tape, film, fiber
Poly(vinyl chloride) (PVC)	$-CH(Cl)-CH_2-$	Siding, pipes, flooring
Polystyrene (PS)	$-CH(C_6H5)-CH_2-$	Foam cups, cutlery
Poly(tetrafluoroethylene) (PTFE)	$-CF_2-CF_2-$	Nonstick surfaces
Poly(vinylidene chloride) (PVDC)	$-CH_2-CCl_2-$	Saran wrap
Polyacrylonitrile (PAN)	$-CH_2-CH(CN)-$	Textile fibers
Poly(methylmethacrylate) (PMMA)	$-CH_2-C(CH_3)(COOCH_3)$	Glazing in skylights, windows, and picture frames. Used also in lighting fixtures
Poly(vinyl acetate) (PVAc)	$-CH_2-CH(COOCH_3)$	Chewing gum and bubble gum base, adhesives

The sequence of reactions is illustrated below in the polymerization of ethylene:
Initiation of reaction by free radical formation

$$R-O-O-R \rightarrow 2Ra \bullet \text{(free radicals)}$$

Propagation of radical chain

$$Ra \bullet + CH_2 = CH_2 \rightarrow RaCH_2CH_2 \bullet$$
$$RaCH_2CH_2 \bullet + CH_2 = CH_2 \rightarrow RaCH_2CH_2CH_2CH_2 \bullet$$

Termination by combination of macromolecular radicals

$$Ra(CH_2)_m \bullet + \bullet (CH_2)_n Ra \rightarrow Ra(CH_2)_m (CH_2)_n Ra$$

The overall reaction is summarized as follows:

$$nCH_2 = CH_2 \rightarrow \left[-CH_2 - CH_2 - \right]_n$$

3.3.2   Condensation or Step Growth Reaction

Condensation reaction is one where two monomers or short functional macromolecule segments combine together to form a long segment, usually with the elimination of a small molecule such as water or alcohol. For this to take place, each reacting segment must have chemically reactive end groups. The reaction of a pair of such difunctional molecules or oligomers yields a polymer by step growth process and is illustrated schematically below:

$$x - A_n - x + y - B_m - y \cdots x - (A_n + B_m) - y + xy$$

The reaction between a diamine and a dicarboxylic acid to yield a polyamide (or a nylon) is an example of a polycondensation. The basic reaction is shown below. The reaction does not stop here as the product too has the same functional groups and can therefore react with other products to yield chains of different lengths. In this type of reaction, all molecular species such as monomers, dimers, trimers, and longer chains can react concurrently, yielding chains with high enough **Mw** values that would make the polymer practically useful:

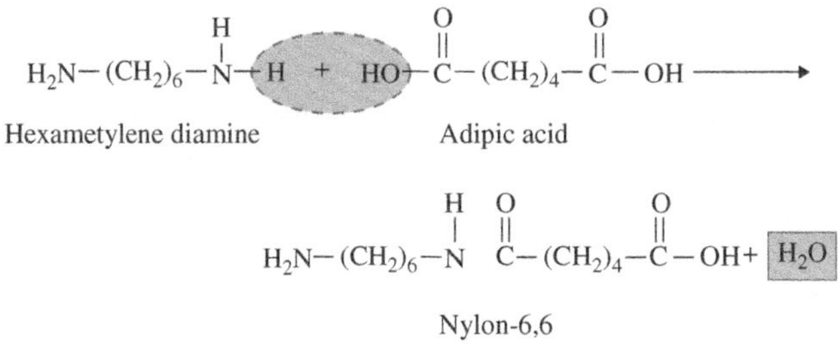

Nylon-6,6

Since these are equilibrium reactions, the water or other small molecular product needs to be removed from the mixture to encourage the reaction to proceed in the forward direction. Another example is the condensation of bisphenol A monomer with phosgene to yield PC plastic:

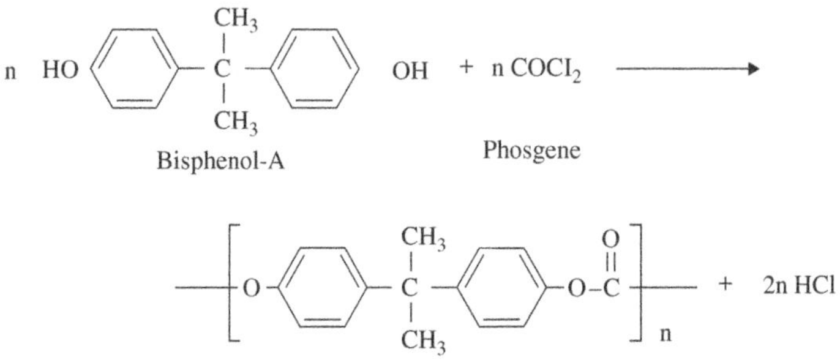

Polycarbonate

High molecular weights are obtained in condensation polymerization but only at very high percent conversions of monomer or where nearly all the functional groups are reacted. High conversion is encouraged by carefully matching the stoichiometry of end groups. A variant of this reaction is when both reactive functional groups are located on the same molecule (such as in the case of hydroxy acids). If one or both of the monomers have more than two functional groups per molecule, then condensation reaction will yield a branched polymer.[10]

---

[10] Phenol-formaldehyde resin, the first all-synthetic plastic (trade name Bakelite) made in 1907, was made via a polycondensation reaction.

Coupling reactions between monomers can also occur without the loss of a small molecule such as water to yield a step growth polymer. For instance, a diisocyanate can couple with a dialcohol to yield a polyurethane as shown below. The $-\overset{O}{\underset{H}{\overset{||}{C}}}-\overset{}{\underset{H}{N}}-$ linkage formed is called a "urethane" group, hence the name polyurethane:

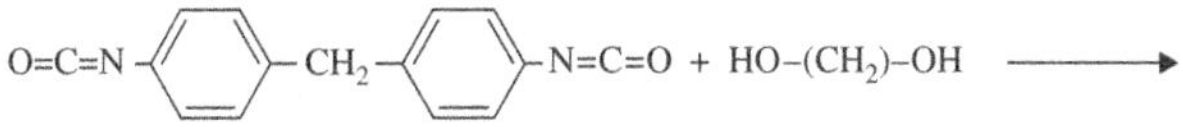

Methylene Diphenyl Diisocyanate   Ethyleneglycol

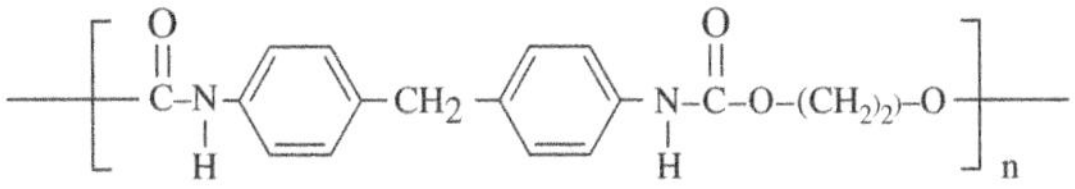

Polyurethane

Common condensation polymers and their structures are listed in Table 3.3.

TABLE 3.3   Some Common Condensation Polymers

Polymer	Repeat unit
Poly(ethylene terephthalate) (PET)	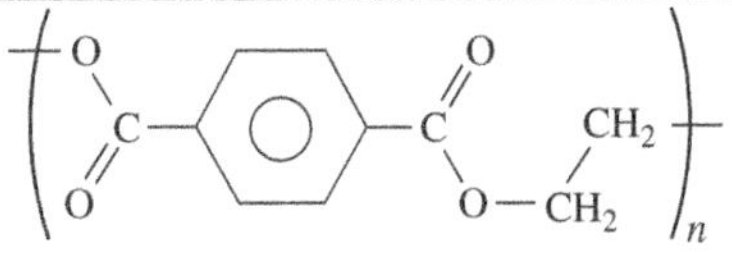
Polycarbonate (bisphenol A polycarbonate) (PC)	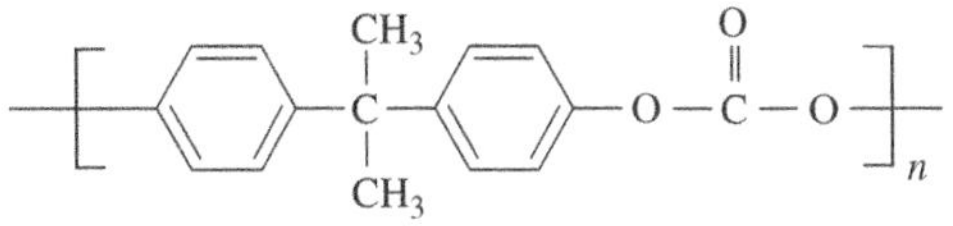
Polyamide (nylon 6) (PA)	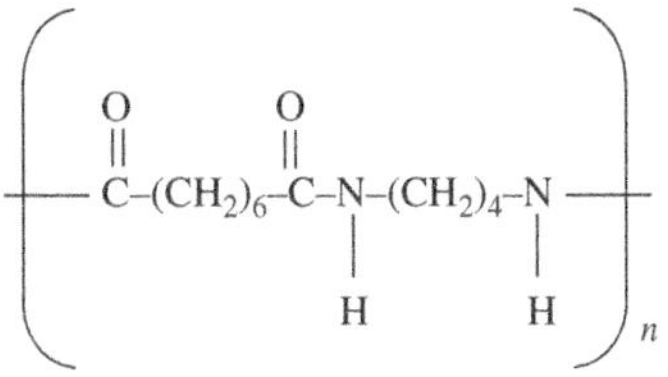
Polyurethane rubber (PU)	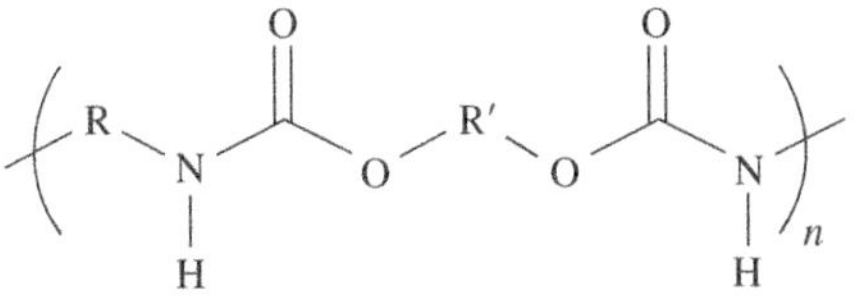
Kevlar	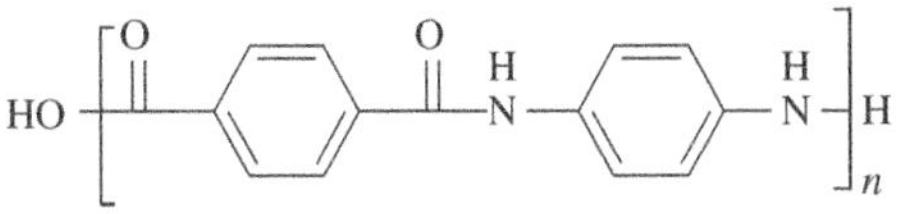

3.3.3 Copolymers

A copolymer is made by polymerizing a mixture of two or more monomers and yields a polymer that incorporates the repeat units of all the monomers used. With a pair of monomers, the ratio of the two repeat units in the polymer is determined by the ratio of the monomers in the feed. As the properties of the polymers are highly dependent on repeat unit composition, the monomer feed unit ratio is closely controlled during reaction. In some instances (such as with ethylene copolymerized with heptene or octene), these can add on branches on the PE chains. This is in fact the case in the synthesis of LLDPE.

Copolymerization allows a convenient means of modifying properties of a homopolymer to suit specific applications. For example, PS is a brittle thermoplastic with poor gas transport properties (important in food packaging applications). Provided the slight loss in transparency can be tolerated, a copolymer of butadiene–styrene with large improvement in both the barrier properties and impact resistance can be used in its place. This grade of PS called high-impact polystyrene (HIPS) has more desirable properties including high-impact resistance compared to the general-purpose polystyrene (GPPS) grade. Acrylonitrile used as a comonomer also yields a copolymer with similarly improved properties.

Whenever two or more monomers are polymerized together, their arrangement along the polymer chain also influences the properties of the polymer. Where the pair of monomers A and B are involved, easiest to envision is a random distribution of the repeat units A and B along the chain. However, it is possible to obtain instead an alternating copolymer with the pair of repeat units A and B arranged alternatively: $(-A-B-A-B-A-B-)_n$. Depending on the monomer ratio and catalysts used in the reaction, it is also possible to have block copolymers with short homopolymer sequences or "blocks" of a single repeat unit on the polymer chain such as $(-A-B-A-A-A-A-A-A-B-B-B-A-B-A-B-)_n$. Blocks often aggregate into domains within the bulk of the polymer to yield various 3D morphologies with consequent interesting and useful properties. Figure 3.8 illustrates the structure of different copolymers.

3.4 TESTING OF POLYMERS

A wide range of plastics test methods are available and routinely used to characterize plastics and rubber. Even a basic introduction to the common tests used is beyond the scope of this chapter. The following introductory discussion is on several selected test methods that will allow the reader to better understand the test data presented in other chapters. With most of these tests, there are an extensive theoretical foundation and numerous practical details on how to carry them out as well as the relevant standards that have been left out. With most tests described here, the experience of the operator in sample preparation, carrying out the tests, and analysis of data is critical to obtain reproducible results.

3.4.1 Tensile Properties

When a plastic film, pipe, or rod is extended along its long axis, it is placed under tensile stress. Plastic films in shopping bags, plastic ropes, tape, and fishing gear routinely undergo this type of deformation in use. Therefore, it is important to test how the plastic behaves under such stress. Tensile testing is designed to provide this information.

In a typical tensile test, the plastic sample in the shape of a rectangular (or dog-bone) piece is held at its extremities in a pair of grips and slowly pulled along its long axis. The deformation (or the increase in length of the sample) and the force F applied are recorded continuously. The strain increases at a prescribed rate during the tensile test. Basic quantities involved in the test are defined as follows (Fig. 3.9):

$$f = \frac{F}{a}$$

$$\varepsilon = \frac{\Delta\ell}{\ell_o} = \frac{(\ell_t - \ell_o)}{\ell_o}$$

where f is stress, F is force, and a is the area of cross section of the plastic sample prior to deformation. The extension ε is defined as the change in length $\Delta\ell$ of the sample to the initial length ℓ_o. Lengths are measured as the distance between the grips. Both f and ε vary continuously during deformation, and the test is completed when the sample snaps. When plastics deform by extension along one axis, its width and thickness decrease as shown in the figure.

The behavior of the material in the test is best understood in terms of the stress–strain curve. Figure 3.10 shows typical stress–strain curves for a dog-bone-shaped (a standard ASTM shape) test piece of a series of glass-bead filled LDPE test pieces

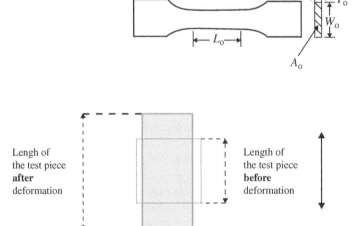

FIGURE 3.9 Upper: Standard dog-bone-shaped test piece used in tensile tests. Lower: Tensile deformation of a rectangular test piece. Notice shrinking of the width. Direction of strain shown by the *double-headed arrow at right*.

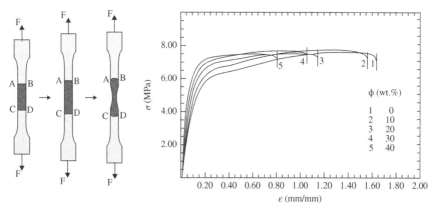

FIGURE 3.10 *Left*: Change in shape of the dog-bone test piece. Source: Reprinted with permission from Coppieters et al. (2011). *Right*: Tensile stress–strain curves for glass bead-filled LDPE at different volume fractions of beads. Source: Reprinted with permission from Liang et al. (1998).

and a schematic for a plastic test-piece undergoing tensile strain illustrating the slow development of a 'neck' region. The value of f at the point of failure is called the tensile strength (MPa), and the strain ε (%) at failure is usually expressed as a percentage and is called the ultimate extensibility or the elongation at break. As the strain is increased, the sample extends elastically (or recoverably) initially. At a higher strain, a maximum force characterized by a maximum in the stress–strain curve called the "yield point" occurs; the corresponding stress is called the yield stress. At or about this strain, the sample develops a "neck" and continues to extend by plastic deformation. The width of the test piece under strain is drastically reduced, and the stress does not increase much on further strain, until it finally breaks. The breaking extension is the "ultimate extensibility (%)," and the corresponding stress is the "tensile strength" (MPa). Typical stress–strain curves shown in Figure 3.10 illustrate the initial linear part of the curve and the stiffening effect of glass bead fillers at different weight fractions on the properties of LDPE.

3.4.2 Thermal Properties: DSC (Differential Scanning Calorimetry)

The range of physical and chemical changes that occur on heating polymers can be monitored using DSC and thermogravimetry (TGA). These thermal techniques are commonly used in plastics industry to characterize plastics.

DSC is generally used to obtain information on properties including (i) the Tg of polymers, (ii) percent crystallinity, (iii) thermal stability or degradation on heating, (iv) cross-linking or thermal curing, and (v) the purity of materials. The main features of a DSC instrument are shown in Figure 3.11. Essentially, it consists of two identical pans (each equipped with sensitive temperature sensors), enclosed in an oven. This arrangement allows the pans to be slowly heated over a range of temperature and the heat flow to each pan at a given temperature to be monitored

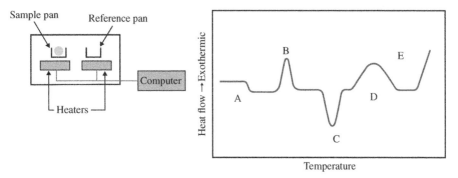

FIGURE 3.11 *Left*: Basic features of a DSC instrument. *Right*: A generalized DSC tracing.

separately. The plastic sample to be analyzed is placed in one of the pans; the difference in heat flow needed to maintain both pans at the same temperature is accurately recorded during the heating process. It is essentially a measurement of the heat capacity of the sample as a function of temperature. The value of ΔT (°C) plotted as a function of temperature T (°C) yields a DSC trace that reflects the thermal characteristics of the material. A generalized DSC trace of a semicrystalline polymer is shown in Figure 3.11 showing the features associated with changes on heating the plastic sample.

The DSC curve shows the crystallization exotherm (B) and a melting endotherm (C) followed by an exothermic decomposition of the plastic (E) at higher temperatures (Bruns and Ezekoye, 2014). At a lower temperature, a first-order step change in the curve (A) is apparent and signifies the Tg of the material (Feng et al., 2013), and in some instances, a transition that signifies cross-linking (D) is also observed.

The melting transitions for blends of PP are shown in Figure 3.12. The area under the DSC tracing of differential heat flow (dH/dT) is the enthalpy change associated with the transitions. This is the sum of heat capacity and the heat flow needed to accommodate the physical/chemical changes undergone by the plastic material:

$$\left\{ \frac{dH}{dt} \right\}_P = C_P \left\{ \frac{dT}{dt} \right\} + f(T,t)$$

The first term on the right-hand side is the change in C_P and the second is the temperature-dependent kinetic response of the plastic. It is the second term that yields peaks (positive or negative) on the baseline of the specific heat C_P versus T tracing in a DSC. Area under these peaks and the temperatures at which they occur can yield information on changes in the polymer responsible for these. The area under the melting peak (J/mol) might be normalized against that for a 100% crystalline form of the polymer (usually estimated) to estimate a value for the percent crystallinity of the sample (D'Aniello et al., 2000). The area of interest is shown in the DSC tracing of PP blends in Figure 3.12.

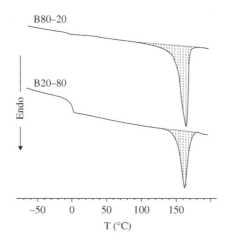

FIGURE 3.12 DSC tracings of two blends of atactic and isotactic PP showing the area under the melting curve. The designations indicate the weight fraction of isotactic and atactic PP in the blend. Source: Reprinted with permission from D'Aniello et al. (2000).

3.4.3 Thermal Properties: TGA

TGA involves accurate monitoring of the mass of a sample as a function of either the temperature, or time at a constant temperature in the isothermal mode. As the sample dries and disintegrates, its mass is reduced. The technique allows the measurement of volatiles and/or the fraction of organic or inorganic filler present in a sample (Hatakeyema et al., 2005). The technique is therefore used for quality control and characterization of compounds that contain known amounts of fillers. It is also used to study the stability of compounded plastics; chemical changes that evolve a product, change the mass of sample that can be monitored as a function of time. A balance with a precision of 0.01–0.001% is used in recent models of the equipment to record the weight change. The temperature can be raised at selected rates up to 1500°C and the samples can be heated in an inert atmosphere, a specific gas or in air. Resolution of the data depends on the scan rate (typically 10–20°C/min rates are used).

3.5 COMMON PLASTICS

Different fossil fuel-derived feedstock used in the manufacture of common plastics are as follows:

Ethylene: PE, PP, PVC, PS, PET, and ABS
Propylene: PP, PS, ABS, PC, and nylons
Benzene: PS, ABS, PC, and nylon
Butadiene: PS, ABS, and nylon
Xylene: PET

3.5.1 Polyethylenes

In the US, ethylene monomer derived from natural gas is primarily used to produce PE though other fossil fuels may also be used for the purpose. The annual world production of PE is approximately 100 MMT.

Polymerizing ethylene in high-pressure (~100,000–350,000 kPa) reactors yields an extensively branched variety of PE, the LDPE; branching is responsible for its characteristic lower density. Essentially, the same polymerization reaction can also be carried out at a considerably lower pressure (~2000 kPa) when Ziegler–Natta catalysts or the more recently developed metallocene catalysts are used. The former catalysts have multiple sites that restrict linking of monomer molecules to a specific regular orientation during polymerization, leading to broad molecular weight distributions, while the latter catalysts yield PE with a relatively narrow distribution. The low-pressure processes yield HDPE with very low levels of chain branching and therefore higher crystallinity and relatively high density. Often, a commoner hexane is used to control branching and therefore the crystallinity of the product.

A third variety, LLDPE, is manufactured using specialized catalyst systems and have a similar molecular structure but with short-chain branching (as opposed to long-chain branching in LDPE). This variety is particularly strong and is often used in the manufacture of thin-walled plastic bags. Newer metallocene family of catalysts yields a grade of LLDPE with more homogeneous molecular structure compared to the regular variety; this mLLDPE can be used to down gauge packaging films as they can be tailored as a very strong and transluscent plastic.

About 40% of PE produced is blow-molded into bottles. Only approximately 15% is used as plastic film. The rest is injection molded or extruded (into pipe products). Both LDPE and LLDPE grades are used predominantly in film applications (~75%), in extrusion coating of paperboard products (<10%), and as cable coating with only a small fraction being injection molded into container. HDPE absorbs very little water and has better specific strength compared to LDPE. Therefore, it is popularly used in molding milk jugs and other containers for packaging liquids including chemicals. Table 3.4 compares the properties of the different grades of PE with PP.

No significant fugitive emission problem is associated with the manufacture of PE or PP, and in any event, the monomers (and the comonomers generally used) are relatively nontoxic. The only exception might be where solvents or diluents and catalysts are used in the manufacture.

TABLE 3.4 Characteristics of Common Classes of Polyethylenes

Property	HDPE	LDPE	LLDPE	VLDPE	PP
Density (g/cm³)	0.94–0.97	0.915–0.935	0.90–0.94	0.905–0.915	0.90
Percent crystallinity (%)[a]	55–77	30–54	22–55	0–22	30–50
Melting temperature (°C)	125–132	98–115	100–125	60–100	175
Shore hardness (type D)	66–73	44–50	55–70	35–55	80–102
Tensile extensibility (%)	10–1500	100–650	100–950	100–600	100–600

[a] Measured calorimetrically. Polyethylene data based on data from Peacock (2000).

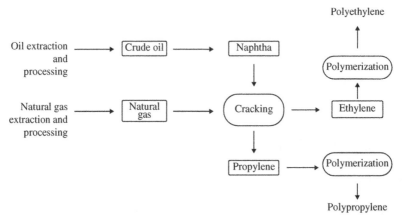

FIGURE 3.13   Flow chart illustrating the manufacture of polyethylenes and polypropylenes.

3.5.2 Polypropylenes

PP is the lightest of common thermoplastics (density $0.905\,g/cm^3$) and has a useful combination of properties that include transparency, high stiffness, good thermal resistance, chemical inertness, and good impact properties. General-purpose PP (with $T_g = -10°C$ and $T_m = 173°C$) is used in a wide range of products including packaging materials (pails, containers bottle caps and film) (~28%), textile fiber (~20%), molded products, housewares, toys, auto parts, and extruded pipe. The annual world production is approximately 52 MMT. Stereoregular grades of PP including syndiotactic PP made using metallocene catalysts, or the m-PP, are commercially available and gaining popularity in high-end applications. The low-pressure polymerization of propylene employs either Ziegler–Natta or metallocenes catalysts. Figure 3.13 shows the main steps involved in the manufacture of PE and PP.

3.5.3 Polystyrene

The highest volume use of PS (including expanded PS foam, EPS) is in food packaging and food service sectors (~37%). PS containers are popularly used for packaging single-serve food items such as yogurt. EPS dominate the market for hot beverage service items and soup bowls and carryout food containers. About 10–11 MMT of the resin is produced annually (2013 statistics), worldwide.

The monomer ethylbenzene is made from petroleum-derived benzene and ethylene. Free radical polymerization is usually carried out in the gas phase with an iron oxide or other metal oxide catalyst yielding PS resin:

$$H_3C - CH_2\ (g) \xrightarrow[\text{Fe}_2\text{O}_3\ \text{Catalyst}]{850\ K} H_2C = CH\ (g) + H_2(g)$$

Ethylbenzene   *Phenylenthene*

The *atactic* grade of PS is a transparent resin with good moisture resistance and moderate chemical resistance. Its brittleness, however, limits its commercial uses. Therefore, in addition to this general-purpose grade, a tougher grade of PS is made by polymerizing styrene with 5–10% poly(buta-1,3-diene) rubber dissolved in the monomer. This tougher product called High impact polystyrene (HIPS) is generally made exclusively by continuous thermal polymerization. Figure 3.14 shows the microstructure of HIPS and compares it with a transmission micrograph of a styrene–butadiene rubber (SBR) copolymer. Both show soft microdomains of rubber dispersed in the styrene matrix. This toughened low-cost plastic, however, is translucent due to light scattering by the rubber microdomains dispersed in the bulk polystyrene matrix. It is best used for structural applications that require low strength, good impact resistance and machinability. Using new metallocene catalysts, stereoregular (*syndiotactic*) grades of polystyrene can be synthesized. This syndiotactic polymer has improved thermal and mechanical properties. Properties of different grades of PS are shown in Table 3.5.

The variety of PS that is ubiquitous is expanded polystyrene (EPS) or PS foam commonly used in cups and food service applications. The expanded grade is manufactured as beads of PS containing pentane, a liquid hydrocarbon at room temperature.

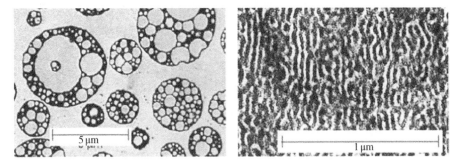

FIGURE 3.14 *Left*: An electron micrograph of a thin section of HIPS showing the rubber microdomains. *Right*: An electron micrograph of a thin section of SBR copolymer. Source: Courtesy of Styrolutions, Styrolux, BASF.

TABLE 3.5 Properties of Different Grades of Polystyrene

	GPSS	HIPS[a]	SPS[b]
Specific gravity	1.04	1.04	1.02
Melt flow (g/10 min)	1–20	2–15	
Tensile strength (psi)	5–7	3–7	1.04
Flexural strengths (psi) × 1000	17–15.0	2–8	11.6
Tensile elongation to break (%)	2–3	15–65	30
Flexural modulus (psi) × 1000	450–500	270–420	465
Vicat softening temperature (°F)	195–228	185–225	

[a] Properties depend on rubber content.
[b] SPS is syndiotactic polystyrene.

When the beads are heated in steam (or "expanded"), the pentane volatilizes to expand the beads into a thermoplastic foam. These foam beads are subsequently packed into molds and fused by heat into the shape desired. The EPS can also be extruded into sheets and thermoformed into products such as meat trays and plates. EPS has good thermal insulation and shock-absorbing properties.

Styrene is copolymerized with butadiene to obtain SBR rubber and with acrylonitrile–butadiene to obtain ABS plastics. ABS is a relatively tougher thermoplastic compared to HIPS. The composition of the latter is typically 60% (w/w) styrene, 25% acrylonitrile, and 15% buta-1,3-diene.

3.5.4 Poly(vinyl chloride)

PVC is a tough and rigid thermoplastic that is the most-used plastic in the United States after PE and PP. It is also the most-used plastic in building applications (extruded profile (~17%), pipe (~24%), and residential siding are popular uses in the US). Plasticized PVC is used in packaging applications, medical devices, and wire and cable insulation and conduit. PVC is a versatile plastic; the polarity of the repeat unit allows a very wide range of additives to be incorporated at very high levels into the polymer. This allows a variety of plastic compounds with different characteristics to be designed. Properly compounded, PVC products are well suited for outdoor applications stabilized.

Ethylene monomer (from natural gas or other petroleum resources) is chlorinated in the presence of an iron salt catalyst to obtain the monomer:

$$\underset{\text{Ethene}}{CH_2 = CH_2(g)} + \underset{\text{Chlorine}}{Cl_2(g)} \xrightarrow{\text{FeCl}_3\text{catalyst}} \underset{\text{1,2 Dichloroethane}}{ClCH_2CH_2Cl(g)}$$

$$\Delta H^\varphi = -178 \text{ kJ / mol}$$

Alternatively, ethylene can be oxychlorinated with HCl in the presence of cupric chloride ($CuCl_2$) impregnated on a porous support such as alumina. Both reactions are exothermic and involve the use of corrosive and toxic gases:

$$CH_2 = CH_2(g) + 2HCl(g) + \tfrac{1}{2}O_2(g) \rightarrow ClCH_2CH_2Cl(g) + H_2O$$

Heating the dichloroethane yields the vinyl chloride monomer with HCl generated as a by-product. Where oxychlorination is used to manufacture the precursor, some of this HCl can be used on-site for the previous reaction. Stripping the HCl as well as unreacted dichloroethane yields a polymerizable grade of monomer:

$$\underset{\text{1,2 Dichloroethane}}{ClCH_2 - CH_2Cl(g)} \xrightarrow{650 \text{ K}} \underset{\text{Chloroethene}}{CH_2 = CHCl(g)} + HCl(g)$$

The monomer is polymerized in aqueous dispersion at 325–350°C at high pressure (~13 atm) to maintain the monomer in liquid phase. In the manufacture of PVC, there

TABLE 3.6 A Comparison of Properties For Polyolefins

Property	Rigid PVC	Soft PVC*	PS	PET
Specific gravity	1.4	1.22	1.04	0.959–0.965
Crystallinity (%)	~10–30	~6		
Tensile strength (psi) × 1000	6–7	1.5–3.5	5–7	5–6
Tensile modulus (psi) × 1000	165–225		25–40	150–160
Extensibility (%)	~15	200–450	2–3	11–1300
Water absorption (% in 24 h)	0.06	–	0.08	0.10

*Plasticized PVC or flexible PVC.

is potential for environmental hazards, and the process has to be therefore carefully controlled.[11] PVC has been under intense attack by environmental groups for decades. The risk[12] of volatile emissions as well as the invariable need to handle and transport large quantities of chlorine to manufacturing plants has contributed to the negative image of PVC. Vinyl chloride monomer is carcinogenic and its release into air is a potential hazard. If the temperature is not carefully controlled, the side reactions that yield by-products such as dioxins can occur. The emission load of polychlorinated dibenzo-p-dioxins and polychlorinated dibenzofurans from the manufacturing facilities to the open environment in the United States was estimated at 12 g International Toxicity Equivalents (Carroll et al., 2001) per year. Also, the chlorine production for PVC manufacture is a particularly energy-intensive operation.[13] There are 17 PVC production facilities in the United States, and these are regulated by the USEPA as to the maximum allowable emissions. Table 3.6 summarizes the properties of PVC in comparison to those of PS and PET.

REFERENCES

Bruns, M.C. and O.A. Ezekoye (2014) Modeling differential scanning calorimetry of thermally degrading thermoplastics. J Anal Appl Pyrol 105, 241–251.

Carroll WF Jr, Berger TC, Borrelli FE, Garrity PJ, Jacobs RA, Ledvina J, Lewis JW, McCreedy RL, Smith TP, Tuhovak DR, Weston AF. Characterization of emissions of dioxins and furans from ethylene dichloride, vinyl chloride monomer and polyvinyl chloride facilities in the United States. Consolidated report. Chemosphere 2001;43:689–700.

[11] In February 2012, the USEPA issued a final rule to update emissions limits for air toxics from PVC production, requiring reduced emissions of harmful toxic air emissions. This rule sets maximum achievable control technology standards for major sources and generally available control technology.

[12] In March 2012, a PVC plant in Geismar, LA, exploded releasing 2500 lbs of HCl, 632 lbs of chlorine, and 239 lbs of VC monomer into the air. In 2004, another PVC plant in Illiopolis, IL, also exploded (four workers were killed). Explosions have been reported in plants in Siberia, Russia (2007), and Northern China (2010), as well.

[13] In a 1993 study, SRI International estimated the global energy cost of about 3000 kW-h/ton of chlorine produced.

Coleman MM, Painter PC. *Fundamentals of Polymer Science: An Introductory Text*. Boca Raton: CRC Press; 1998.

Coppieters S, Cooreman S, Sol H, Van Houtte P, Debruyne D. Identification of the post-necking hardening behaviour of sheet metal by comparison of the internal and external work in the necking zone. J Mater Process Technol 2011;211 (30):545–552.

D'Aniello C, Guadagno L, Gorrasi G, Vittoria V. Influence of the crystallinity on the transport properties of isotactic polypropylene. Polymer 2000;41 (7):2515–2519.

Feng L, Bian X, Li G, Chen Z, Cui Y, Chen X. Determination of ultra-low glass transition temperature via differential scanning calorimetry. Polymer Test 2013;32 (8):1368–1372.

Flory PJ. On the morphology of the crystalline state in polymers. J Am Chem Soc 1962;84 (15):2857–2867.

Hatakeyema H, Tanamachi N, Matsumura H, Hirose S, Hatakeyama T. Bio-based poly-urethane composite foams with inorganic fillers studied by thermogravimetry. Thermochim Acta 2005;431 (1–2):155–160.

Kawakatsu, T. (2004) *Statistical Physics of Polymers: An Introduction*. Heidelberg: Springer.

Kiriy A, Gorodyska G, Minko S, Jaeger W, Štěpánek P, Stamm M. Cascade of coil-globule conformational transitions of single flexible polyelectrolyte molecules in poor solvent. J Am Chem Soc 2002;124 (45):13454–13462.

Liang JZ, Li RKY, Tjong SC. Morphology and tensile properties of glass bead filled low density polyethylene composites: material properties. Polymer Test 1998;16 (6):529–548.

Mouras R, Kilgour A, Elfick A. On the nanotribology of polyethylene single crystals. Wear 2011;270 (9–10):622–627.

Painter PC, Coleman MM. *Fundamentals of Polymer Science: An Introductory Text*. 2nd ed. Boca Raton: CRC Press; 1997.

Peacock A. *Handbook of Polyethylene: Structures: Properties, and Applications*. Boca Raton: CRC Press; 2000.

Mishra M, Yagc Y. *Handbook of Vinyl Polymers: Radical Polymerization, Process, and Technology*. 2nd ed. Boca Raton: CRC Press; 2009.

Rubinstein M, Colby RH. *Polymer Physics*. Oxford: Oxford University Press; 2003.

Rudin A, Choi P. *The Elements of Polymer Science & Engineering*. Watham: Academic Press; 2012.

Smith P, Lemstra PJ, Pijpers JPL. Tensile strength of highly oriented polyethylene. II Effect of molecular weight distribution. J Polymer Sci Polym Phys Ed 1982;20 (12):2229–2241.

Stevens PS. *Polymer Chemistry: An Introduction*. 3rd ed. New York: Oxford Press; 1999. p 234–235.

Strobl GR. *The Physics of Polymers: Concepts for Understanding Their Structures and Behavior*. Heidelberg: Springer; 2007.

Strobl GR, Schneider MJ, Voigt-Martin IG. Model of partial crystallization and melting derived from small-angle X-ray scattering and electron microscopic studies on low-density polyethylene. J Polymer Sci B 1980;18 (6):1361–1381.

Subramanian PM. Plastics recycling and waste management in the US. Resour Conserv Recycl 2000;28 (3–4):253–263.

Chang, L. (2011) Tacticity in vinyl polymers. In: Woo EM. Encyclopedia of Polymer Science. doi:10.1002/0471440264.pst363.

Young RJ, Lovell PA. *Introduction to Polymers*. Boca Raton: CRC Press; 2011.

4

PLASTIC PRODUCTS

There are over 80,000 different plastic compounds or mixed plastic formulations available commercially. These are based on about 20 chemically distinct classes of polymers,[1] such as polyethylenes (PE), polystyrenes (PS), and polypropylenes (PP). Most of the common plastics produced in high volume, called commodity plastics, are based on only about 5–6 of such classes. A single class of polymer such as PP, however, is available in a wide range of different *grades* with identical repeat-unit chemistry but with widely different molecular weights, chain branching, degrees of crystallinity, stereoregularity, and specific gravity.

The average molecular weight of the polymer is the single most important metric in specifying plastics for a given application. For instance, PE with an average molecular weight[2] of $\mathbf{Mn} \sim 1000$ g/mol with an average of only 40 repeat units per chain molecule is a soft waxy material. But ultrahigh-molecular-weight polyethylenes (UHMWPE) with $\mathbf{Mn} \sim 2$–5 million g/mol are very strong materials[3] used as gel-spun fibers in bulletproof vests and helmets. Yet, both are polyethylenes with the same repeat units. PE with the same $\mathbf{Mn}$ but with different molecular weight

---

[1] We use the term "polymers" and "plastics" interchangeably in this discussion for convenience. Strictly speaking, plastics are a subset of the larger group of chemical compounds called polymers. Polymers include all long-chain molecules with repeat units in their chemical structure and include rubbery materials as well as fibers that are not considered plastics.

[2] The average value of the molecular weight is denoted by the bold letters ($\mathbf{Mn}$ or $\mathbf{Mw}$).

[3] The tensile strength of UHMWPE fiber (Spectra 1000) is in the range of 2.9–3.7 GPa. This compares well with that of high-carbon steel (1.2 GPa). Multiwalled carbon nanotubes, however, have even higher tensile strength of over 60 GPa!

Plastics and Environmental Sustainability, First Edition. Anthony L. Andrady.
© 2015 John Wiley & Sons, Inc. Published 2015 by John Wiley & Sons, Inc.

TABLE 4.1   Choices Available in Selecting a Polystyrene Resin in the US Market[a]

All polymers		Choices
Select polymer class	LDPE, HDPE, PP, PC, **PS**, PMMA, SAN, etc.	1 class
Select grade of PS	GPPS, **HIPS**,[b] EPS, etc.	7 grades
Select manufacturer of HIPS	Availability of resin in US marketplace	~65 manufacturers
Select a resin types within HIPS	High-impact-grade (HIPS) resin choices	~400 types (all manufacturers)
Preformulated compounds of selected HIPS	Polystyrene compound ready to be processed into products	~700

[a]Numbers based on data from IDES database (UL IDES Prospector Materials Database).
[b]High-impact grade of PS.

distributions or polydispersity indices (PI) can have very different properties. In highly oriented fibers of PE, the tensile strength doubled (Smith et al., 1982) when the PI was reduced from 8 to 1.1 at the constant $Mw \sim 10^5$ g/mol. The same grade of the same variety of plastic resin is manufactured by a host of different companies worldwide. There will be subtle changes in properties in resins with similar specifications but manufactured by different companies.

Plastic resins are rarely used in virgin form but are intimately mixed with several chemicals (called additives) before they can be formed into useful plastic products. Processing plastics requires them to be subjected to harsh mechanical and thermal treatment in processing machinery. Some of the additives incorporated reduce the amount of thermal degradation or breakdown of the resin during processing. Other additives ensure that the resin is imparted with specific properties such as high strength or long-term durability required in the intended product. Popular formulations or "compounds," especially those intended for common high-volume products, are often preformulated with additives already mixed in. Yet, even an off-the-shelf plastic formulation might be further modified by the processors by mixing in even more additives to obtain the unique mix of properties that best serves the application.

This diversity presents a bewildering array of choices to a designer of plastic products. Table 4.1 illustrates the diverse choices available to a designer looking for a PS for a product application. First, he/she must select a grade and then a manufacturer. In each step shown in the table, available choices expand and the designer will select a low-cost formulation that has the required functionality. The number of available choices will also include various copolymers of styrene as well as blends of PS with other polymers (not included in this estimate of choices).

4.1   PLASTICS: THE MIRACLE MATERIAL

Most plastics used in high volume today were in fact synthesized and commercialized in the 1950s. Figure 4.1 gives a timeline for the development of common plastics in use today. Plastics are truly unique materials destined to replace most of the

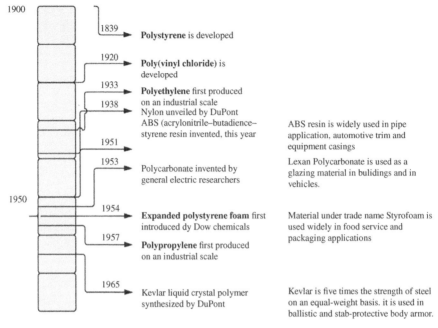

FIGURE 4.1 The timeline for development of the common classes of thermoplastic polymers.

conventional materials in building, transportation, and packaging applications. Metal and asbestos-cement pipes have been replaced by either poly(vinyl chloride) [PVC] or polybutene for water/sewage pipes in construction. Wood-based siding, window frames, and door frames have for the most part been replaced by PVC alternatives. In the packaging sector, glass, metal, and paperboard continue to be replaced by plastic film, while natural-fiber textiles are being replaced by synthetic-fiber substitutes. Plastic composite is becoming a popular choice in transportation applications as well.

Growing popularity of plastics as a material is reflected in the exponential growth of its global production volume (see Fig. 4.2) over its short history as a commercially available material. 280 million metric tons of plastic resin was manufactured globally in 2012, and the annual trend is for this to continue into the future. Thermoplastic resins make up over 90% of this production. Thermoplastics is a general term used for plastics that can be softened into a viscous liquid melt by heating and can therefore be remolded again into the same or a different shape. (This is why postconsumer thermoplastics can be readily recycled into other products.)

Five classes of thermoplastics (see Fig. 4.2) dominate the global production of resins. These are PE, PP, PS, PVC, and poly(ethylene terephthalate) [PET]. The popularity of these five classes of plastics is mostly due to their extensive use in packaging and building applications. PE is the highest-volume plastic resin

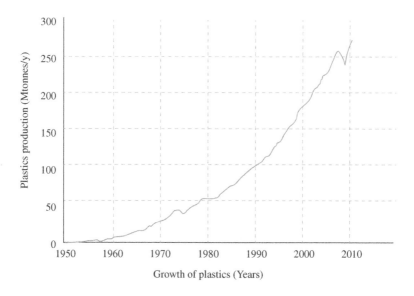

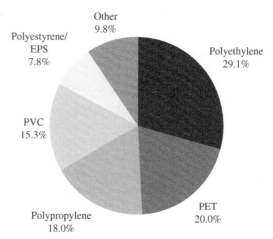

FIGURE 4.2   Upper: world plastic production in recent years. Lower: pie diagram of world thermoplastic resin capacity 2008. Source: Plastics News (2010).

produced and used in the world. Nearly half of it is devoted to making plastic films used in carrier bags, sandwich bags, freezer bags, most cling wrap, and agricultural mulch film.

Not included in Figure 4.2 are thermoset plastics that are extensively used as glass-fiber- or carbon-fiber-filled materials in building and transportation. For instance, watercraft or aircraft structures and in wind turbine blades in the energy sector are fabricated out of lightweight composites or thermoset plastics filled with reinforcing fillers. Thermoset plastics cannot be melted and reshaped into

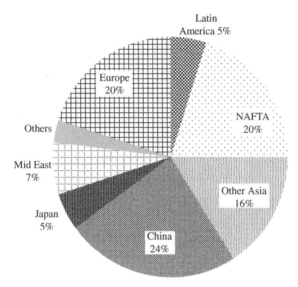

FIGURE 4.3 Plastic resin production in different regions of the world.

other products. Asian composite capacity will outpace that of the United States and Europe by 2015 and will account for 43% global production (Toloken, 2012).

The primary geographic regions of resin production are in North America, Europe, and Asia (including Japan) (Fig. 4.3). Asia is the leading producer presently, accounting for 35% of the production with only 25% being produced in North America. This rapid growth in Asian production suggests a trend towards relocating manufacturing to Asia, away from the NAFTA and European regions to continue. The largest variable costs in plastic processing plants are materials, direct labor, and energy. The adoption of best practices seeks to reduce direct labor costs in Western plants[4] (Kent, 2008), while in Asia the labor costs are already relatively low and the regulatory constraints tend to be less stringent.

Given the expected rapid growth in plastics as a material based on extrapolating the curve in Figure 4.1, can future shortage of raw materials for plastic production be expected? Adjusting for population growth, the global demand will likely rise to over 800 MMT by 2050. That would amount to a significant fraction of the fossil fuel production, and using oil as the feedstock will have to give way to alternatives such as coal and natural gas available in good supply in the medium term. The use of natural raw materials to synthesize plastics (bio-based plastics) is emerging, but the technology needs to dramatically advance to meet the demand in a practical timescale. Bio-based plastics will have an important role to play in future not only because of these resource limitations but also because of sustainability considerations.

---

[4] Direct labor in plastic plants has reduced from an average of 25% of manufacturing costs in 1960 to only 10% today.

4.2   PLASTIC PRODUCTION, USE, AND DISPOSAL

Plastics industry is the third largest manufacturing industry in the United States. While infrastructure, organization, and management styles in the industry can be different worldwide, the basic steps involved in production are the same. These steps can be thought of in terms of three phases in the life cycle of plastic products: production, use, and disposal. This last phase is an equally important aspect of the industry. Concerns on the growing fraction of plastics in the municipal solid waste (MSW) stream and in urban litter has recently promoted the notion that the manufacturer's environmental responsibility and stewardship does not end at the gate (or when the product is delivered to the consumer). An integral part of the design process is therefore to ensure that viable waste management routes exist for post-consumer disposal of the product as well. Educating the consumer as to what these options might be is also increasingly expected of the manufacturer. Volunteer guidelines adopted by industry members, the publicization of the issue by watchdog groups, and moves to regulate plastics use by the governments[5] in recent years have highlighted the public concern on plastic waste management.

Figure 4.4 shows the activities that fall into each of the three phases in the life cycle. Sustainability considerations suggest that industry pay equal attention to all three and streamline their operations to be consistent with identified environmental goals to ensure that it survives robust and profitable in the future. The plastics industry has the obligation to make their entire operation environmentally benign and sustainable in the long run. The broad sustainable development guidelines within each phase can be derived using the sustainability discussion developed in Chapter 2 and might be summarized as follows:

Sustainability considerations in the production phase

1. Minimize the use of energy, especially fossil fuel-based energy, in manufacturing resins, fabrication of products, transportation, and disposal of plastics waste.
2. Develop and use green alternative monomers (especially renewable materials) in place of fossil fuel-derived monomers wherever technically and economically feasible.
3. Minimize the residual monomer content in virgin resin pellets reaching the processors and ending up in plastic products.
4. Continually research appropriate chemistries for greener alternatives for additives identified as posing a potential health hazard.
5. Ensure minimum waste of all materials during manufacture.
6. Guard against polluting the air, water, and soil by using the best available industrial practices. This also includes minimizing incidental loss of virgin plastic resin pellets during transport.

---

[5]Some local governments and even countries have adopted laws that limit plastics use in specific applications. However, it is also not uncommon for the regulations to be so lax that it is environmentally counterproductive but might be profitable.

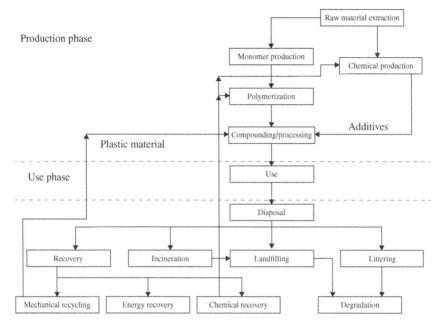

Production phase

Raw material extraction

Monomer production

Chemical production

Polymerization

Additives

Compounding/processing

Plastic material

Use phase

Use

Disposal

Recovery   Incineration   Landfilling   Littering

Mechanical recycling   Energy recovery   Chemical recovery   Degradation

Disposal phase

FIGURE 4.4   A generalized flow diagram of the plastics industry showing the three phases of activity. Source: Modified from Lithner (2011).

7. Study and research potential health hazards and ecological problems associated with plastics additives popularly used. Communicate these potential hazards and what steps are being taken to minimize them, to the consumers.

8. Design plastic products that fully exploit the unique characteristics of the material.

Use phase

(a) Where possible to do so safely, design plastic products to be reusable to ensure that consumers derive the full value of the material.

(b) Encourage proper disposal of plastic products to avoid littering especially in the marine environment where litter management options are not readily available.

(c) Assess the likelihood of release of residual monomer and additives, from plastic packaging especially into food and beverage.

(d) Inform consumers about the positive as well as negative environmental attributes of the product during use and the industry efforts at correcting perceived problems.

Waste disposal

(a) Encourage reuse, recycling, and resource or energy recovery with postconsumer plastic waste to cost-effectively extend the value of the product beyond its immediate service life.

(b) Promote and support subsidies and incentives to encourage recovery of discarded plastics in the environment.

(c) Adapt a whole-environment approach to waste management where the selected options do not shift the pollution burden from one medium to another less visible one or to a different environmental compartment.

(d) Research emerging and novel approaches to recycling and waste management that are particularly well suited for plastics.

4.2.1 From Resin to Products

Production of a plastic product (the first of these phases) involves four separate segments of the polymer industry: (1) the extraction of raw materials (oil, coal, natural gas) and refinement/processing that yields monomers (2) polymerization of this monomer into a plastic resin that is made available as resin pellets or powder, (3) intimate mixing of the plastic resin with fillers and additives, and (4) forming (or shaping) this mix to produce a useful product. In the flow chart (Fig. 4.4), compounding and processing are combined into a single box as both operations are often carried out by a single company.

4.2.1.1 Resin Manufacture   Plastic resin is manufactured primarily from crude oil or natural gas (though other fossil fuels and biomass might also be used for the purpose). Refining crude oil yields chemicals that are converted into monomers, the basic building blocks of polymers. These are reacted or polymerized in the presence of appropriate catalysts to yield plastic resin. A more detailed discussion of resin manufacture follows later in this chapter. The end product of this step will be the virgin plastic resin in the form of prils or powder.

4.2.1.2 Compounding   Compounding is the intimate mixing of plastic resin in the form of pellets or powder with a variety of additives and fillers. The resin is melted and the melt is mixed with the additives to ensure their even dispersion in the plastic matrix. Given the range of available additives, thousands of different formulations or "compounds" of plastic are theoretically possible. Compounders are specialists in designing and mixing resin formulations into pellets (or dry blended into powders) for use by the processors. Heavy equipment such as compounding extruders, roll mills, and internal mixers are used to carry out this operation.

Though not as popular in the 1980s, the custom compounder segment in the US plastics industry is still viable. But compounding is increasingly undertaken by resin manufacturers or more commonly by plastic processors. This often results in cost savings (and even environmental benefits) as the resin does not have to be transported. Pre-made formulations or "compounds" are commercially available for processors to use. Some of these are available as "concentrates" or master-batches where the additives and fillers are incorporated at levels that are much higher than that typically used in the final product. The processer generally mixes an appropriate ration of the masterbatch with the virgin plastic pellets, remelts to mix them, and pelletizes the mix prior to processing the mix into a product. The mixing is typically carried out in a compounding extruder or in an internal mixer. The scope of this work does not permit any substantial treatment of these steps or even an adequate technical discussion on the topic. Only basic information that allows the reader to appreciate the content in the following chapters is included in the following section. Several excellent books on the topic are available (Manas-Zloczower, 2009; Todd, 1998).

4.2.1.3 Processing into Product Plastics industry uses several different processing techniques to convert plastic resin into different products. These are summarized in Table 4.2 (with those for thermoset plastics summarized in Table 4.3). The most used of these by far are injection molding and extrusion, each accounting for about 1/3 of the global thermoplastic processing capacity. Other less common techniques are blow molding of bottles, film blowing, and coating of paper goods. Each of these techniques requires dedicated machinery that automates the process. In general, all processing techniques heat the plastic pellets (which are precompounded) and either force it through a specially designed die (extrusion) or inject it into a cavity mold (as with injection molding and blow molding). Naturally, these are energy-intensive processes as the polymer needs to be softened and the viscous melt worked mechanically to plasticize it. The mechanical forcing of the melt is usually via a rotating screw system, and during passage an enormous amount of frictional heat is generated that further heats up the melt. Thermal stabilizers are generally used as additives to minimize the heat-induced degradation of resin during processing. Either air or water is used to cool the equipment and to solidify the plastic shape created.

4.3 PROCESSING METHODS FOR COMMON THERMOPLASTICS

4.3.1 Injection Molding

This is the most popular processing technique employed by industry and is similar to metal die casting. Injection molding machines essentially consists of two parts: a heated barrel and a cooled two-part cavity mold that opens to allow removal of the product. The barrel carries a long reciprocating screw that transfers the plastic pellets from the hopper at the base of the screw forward toward the mold. The barrel and screw are heated to melt the plastic resin into a highly viscous liquid. A large amount of mechanical power is needed to melt, transfer, and compress the plastic in the barrel. Once sufficient quantity of melted plastic (or the "shot") is transported into the nozzle area in front of the screw, the screw itself acts like a piston to ram it into the cavity mold. The ramming pressure is held just long enough for the plastic melt in the mold to consolidate, cool, and solidify before the screw now rotates back to

TABLE 4.2 Commonly Used Processing Techniques for thermoplastics

Processing technique	Plastic resin	Typical products	Base technology
Film blowing (extrusion blowing)	HDPE, LDPE, and LLDPE	Shrink film, stretch film, bag film or container liners, greenhouse film, garment bags, and garbage can liners	Melt extrusion and air blowing
Blow molding	HDPE, LLDPE, PP, PET, PVC	Bottles and containers, fuel tanks, air ducts, garden supplies	Melt extrusion/molding
Profile extrusion	HDPE (rigid) PVC	Plastic pipes, window frames, drainage tubing, wastewater pipes, agricultural tubing, conduit/cable protector, guttering, plastic siding, fascia and soffit, garden fence posts and decking	Melt extrusion
Injection blow molding	LDPE, HDPE, LLDPE, PP, PET, PVC	Plastic bottles and containers	Melt injection/blowing
Injection molding	ABS, nylon (PA), PP, and PS Polycarbonate (PC)	Power-tool housing, telephone handsets, TV cabinets, computer housing, DVDs, automotive bumpers, dashboards, syringes, kitchenware, cutlery, crates, bottle lids/closures	Melt injection Variations include gas-assisted injection molding and stretch injection molding
Expansion/molding/extrusion	PS	Expanded polystyrene products, hot beverage cups, floats, insulating tiles. Meat trays, snack boxes for fast food	Steam expansion/molding Steam expansion/extrusion
Rotational molding	LDPE, LLDPE, PP, PVC	Rainwater tanks, slides and climbing frames, diesel fuel tanks, children's playhouses, traffic cones, canoes and kayaks and pallets	
Thermoforming and vacuum forming	PE, PS, PET.PVC, PC, ABS, PMMA, PVC, and EVOH	Meat trays, microwave trays, frozen food containers, ice cream tubs, delicatessen tubs, bakery packaging, vending beverage cups	Melting followed by forming at ambient pressure

Source: From http://www.bpf.co.uk/plastipedia/processes/default.aspx#blownfilm

TABLE 4.3 Commonly Used Processing Techniques for thermoset Materials

Processing technique	Plastic resins (a variety of reinforcement is used in all these processing techniques including polyester, vinylester, epoxy, and phenolic and methyl methacylates)	Typical products	Base technology
Pultrusion	Unsaturated polyester, polyurethane, vinylester, and epoxy (recently, some thermoplastics such as poly(ethylene terephthalate) (PET) has been successfully used in pultrusion)	Round rods, rectangles, squares, "i" sections, channels, dog bone profiles, corner profiles, hollow sections	The resin impregnation of roving and subsequent cure
Resin transfer molding	Unsaturated polyester, vinylester, epoxy,	Products with a complex shape where wall thickness is critical and yields a smooth finish on both sides	Thermoset resin impregnation of fiber mats and cure
Sheet molding compound (SMC)	Same as above	Sheets and laminates	Cut short fiber embedded in resin layer and cured
Hand lay-up and spray lay-up	Same as above	Used in molding large complicated shaped objects including boat hulls, large tanks, and hot tubs	Laying resin impregnated reinforcing fabric by hand on a mold and end up with a gel coat of resin
Compression molding	Same as above	Blower or fan blades, heat-resistant knobs, electrical boxes, and parts in industrial machinery	The blend placed in mold and heated under pressure within a steam-heated press
Filament winding			A continuous yarn or fiber that is impregnated with resin/catalyst mix is wound around a tool mandrel and cured

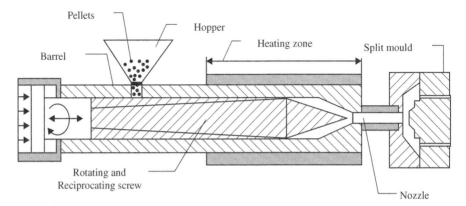

FIGURE 4.5 Schematic diagram of an injection molding machine showing the reciprocating screw and different heating zones. Source: Reproduced with permission from Bull and Zhou (2001).

FIGURE 4.6 An injection molding machine and examples of molded products. Source: Reproduced with permission from Styrolution, Styrolux, BASF.

position itself for another cycle. The mold is rapidly cooled (usually with chilled water) to solidify the melt into a product. When the mold opens steel pins built into it push out the formed now cold part. The mold recloses and snugly fits against the gate through which the melt is delivered, in anticipation of the next charge of melt to be delivered into it.

Injection molding machines are rated according to the clamping force available to keep the mold closed against the ramming pressure of the melt driven by the screw mechanism. Figure 4.5 illustrates the basic parts of an injection molding machine, and Figure 4.6 shows an actual machine and some typical molded products. With multicavity molds, some of the melt is inevitably used up to fill the runner areas of the mold. However, after recovering the product, the runner waste is recycled.

Injection molding is estimated to typically require 1.3–1.6 kW/kg of moldings; however, less than 10% of this energy is input into the plastic resin and most being used to operate the equipment (British Plastics Federation, 1999). More than half the energy is spent during the so-called "recovery" step where the screw rotates to create a fresh shot of melt to be injected (Franklin Associates, 2011). Injection, mold release, and heaters each account for about 10–15% of the energy; the heating energy is low because the frictional heat contributes significantly to melting.

4.3.2 Extrusion

Extruders are designed to heat, melt, and homogenize the plastic pellets or powder and to transport the melt forward to eventually force it through an open die to obtain a long, continuous shape such as plastic pipe. The equipment consists of a long heated barrel (similar to that in injection molding machines), but the nozzle at its terminus forces the melt through a die. A single screw (or a reciprocating pair of screws) in the barrel moves the partially melted plastic forward toward the die. However, unlike with injection molding, the screw does not act like a piston. The process is more akin to a paste being pressed out of a tube to obtain a cylindrical extrudate. Extruded shape that is produced in a continuous profile is rapidly cooled to ensure that it is not deformed and is cut into suitable lengths. Plastic tubing, plastic sheets, window/door frames, molding used in building industry, and all thermoplastic pipes are extruded products. Figure 4.7 shows the basic features of a single-screw extruder and a sheet extrusion die in cross section. However, twin-screw extruders are also used. Naturally, the throughput of extrudate from the latter will be higher compared to single-screw extruder.

The engineering design of the screw can be complex. Different regions of the screw perform slightly different functions and are therefore designed with appropriate pitch and depth. At least three basic zones can be identified: a feed zone, a compression or compaction zone, and a metering zone. In the feed zone, the plastic pellets begin to melt, and the material is a mix of partially melted pellets and liquid melt. By the time the melt is in the compression zone, it is homogeneous and has compacted during its travel through the other zones. The metering zone of the screw ensures delivery of homogenized melt to the die at a constant rate and pressure.

An extruder can also be used with a cylindrical annular die to extrude a tube of thin-walled plastic. The tube is not allowed to collapse by blowing air through the mandrel of die to balloon it out until it cools. The collapsed, continuous tube of film can then be heat sealed and cut into plastic bags. This process is commonly referred to as "film blowing" and used extensively to produce plastic bag stock and other plastic films. Several extruders working in concert can be used to deliver several layers of melted resin to a single complex die to fabricate multilayered plastic laminates. Each layer performs a different function such as enhancing barrier properties (excluding oxygen and moisture), providing good mechanical strength, or even acting as a "tie layer" that holds two incompatible layers of plastics together. Multilayered plastic films can be produced in a film blowing operation as well as in cast film mode.

4.3.3 Blow Molding

This process borrowed from glass technology is used extensively to fabricate plastic bottles. Initially, a short thick length of polymer tubing (called a parison) is extruded through an annular die into an open two-piece mold. Once the mold closes, air is

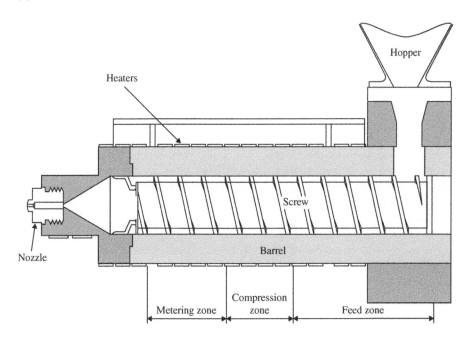

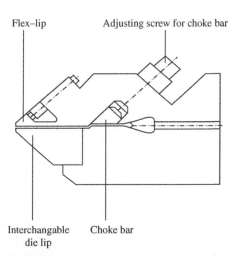

FIGURE 4.7 Upper: schematic diagram of a single-screw extruder. Lower: a sheet extrusion die for plastics. Source: Reproduced with permission from Styrolution, Styrolux, BASF.

injected through the parison into the mold to press the molten polymer against the walls of the cavity. The polymer in mold is cooled, generally using chilled water. Figure 4.8 illustrates this process for molding a bottle. While the mechanics of the process are simple, the pressure, temperature, and parison design have to be carefully controlled to obtain a consistent quality of product.

Table 4.4 compares the three popular processing methods in terms of energy demand by selected aspects of the process and underlines the energy use by auxiliary

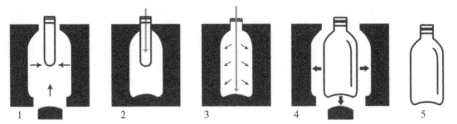

FIGURE 4.8 A diagram of the bottle blow molding process. 1. Heated parison. 2. Mold closing. 3. Blowing air into mold 4. Cooling and opening mold. 5. Molded bottle.

TABLE 4.4 Relative Energy intensity of Selected Plastic Processing Techniques

Processing method	Blend	Mix	Expand	Heat mold	Cool mold	Cure
Injection molding	+++	++		++	++	
Extrusion	++++	++			++	
Film blowing	++++	++	+++		+++	
Compression molding		+	+	+++		+++++
Thermoforming	+		+	+++	++	
Calendaring	+++	+			+	
Rotational molding	++	+	+	++	++	
Foam blowing	+	+	++			++
Unsaturated polyester	*	+				+++

Source: Based on data from the British Plastics Federation (1999).

TABLE 4.5 Estimate of Plant Energy Distribution for Three Plastic Processes

Operation	Extrusion	Injection molding	Blow molding
Plant	3	10	12
Barrel heating	4	19	6
Extruder drive	41	18	38
Auxiliary equipment[a]	42	17	46
Mold clamping		36	

Source: Based on data from Guide to Energy Efficiency Opportunities in the Canadian Plastics Processing Industry (2007).
[a]Includes dryers, coolers, granulators, compressed air generators, and material handling.

(usually downstream) equipment in nearly all the processes. Table 4.5 summarizes approximately energy demand by different aspects of the basic process itself. In injection molding, extrusion, and film blowing, the blending of the viscous mass of plastic takes up most of the energy.

4.4 THE ENVIRONMENTAL FOOTPRINT OF PLASTICS

Does the manufacture and processing of plastics contribute significantly to energy use and climate change? Do the societal benefits derived from plastics justify this footprint?

4.4.1   Energy Considerations in Resin Manufacture

In the United States, liquid petroleum gases (LPG), natural gas liquids (NGL), and natural gas are used to produce plastics.[6] In 2010, the US plastic production accounted for only about 2.7% of the total petroleum and 1.7% of the natural gas consumed. The electricity used was estimated at a further 1% of the consumption. Then estimating conservatively, only about 5.4% of petroleum resources were devoted to manufacturing plastics[7] (US Energy Information Administration, 2013). The operational or process energy may add another approximately 3–4% to this, bringing the total fossil fuel investment in plastics to about 8.6%.

The question is whether the improvements in human condition (convenience, comfort, and safety) as well as energy savings due to use of plastics (see Chapter 5) justify the investment of 8–9% of the crude oil resources in the plastics industry. Fossil fuels are used efficiently in manufacturing plastics with little waste. In the United States, over 70% of crude oil is used for transportation (in engines that waste 2/3 of the energy as heat!),[8] which is also a highly polluting application. Indispensability of reliable transportation in modern life has allowed this energy cost to be accommodated. The 8–9% investment in fossil fuels provides an incredibly wide range of useful items that we simply cannot avoid coming into contact with in our daily lives. Plastics in some instances (such as in residential insulation or aircraft construction) save energy and in applications such as wind turbines even help generate green energy. The reader is invited to review some of these benefits in Chapter 5.

Embodied energy (EE) in a material or product is all the energy directly or indirectly used in the creation of a unit mass (for instance, a metric ton) of it. This would include the energy used in extraction of raw materials, their transportation to a plant, processing in the plant, and maintenance energy for dedicated plant.[9] The metric therefore depends on the boundary conditions used in the definition and the mix of energy used (therefore the location of manufacture). When it is taken to mean all energy from extraction of raw materials until the product leaves the manufacturer's location, it is called "cradle-to-gate" energy or the "front-end" energy investment in producing the material. It does not include the energy cost of using and disposal of the material or product. The values of EE reported vary because of how the quantity is calculated;

$$\text{Embodied energy}\,(\text{Em}) = \frac{\sum \text{energy input in production per unit time}}{\text{mass of product generated per unit time}}$$

[6]LPG is a by-product of the petroleum refining process, and NGLs are removed from natural gas before it is used as fuel gas in transmission lines. Coal can also be used to make plastics; the first nylon made was based on coal feedstock.

[7]According to 2006 data by the US Energy Information Administration, 4.6% of petroleum resources were used by plastics industry.

[8]Percentage of the thermal energy from burning fuel converted into mechanical work in an internal combustion engine is around 26%. However, the actual mechanical efficiency is lower as some of the energy is wasted in overcoming frictional forces.

[9]This is why theoretical calculations based on thermodynamic considerations are not that useful in environmental assessments. All processes are inefficient and produce waste, and the energy costs depend on the type of energy used.

One criterion in sustainable material selection is to select one with the least EE that also meets the performance requirements for a given application. When the raw material itself is an energetic material (such as oil, coal, or gas) with fuel value, EE also includes the raw-material energy in addition to the energy expended in manufacturing and transportation. The EE referred to here is that associated with converting feedstock materials into plastic resin pellets and not that associated with manufacturing a product. In a subsequent step where the resin pellets are converted into a product such as a bottle or a bag, additional energy has to be expended. This latter energy will be referred to as "process energy," for clarity.

Although minimizing the EE is a sustainable strategy, at least two other factors must be taken into consideration: (1) the complete service life of the material and (2) energy cost per functional unit of the material. There are instances where a material with a higher EE is justified when the life cycle costs that also include maintenance and replacement costs are taken into account. For instance, an aluminum window with an EE ~ 5500 (MJ/m²) and a carbon emission (GWP) 279 (kg/m²) will have several times the service lifetime of a comparable wooden window with the lower EE of approximately 230–490 (MJ/m²) and a GWP of only 12–25 (kg/m²) (Hammond and Jones, 2008a). Similarly, a material with a higher EE might be justified when a smaller quantity of it can be used to provide the same functionality as the material it replaces. In any event as discussed in Chapter 2, other criteria such as associated externalities including the embedded carbon emissions and reusability of materials need to be considered.

Estimates of EE for different plastics (US data) are shown in Figure 4.9. The estimates for common plastics are not that different from each other and generally amount to about 75–100 GJ/ton of resin manufactured.[10] This value being dependent on the specific sources of energy used in manufacture and on the technology used can be somewhat different in other parts of the world. For instance, the comparable energy calculations in the PlasticsEurope database are 4–13% lower for common plastics such as PE, PP, PVC, and PS. The cradle-to-gate values for EE in the figure for neat plastics resins will change if they are compounded with additives.

Embodied energy in plastic material can be compared to that for other materials[11] as shown in Table 4.6. Plastics are more energy intensive to produce than most materials of construction in the table except for aluminum. Like aluminum, plastics also have to be removed from a finite in-ground resource pool and have to be refined before use; both can be recycled as well.

Fossil fuel resources are used in plastics manufacture both as a raw material and also directly or indirectly as a source of process energy (for instance, to operate injection molding machines). Figure 4.10 compares the percentage of energy in each of these categories and the small amount of energy used in transportation of raw materials in the manufacture of plastic resins. Similar data for thermoset polymer (polyurethane (PU)) are also included in Figure 4.10. For PU, the process energy spent in multistep synthesis of precursor is larger than that for thermoplastics. With most common polymers (other than PVC), more than half the fossil fuel cost is in raw materials.

---

[10] By comparison, the energy in crude oil is about 47 GJ/ton.
[11] The value for glass is from Joshi et al. (2004).

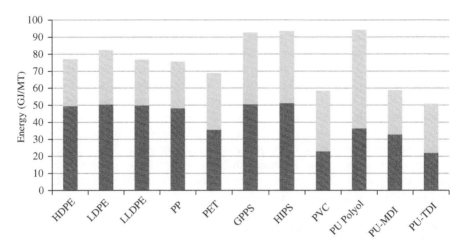

FIGURE 4.9 Embodied energy for selected classes of plastic resin. The top part of each bar is for manufacturing energy (including recovered energy), and the bottom part is for material energy. Source: Based on data from Franklin Associates (2010).

TABLE 4.6   Comparison of EE Values and Carbon Emissions for Different Building Materials

Material	Embodied energy (MJ/kg)	Carbon emission (kg of CO_2/kg)	Density (kg/m^3)
Steel	24.4	1.77	7850
Aluminum	218	11.46	2710
Recycled aluminum	28.8	1.69	7850
Concrete	0.95	0.13	2710
Copper	40–55	2.19–3.83	9000
Glass	15	0.85	2500
Bitumen	47	0.48	1200–1360
Ceramic	10	0.65	5600
HDPE	**76.7**	**1.6**	**~970**
LDPE	**78.1**	**1.7**	**~910**
PP (oriented film)	**99.2**	**2.7**	**~900**
Polycarbonate	**112.9**	**6.0**	**~1300**
GPPS	**86.4**	**2.7**	**~1000**
HIPS	**87.4**	**2.8**	**~1000**
PVC	**77.2**	**2.41**	**1390**

Source: Data of Hammond and Jones (2008b).

Processing is an energy-intensive operation as the resin has to be plasticated and melted prior to forming. The approximate energy demands by various stages in common processing techniques are as follows:

Extrusion: Drive mechanism (41%), process chilling (19%), barrel heating (7%), resin drying (7%), and granulator (6%)

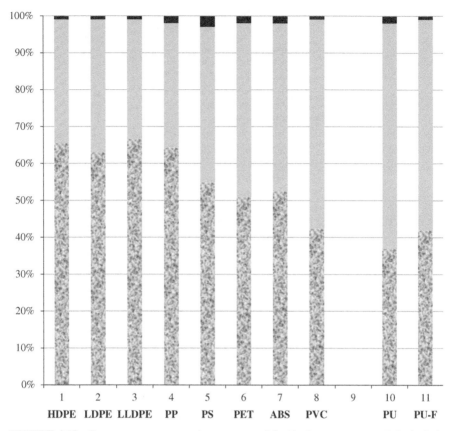

FIGURE 4.10 Percentage energy used as raw materials (the lower segment of the bar), in manufacturing operations (middle, grey segment), and in transportation of raw materials (upper black segment) in the manufacture of different plastic resins in the United States. Source: Based on data from Franklin Associates (2011).

Injection molding: Mold clamping (36%), barrel heating (19%), drive mechanism (18%), resin drying (6%), and process chilling (4%)

Blow molding: Drive mechanism (38%), process cooling (19%), resin drying (7%), granulator (6%), and finishing equipment (7%)

Significant energy savings can be accrued in the fabrication or processing of plastics by making fairly modest process modifications. Table 4.7 summarizes such improvements in processing that deliver substantial energy savings.

4.4.2 Atmospheric Emissions from Plastics Industry

Both globally and in the US, greenhouse gases (GHG) released to the atmosphere are dominated by CO_2 from combustion of fossil fuels. Most of this carbon load emitted

TABLE 4.7   Energy-saving Opportunities in Plastics Processing

Process	Improvement	Energy saving (%)
Extruder drive system	Improve drive mechanism by matching size and speed of motor. Investigate high-efficiency motors	20
Extruder barrel heating	Insulate barrel in extruders or molding machines	15
Mold closing and clamping system	Use variable hydraulic power to match load requirements. Explore the use of variable speed drives, variable displacement pumps, accumulators, and control systems	45
Hydraulic system	Use a single centralized hydraulic power system to supply a group of machines	50
Compressors	Use compressors of proper size that are well maintained and "staged"	20

Source: From the Canadian Plastics Industry Association (2007).

is derived from burning oil (43%),[12] coal (36%), and natural gas (20%). The annual US emissions of about 30 MMT of CO_2 (based on 2007 data) amount to about 20% of global emissions. Industry is the third largest emitter of carbon (20%) after electricity production (41%) and transportation (22%).

The plastics industry also contributes to air emissions of both CO_2 and other pollutant gases. In addition, certain plastic processes do emit volatile organic chemicals (VOCs) into the environment. Composite technologies using urethane chemistry and unsaturated polyester prepolymers are probably the worst in this regard. Foamed plastics (PE foam, expanded polystyrene [EPS], as well as PVC foam) also contribute to this problem (Canadian Industry Program for Energy Conservation, 2007). A recent European study (Pilz et al., 2010) estimated the plastics industry contribution to the overall carbon footprint to be only about 1.3%. The corresponding estimate for leisure and recreational activity, for instance, was estimated to be 18%.

Another approach to assessing the relative impact of an industry sector is to assign a monetary value to the natural capital consumed and damage incurred by different industries. This aggregate environmental cost can then be expressed as a percentage of the revenue generated by the sector (which is a measure of "benefits") from a given application sector. This is called its impact ratio (IR). Where the social cost or abatement cost per unit of revenue generated is much greater than 1, the activities are deemed to have a high environmental cost. Figure 4.11 compares the value of IR (%) for the plastics industry with those for several other

[12]Data is for 2010 and global CO_2 emissions. Source *"CO_2 Emission from Fuel Combustion." (International Energy Agency, Paris, France).*

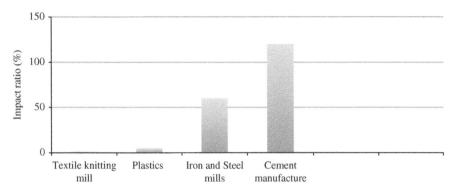

FIGURE 4.11 Total direct environmental damage as a percentage of revenue for several selected industries. Source: Based on data from Trucost Plc. (2013).

industries[13] based on one recent study (Trucost Plc., 2013). Serious biases, approximations, and omissions will be inherent in any such effort. Uncertainties in valuation of environmental services, for instance, must have been quite large, and all true costs are rarely reflected in market costs. Even taking these limitations into account, the plastics industry was estimated to be far less damaging compared to other selected industries, at least qualitatively.

However, as already alluded to in the introduction, the plastics industry constitutes only a very small fraction of total industry. As with the associated energy costs, the question is if the environmental "cost" is excessive compared to the benefits afforded by plastics. This is discussed in Chapter 5.

4.5 PLASTICS ADDITIVES

To ensure the desired properties and for facile processing of the material, additives are mixed in with the resin during compounding. Some of these are inert fillers or recycled plastic used to bring down the cost of the manufactured product, while others are used to ensure specific functionalities. Table 4.8 indicates the ranges of levels at which these additives are typically used. The exact level depends on the grade of resin as well as on the properties demanded of the compound. The main classes of additives in plastics are shown in Figure 4.12:

(a) Stabilizers that protect plastic from thermal degradation during processing and light-induced thermo-oxidative and biological damage during use
(b) Reinforcing fillers and plasticizers that improve the mechanical properties such as the modulus of plastic material, to match the service requirements
(c) Processing aids that make the processing of the material easier
(d) Inert fillers and extenders

[13] The value is for plastic resin manufacturing operation only. For extraction of petroleum, the IR ~ 80%.

TABLE 4.8   Levels of Common Additives Used in Common Plastics

Additive	Percentage by weight of resin typically used	Chemical nature	Examples of use
Plasticizers	10–70%	Phthalates and chlorinated paraffins	Plasticized PVC products
Flame retardants [F]	12–18%	Brominated or chlorinated compounds	Electronic equipment casing, insulating foam
UV stabilizer antioxidants	0.05–3	Hindered amines, alkylphenols, metal compounds	Plastics used outdoors such as greenhouse films or stadium furniture
Heat stabilizer	0.05–3	Cadmium and lead compounds	Compounds intended for extrusion or molding
Slip agents	0.1–3		
Lubricants	0.1–3		
Antistatics	0.1–1		
Curing agent	0.1–2		
Foaming agent	—		
Biocides	0.001–1		
Soluble dye	0.25–5		
Organic pigments	0.001–2.5		
Inorganic pigment	0.01–10	Zinc sulfide, zinc oxide, iron oxide, titanium dioxide	
Fillers (reinforcing and nonreinforcing)	Up to 50%	Calcium carbonate, talk, clay, zinc oxide, glimmer, wood powder	

Sustainable products need to be safe to use. Safety of additives that can leach out of food-contact plastics has drawn considerable public attention. Endocrine disruptor effects (see Chapter 7) and toxicity often attributed to "plastic" products (Chapter 8) are often due to additives such as phthalate plasticizers and brominated flame retardants. What is not sustainable is to keep using additives that are evidently hazardous (and those with unknown ecological impacts) simply because they have superior performance and are economically the best choice. Alternative additives that not only match or exceed the performance of conventional ones but do not pose a threat to the biosphere including its human occupants, need to be developed as replacements for the ones in question.

The complexity of designing a formulation with several additives is illustrated by the following recipes. Additives have to be mixed intimately, usually in a compounding extruder, and pelletized prior to processing.

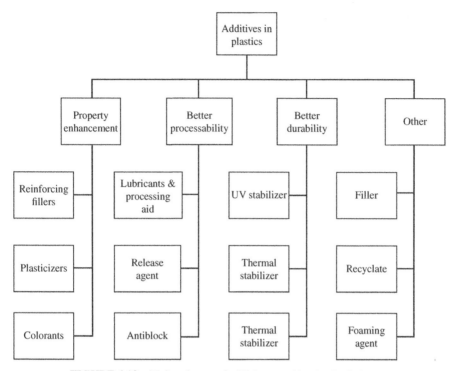

FIGURE 4.12 Major classes of additives used in plastics industry.

Ingredient	Amount (phr)	Function
A rigid PVC extrusion compound		
PVC resin	100	Base plastic
Methyl tin stabilizer	1.25	Thermal stabilizer
Calcium stearate	1.0	Lubricant (modifies viscosity)
Paraffin wax	0.15	Lubricant
Acrylic modifier	4.5	Increases impact resistance
Rutile titania	8	Opacifier (light stabilizer)
Ground $CaCO_3$	10	Inert filler
Polypropylene molding compound		
PP resin	100	Base plastic
Impact modifier	18.4	Increases impact resistance
Carbon black masterbatch	3.06	Carbon black (30%) reinforcing filler
Talc	30.6	Reinforcing filler
Calcium stearate	0.15	Lubricant
Stabilizer	0.07	Stabilizer
Ground $CaCO_3$	10	Inert filler

4.5.1 Fillers for Plastics

Fillers are low-cost, mostly inorganic, powders that are used in plastic formulations to either reduce the amount of plastic resin used and hence the cost (called extender fillers) or to improve performance of the compound. Carbon, glass fiber, mica, talc, glass, titania, clays, and silica are well-known reinforcing fillers used to increase the modulus and strength of plastics. With these, the amount, particle size (and its distribution), particle shape, and compatibility with the polymer matrix determine the level of reinforcement. In general, the improvement obtained is proportional to the volume fraction of the filler used. Maximum performance of a functional filler at a given volume fraction is fully expressed only if it is well dispersed in the plastic matrix. Therefore, the processing method also influences filler performance. Chemical modification (or functionalization) of the filler surface and the use of a dispersants are common methods to ensure improved compatibility between phases and therefore superior reinforcement in filled plastics. In both instances, the surface of filler particle is covered with a hydrophobic layer that has better compatibility with the polymer matrix.

With emerging nanoscale fillers where the particle sizes are in the 100s of nanometers, the specific surface area of filler particles can be orders of magnitude larger than that for conventional fillers. As fractional volume of matrix–filler interphase determines the level of reinforcement achieved in the composite, nanofillers can achieve superior properties at relatively low volume fractions, provided they are adequately dispersed in the base polymer.

4.5.2 Plasticizers in PVC

Plasticizers are commonly used in plastics technology and especially in PVC, a plastic produced in high volume. They are used at high volume fractions; phthalate levels can be as high as 60–70% by weight in some "soft" or flexible PVC products (Rudel and Perovich, 2009). Products that consumers are routinely exposed to include shower curtains, toys, floor tiles, containers, sealants, and automotive interior trimming that are made from plasticized PVC. Phthalates are the most commonly used plasticizer in PVC, though others such as epoxides, trimellitates, and phosphates can also be used (phthalates are discussed in Chapter 7). Of the 6.4 MT of plasticizer globally produced in 2011, for instance, 87% were phthalates. Plasticizers also yield the soft, pliable feel to some types of "cling" films (Sablani and Rahman, 2007).[14]

Molecules such as phthalates have polar structural features that can associate reversibly with the polar structural units in PVC.[15] The plasticizer molecules are able to readily mix with the long-chain polymer molecules weakening the polymer–polymer attractive forces (van der Waals forces) that impart rigidity to the material. In so doing, the plasticizers increase the "free volume" or the interchain space available in the plastic, making it easier for segments of the long chains to flex and slide past each other easily. These molecular-level changes are reflected in corresponding

[14] All varieties of "cling" films do not contain plasticizers; some are based on polyethylene.
[15] Good plasticizer molecules have a balance between polar and nonpolar groups, the latter allowing self-association or "pooling" within the polymer matrix.

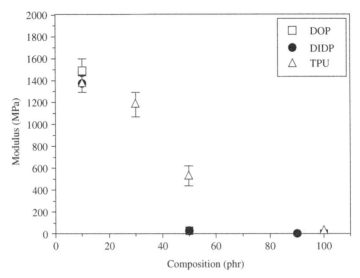

FIGURE 4.13 Dependence of the modulus of PVC on plasticizer content. DODP and DIDP are types of phthalates TPU is a thermoplastic PU. Source: Reproduced with permission from Pita et al. (2002).

changes in lowering of the modulus (or increasing flexibility) as well as the glass transition temperature Tg (C) of the plastic material (Fig. 4.13).

Most plasticizers including phthalates are not covalently linked (bonded or grafted) to PVC molecules and are free to leach out of the plastic product.[16] With plasticized packaging films in direct contact with food, leaching of plasticizer is a concern.

4.6 BIOPOLYMER OR BIO-DERIVED PLASTICS

A *biopolymer or bioplastic*[17] is a polymer produced by a living organism. Cellulose is the most common biopolymer constituting about a third of all plant biomass. All living things produce other biopolymers such as polynucleotides and proteins. An example of a class of useful biopolymers are the so-called bacterial polyesters that are harvested from bacteria grown under specific conditions.

A *bio-derived plastic* on the other hand is biopolymer that is chemically modified to improve its properties. Cellulose from plants can be acetylated to yield cellulose acetate, the bio-derived plastic used in cigarette filters. It can also be xanthated and extruded into cellophane (or Rayon), a bio-derived plastic. Chitin, a biopolymer from crab shells, can be processed (by converting the amide functionalities into amine functionalities) into its amine analog to obtain a bio-derived plastic, chitosan.

[16] There are a few examples of plasticizers such as epoxy plasticizers in PVC that are covalently linked to the polymer chain.

[17] Again, we use the terms biopolymer and bioplastics interchangeably here. Plastics are a subset of polymers as already pointed out in a previous chapter.

Chitosan   Chitin   Cellulose

A *bio-based* polymer on the other hand is a man-made polymer where the starting materials or raw materials (but not the polymer itself) are derived from living organisms (generally plants). These renewable feedstocks are used to make polymers varieties that are identical in chemistry to conventional fossil fuel-based plastics.

Despite the fact that they are chemically identical, it is possible to analytically identify bio-based plastics apart from identical fossil fuel-based plastics. Carbon dioxide in air has two isotopic forms of carbon, $^{12}CO_2$ and $^{14}CO_2$, in equilibrium with each other, and plastics derived from plant sources will therefore have both these isotopes. Any ^{14}C isotopic carbon in fossil fuel deposits, however, had long decayed over the millions of years since their formation (half-life of ^{14}C is 5730 years) into ^{12}C and fossil fuels, and plastics made from them do not have significant levels of ^{14}C. Burning a sample of plastic and measuring the ratio $\{^{14}C/^{12}C\}$ in the CO_2 using liquid scintillation counting or isotope-ratio mass spectrometry therefore provide a test as to the origin of the plastic (ASTM D 6866).

The distinction between bio-based and conventional plastics is based on the source of feedstock and should not be confused with "biodegradable polymers" (discussed in Chapter 6). Biodegradability is merely a property or functionality of a plastic material. Both bio-based and conventional polymers can be biodegradable or nonbiodegradable in the environment as illustrated in Table 4.9. As "nonbiodegradable" is strictly a misnomer (see Chapter 6), the term "durable" is used in its place.

Using renewable feedstock to make plastics is a key dictum in environmental sustainability. An abundance of biomass that can be used as raw material is available. Of the 170 billion tons of biomass produced annually by nature, less than 4% is used by humans (mostly for food and wood-based industries (Thoen and Busch, 2006)). The first bio-derived plastic, celluloid, was invented back in 1860 followed by a few others in the 1940s. But with the discovery of oil, these inventions were never developed into commercial scale. With future shortage of fossil fuels, the time is ripe to exploit bio-based and bio-derived technologies (Momani, 2009).

The savings in primary energy and avoided CO_2 emissions in using bio-based feedstock compared to conventional petroleum feedstock are significant. For instance, the GHG emission for conventional PET in a cradle-to-grave LCA estimate was 3.36 (kg equiv. CO_2/kg plastic). The corresponding number for bio-based PET was 2.34–2.67 (kg equiv. CO_2/kg plastic), depending on feedstock used (Shen et al., 2012).

TABLE 4.9 Selected Examples of the Three Classes of Plastics

	Durable	Readily biodegradable
Conventional	PE, PP, PS, PET, PVC	Poly(caprolactone), poly(butylene succinate/adipate), copolymers of poly(butylene succinate)
Bio-derived	Cellulose esters, cellulose ethers	Rayon, cellophane, chitosan, gelatin, gluten (wheat), zein (corn), pectin
Bio-based	Bio-PET, bio-PE, biopolyamide 11, biopolyamide 610	Poly(lactic acid)—compostable
Biopolymers	Lignin, humus	Cellulose, starch, chitin, pullulan, zein

4.6.1 Bio-Based Plastics and Sustainability

Responsible use of non-renewable resources is one of the three emphasis areas in the sustainable growth (Chapter 2). Any technologies that help conserve fossil fuel reserves help achieve that goal. Using renewable bioresources to make plastics, especially types of plastics used in high volume, is therefore a sustainable move by the industry (Philp et al., 2013; Reddy et al., 2013). Biomass raw materials can yield the same basic chemical intermediates that are derived from fossil fuel raw materials, making bio-based synthesis of conventional and novel plastics a possibility. The option is particularly attractive where the embodied energy as well as the carbon emissions associated with manufacture of the bio-derived variety is at or below that of conventional plastic material.

Figure 4.14 compares the conventional and bio-based pathways for thermoplastics PS and PET and poly(lactic acid) (PLA). The renewable pathways rely on fermentation and simple reactions, while the fossil fuel materials have to be cracked and refined to obtain the same basic chemicals.

For instance, the popular PLA and PHA (bacterial biopolymer), both in commercial production, might be considered. Numerous LCA studies have been carried out on these (Bohlmann, 2004; Groot, 2010; James and Grant, 2005; Madival et al., 2009; Shen and Patel, 2008; Vink et al., 2003, 2004, 2007). Figure 4.15 is the result of a meta-analysis of different LCA studies where these two polymers were compared to common plastics. Petroleum-based plastics referred to are LDPE, HDPE, PP, PET, PS, and PC. Not surprisingly, different scopes and boundaries of these cradle-to-gate studies yield very different results. But the trend is clear; both PLA and PHA are superior to conventional thermoplastics in terms of their GWP. Savings of fossil fuel by as much as 68% when using bio-based replacement resins have been estimated by the producers.[18]

However, the externalities associated with bio-based plastic production (that includes intensive agriculture to produce the biomass) can exceed those for conventional plastics. Pesticide/fertilizer pollution from agriculture is already widespread and can affect human populations. For instance, Hottle et al. (2013) report

[18] NatureWorks LLC, NatureWorks LLC Announces World's First Greenhouse-Gas-Neutral Polymer. 2005.

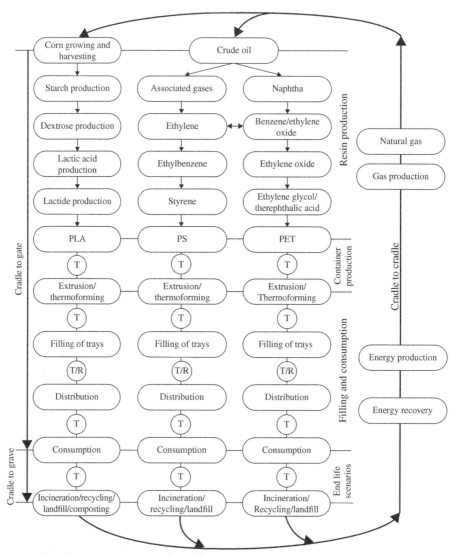

FIGURE 4.14 Basic pathways to derive chemical feedstocks from renewable and fossil fuel raw materials. Source: Reprinted with permission from Madival et al. (2009).

the acidification potential, ecotoxicity, eutrophication, ozone depletion, and PM 2.5 particulates to be significantly higher for PLA compared to conventional PE and PET. The overall potential for human health impacts is also lower for PLA compared to PET (but still higher than for PE or PP). Similar results have been reported for PHA. With bio-derived polymers that are biodegradable (such as PHA and PLA), an additional consideration is the potential negative impact these have on material recycling.

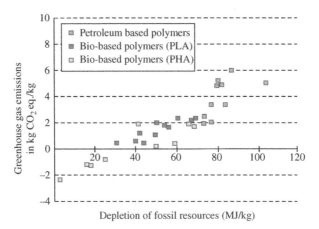

Depletion of fossil resources (MJ/kg)

FIGURE 4.15 A comparison of fossil resources and carbon footprint of conventional plastics with PLA and PHA. Source: Reproduced with permission from Nova Institute GmbH, Huerth/Germany (2012). Downloaded from http://www.bio-based.eu/ecology/

4.6.2 Emerging Bio-Based Plastics

Presently, only about 1% of plastics produced globally are bio-based, with a capacity of only 1.4 MMT in 2012. This is expected to rapidly increase with ramped-up production in Asia and South America to >6 MMT by 2017 (Institute for Bioplastics and Biocomposites, 2012). The biomass used can be food crops (corn and soy), nonfood crops (switchgrass), or agricultural waste. Using the first two categories that include corn, sugarcane, sugar beet, potato, cassava, rice, wheat, and sweet potato, will directly or indirectly compete with land use for food production. This is a valid concern on bio-based plastics technology.[19] While the percentage of arable land needed to support this growth by 2016 is claimed to be less than 0.025% of the global arable land (Institute for Bioplastics and Biocomposites, 2012), adverse regional impacts of such land and water use in future cannot be ruled out. Clearly, alternative nonfood sources of biomass need to be found to support the growth of this plastic. Efforts are being made to convert agricultural waste into PLA and other bio-based plastics (Kim and Dale, 2004). A second generation biomass feedstock that is independent of food crops, perhaps based on agricultural waste, would lend considerable impetus to future growth in bioplastics.

Table 4.10 lists the six highest-volume bioplastics that can be expected in the near future. The leading position of PE is to be expected as it is the most processed resin in any event, and a "drop-in" bio-version of the same resin is particularly easy to integrate into existing manufacturing lines. The PET capacity will be supported by the demand of the resin for beverage bottles.

---

[19] Using agricultural subsidies (as with gasohol production for automobile fuel market) is not a good strategy to expand bio-based plastic production. Instead, the as yet undeveloped options of using agricultural waste, weeds, and algae as biomass feedstock, that will not interfere with the food supply need to be encouraged instead.

TABLE 4.10   Highest-capacity Bio-based Plastics by 2015

Plastic	Percentage[a]	Comments
Biopolyethylene	26	HDPE already in production (Braskern) and used in US packaging market For example, single-serve fruit juice (Odwalla) and yoghurt drink (Dannon) and other beverage (Tetra Pak) packaged in bio-based HDPE
Biopoly(ethylene terephthalate)[b]	17	Bio-based PET beverage bottles (used by Coca-Cola and Pepsi-Cola) already in the market
Poly(lactic acid)	13	Manufactured in the United States by NatureWorks and already in the market in food service applications
Poly(hydroxyalkanoates) (PHA)	9	Compostable agricultural products
Biopolyesters	8	
Biopoly(vinyl chloride)	7	The ethylene for monomer is obtained from bio-based ethanol
Bionylons	5	

Source: From European Bioplastics/University of Applied Sciences and Arts Hanover (2012).
[a] Percentage of the total global production capacity.
[b] The estimate takes into account partly bio-based PET as well.

4.6.2.1 Bio-PE   Bio-PE is made from ethylene monomer derived from dehydration of ethanol made by fermentation of corn or molasses. Both ethylene and ethylene glycol (EG) can also be produced by cracking bionaphtha. With sugarcane, feedstock a hectare of land yields about three tons of bio-PE. Coupled with fossil fuel resource conservation, the use of bio-based feedstock also results in substantial reductions in carbon emissions. For instance, sugarcane-derived PP has a net negative carbon emission (or a sequestration) of 2.3 kg of CO_2/kg of bio-PP produced. This compares with the emission of 3 kg/kg of PP produced, resulting in a net approximately 5 kg carbon emission avoided per kg bio-PP (Godall, 2011). Reliance on agricultural biomass as a raw material is the main limiting factor for this technology. However, improving the process to use lignocellulosic waste materials can address this problem.

4.6.2.2 Bio-PET   Popularity of bio-PET is due to its successful application in beverage bottles. Polyester is made by polycondensation of EG with terephthalic acid [TPA]. Bio-based EG is made from molasses, sugarcane, switchgrass, and bagasse via their fermentation into ethanol. Using this bio-EG with fossil fuel-based TPA, a PET that is partially bio-based is obtained. The "PlantBottle" by Coca-Cola in 2009 is made using this partly (22.5%) bio-based resin. The Bio-PET is recyclable, can be processed without variance from regular PET, performs the same in products and reduces carbon emissions of 8–11% as well, according to the manufacturer, compared to the fossil fuel-based PET.[20]

Non-renewable   Renewable
feedstock   biomass feedstock

A fully bio-based polyester beverage bottle was subsequently developed (lab scale) by PepsiCo. Coca-Cola is also pursuing a 100% bio-based bottle. TPA can also be derived from bio-derived intermediates[21] such as bio-xylene or from bio-derived 5-hydroxymethylfurfural or furandicarboxylic acid [FDCA] from fructose via alkoxymethyl fufural. Efforts to synthesize bio-TPA via the muconic acid route without going via *p*-xylene are also under development.[22]

Another 100% bio-based alternative polyester well suited for beverage bottles and likely to be economical compared to PET is poly(ethylene furanoate) (PEF) (Byun and Kim, 2014). Biomass-derived saccharides (such as sucrose) can be converted into alkoxymethyl furfural, which is oxidized into FDCA. The FDCA can be condensed with EG to yield PEF polymer with a higher Tg (86°C), a lower Tm (235°C), and a higher modulus compared to PET. Its barrier properties are also reported to be superior to those of PET.[23]

4.6.2.3 PLA The popular bio-derived plastic PLA is made from lactic acid obtained by fermenting starch into glucose and then into lactic acid (Bordes et al., 2009; Jem et al., 2010; Okada, 2002) (see Fig. 4.14). The lactic acid as with most bioproducts is in the L form and can be condensed into a L-lactic acid polymer. However, if a mix of D- and L-form monomers are used, the properties of the plastic depend strongly on the ratio of the stereoisomers (or D:L) in the

---

[20] The detailed LCA reports are as yet not publicly available for review.

[21] See, for instance, the Japanese patent JP2011219736 (Toray Ind. Inc. 2011) and the Coca-Cola patent and WO 2009120457 A2 2009 Coca-Cola Co 2009.

[22] Amyris Inc., Emeryville, Calif. (www.amyris.com)

[23] Oxygen barrier is six times and CO_2 is three times better than for PET.

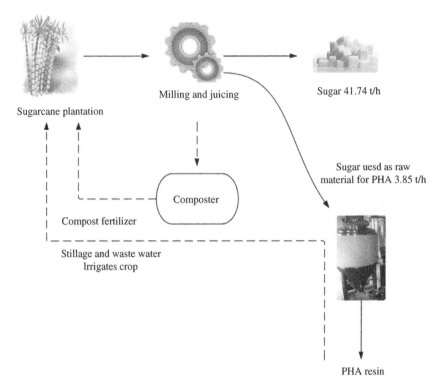

FIGURE 4.16 Schematic of PHA production facility illustrating the recycling of solid and water waste into sugarcane field. Source: Based on information from Nonato et al. (2001).

polymer (Garlotta, 2001). Two routes to a high-molecular-weight PLA are shown in the figure.

PLA is a semicrystalline plastic that can be injection molded, extruded or blown into film, spun into fibers,[24] blow molded, extruded, and expansion molded. The somewhat lower thermal stability and impact strength (Lim et al., 2008) as well as the moderate oxygen/water permeability (van Tuil et al., 2000) may limit its use in some applications despite its low cost (Lunt, 1998). Compostability of PLA is emphasized in marketing PLA products; even though the product is compostable, it does not readily biodegrade under ambient conditions or under anaerobic conditions (Integrated Waste Management Board, 2007) (see also Chapter 6).

But PLA is not suited for all applications: a drawback, for instance, is that in unblended form, it softens at approximately 60°C. The biobased polymer is currently manufactured by NatureWorks LLC (United States) and Hycail (Netherlands), Mitsui Chemicals, and Toyota (Japan).

---

[24] Even a PLA-based fabric, Lactron, that is softer than cotton or polyester, has been developed in Japan by the Japanese company Kanebo Gohsen Ltd.

4.6.2.4 Poly(Hydroxyalkanoates) Unlike bio-based PE, PET, and PLA, the poly(hydroxyalkanoates) (PHA) are bioplastics synthesized by bacteria. It was the first bacterial polymer to be harvested commercially. PHAs are deposited within the bacterial cells of many species as a lipoic material (Burdon, 1946). It is also unusual in that PHAs though hydrophobic still rapidly biodegrade in the environment. All bacterial polymers are not necessarily biodegradable (Steinbüchel, 2005); PHA's biodegradability is attributed to its saturated polyester chemical structure.

Bacteria grown under specific conditions of low limiting nutrients (N, P, and S) generate PHAs as storage materials in high yield. A variety of substrates can support the growth of PHA-producing bacteria. These include milk whey (Nath et al., 2008), molasses (sugarcane or beet) (Solaiman et al., 2006), and glycerol (Ibrahim and Steinbüchel, 2009). Over 250 bacterial species produce PHAs, and some of the best species (*Cupriavidus necator*) can accumulate the polymer up to 80% of its dry cell weight. In addition to naturally occurring species, genetically manipulated recombinant strains (mainly of *E. coli* modified to increase PHA yield and bioreactor expression) have been used for PHA production (Keshavarz and Roy, 2010). The schematic in Figure 4.16 illustrates the process for a specific pilot plant in a sugarcane factory in Brazil (Nonato et al., 2001) set up to produce poly(hydroxybutyrate) [PHB], the first type of PHA to be discovered. This example is particularly interesting because of the sustainability aspects of using bagasse to produce energy to run the operation and the near closed-loop recycling of the waste in the plant.

PHAs are manufactured by Telles (United States), Biomer (Germany), Mitsubishi Gas and Kaneka (Japan), PHB Industrial S/A (Brazil), and Metabolix (United States).

The chemical structures of PHB and its copolymer PHBV are shown below, and the polymer has an average molecular weight of **Mn** ~ 500,000 to a million g/mol. Common copolymers[25] typically have high (80–95%) of the HB monomer. Because of high stereoregularity, PHB polymer has a high degree of crystallinity (~80%) and a high Tm of 179°C. It is a plastic that matches the properties of popular thermoplastics such as polystyrene or polypropylene and can therefore be processed in conventional equipment but is still biodegradable in the environment. The copolymers have reduced crystallinity and Tm values that range from 50 to 180°C depending on composition.

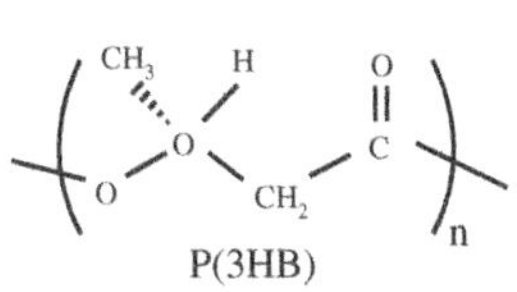

P(3HB)

Poly(3-hydroxy butyrate)

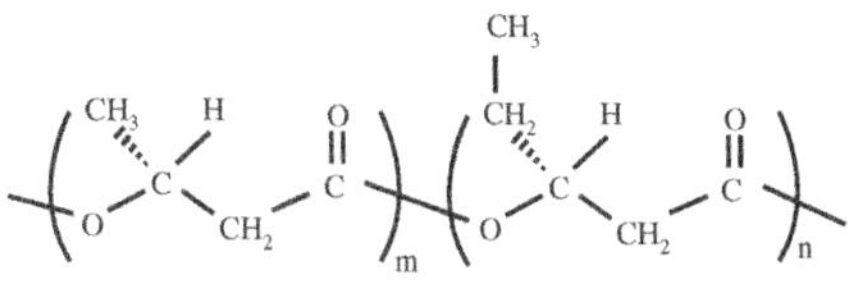

P(3HB-*co*-3HV)

Poly(3hydroxy butyrate-*co*-3 hydroxy valerate)

[25] Poly(3-hydroxybutyrate-*co*-3-hydroxyvalerate), PHBV, poly(3-hydroxybutyrate-*co*-4-hydroxybutyrate): poly(3-hydroxyoctanoate-*co*-hydroxyhexanoate): poly(3-hydroxyoctanoate-*co*-hydroxyhexanoate).

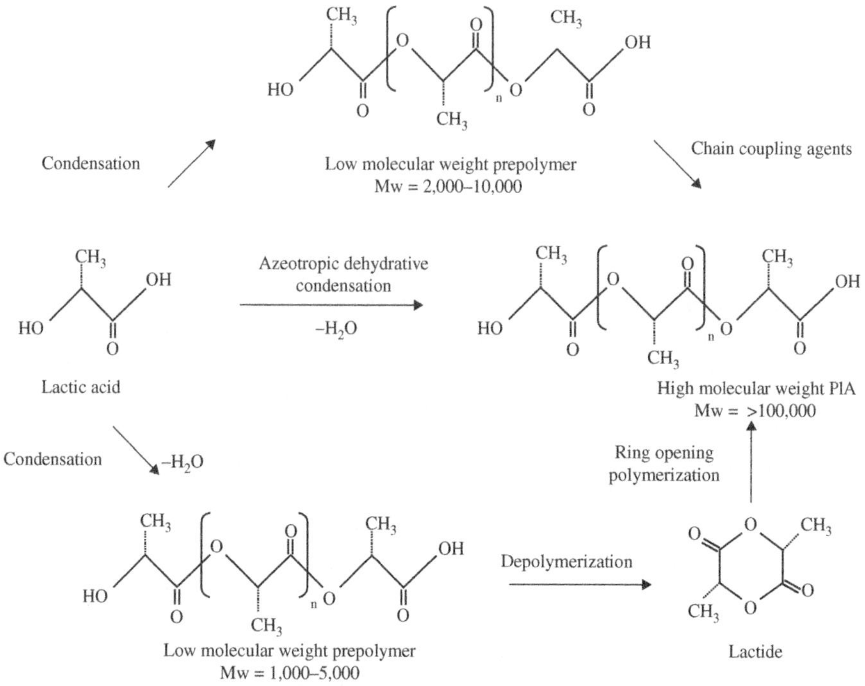

FIGURE 4.17 Schematic diagram of poly(lactic acid) manufacture from L-lactic acid. Reproduced with permission from (Gupta et al. 2007).

A copolymer of P3HB with valeric acid, P(3HB-*co*-3HV), with the structures shown above, can also be bacterially produced and available commercially. In the United States, PHB is manufactured by Telles (ADM/Metabolix) and Meridian Inc. PHBV is manufactured by Metabolix Inc.

4.6.2.5 Bio-Based Thermosets: PU PUs are condensation polymers of a bifunctional or multifunctional triol (polyetherol) with a polyisocyanate. Research efforts at replacing the synthetic polyol with one derived from castor oil are in progress (Lin et al., 2013; Zhang et al., 2014). Commercial polyols partially based on castor oil are already available,[26] and when used as a drop-in replacement in conventional PU, foam slab stock production results in a product that is ~25% bio-based.

REFERENCES

Bohlmann GM. Biodegradable packaging life-cycle assessment. Environ Prog 2004; 23:342–346.

---

[26] BASF markets Lupranol Balance 50, which is a polyol partially derived from castor oil.

Bordes P, Pollet E, Avérous L. Nano-biocomposites: biodegradable polyester/nanoclay systems. Prog Polym Sci 2009;34:125–155.

British Plastics Federation. *Energy in Plastics Processing—A Practical Guide*. London: Crown Copyright; 1999. *Good Practice Guide 292*; p 18.

Bull SJ, Zhou Q. A simulation test for wear in injection moulding machines. Wear 2001;249 (5–6):372–378.

Burdon KL. Fatty material in bacteria and fungi revealed by staining dried, fixed slide preparations. J Bacteriol 1946;52:665–678.

Byun Y, Kim YT. Chapter 15—Utilization of bioplastics for food packaging industry. In: Han JH, editor. *Innovations in Food Packaging*. 2nd ed. Volume 2014, San Diego: Academic Press (Elsevier); 2014. p 369–390.

Canadian Industry Program for Energy Conservation [CIPEC]. *Guide to Energy Efficiency Opportunities in the Canadian Plastics Processing Industry*. Ottawa: Her Majesty the Queen in Right of Canada; 2007.

Canadian Plastics Industry Association. *Guide to Energy Efficiency Opportunities in the Canadian Plastics Processing Industry*. Ottawa: Her Majesty the Queen in Right of Canada; 2007.

European Bioplastics/University of Applied Sciences and Arts Hanover. Why green plastics are here to stay. Compounding World 2012; (June Issue):45.

Franklin Associates. Cradle to gate lifecycle inventory of nine plastics resins and four polyurethane precursors. Report prepared for the Plastics Division of the American Chemical Council. Prairie Village: Her Majesty the Queen in Right of Canada; 2010.

Franklin Associates. *Life Cycle Inventory of Plastic Fabrication Processes: Injection Molding and Thermoforming*. Kansas: Prairie Village; 2011.

Garlotta D. A literature review of poly (lactic acid). J Polym Environ 2001;9 (2):63–84.

Godall C Bioplastics: An important component of global sustainability. A White Paper; 2011. Downloaded from http://www.carboncommentary.com/blog/2011/09/02/bioplastics-an-important-component-of-global-sustainability?rq=bioplastics. Accessed Oct 8, 2014.

Groot W, Borén T. Life cycle assessment of the manufacture of lactide and PLA biopolymers from sugarcane in Thailand. Int J Life Cycle Assess 2010;15:970e84.

Hammond GP, Jones CI. Embodied energy and carbon in construction materials. Proc ICE Energy 2008a;161 (2):87–98.

Hammond GP, Jones CI. Embodied energy and carbon emissions are from the Inventory of Carbon and Energy. Version 1.6a. Bath: University of Bath; 2008b.

Hottle TA, Bilec MM, Landis AE. Sustainability assessments of bio-based polymers. Polym Degrad Stab 2013;98 (2013):1898–1907.

Ibrahim MHA, Steinbüchel A. Poly(3-Hydroxybutyrate) production from glycerol by Zobellella denitrificans MW1 via high-cell-density fed-batch fermentation and simplified solvent extraction. Appl Environ Microbiol 2009;75 (19):6222–6231. 0099–2240.

Institute for Bioplastics and Biocomposites. *European Bioplastics*. Berlin; 2012. Available at http://en.european-bioplastics.org/wp-content/uploads/2013/publications/EuBP_FactsFigures_bioplastics_2013.pdf. Accessed Sep 30, 2014.

Integrated Waste Management Board. Performance evaluation of environmentally degradable plastic packaging and disposable food service ware—Final report. Integrated Waste Management BoardSacramento2007. Downloaded from http://www.calrecycle.ca.gov/Publications/Documents/Plastics/43208001.pdf. Accessed Oct 8, 2014.

James K, Grant T. *LCA of Degradable Plastic Bags*. Melbourne: RMIT (Royal Melbourne Institute of Technology) University; 2005.

Jem K, van der Pol J, de Vos S. Microbial lactic acid. Its polymer poly(lactic acid) and their industrial applications. In: Chen GQ, editor. *Plastics from Bacteria-Natural Functions Applications*. Heidelberg: Springer-Verlag; 2010. p 323–346.

Joshi SV, Drzal LT, Mohanty AK, Arora S. Are natural fiber composites environmentally superior to glass fiber reinforced composites? Compos Part A 2004;5 (3):371–376.

Kent R. Energy management in plastics processing. 2008. 265pp. www.pidbooks.com.

Keshavarz T, Roy I. Polyhydroxyalkanoates: bioplastics with a green agenda. Curr Opin Microbiol 2010;13 (3):321–326. 1369–5274.

Kim S, Dale BE. Global potential bioethanol production from wasted crops and crop residues. Biomass Bioenergy 2004;26:361.

Lim LT, Auras R, Rubino M. Processing technologies for poly(lactic acid). Prog Polym Sci 2008;33:820–852.

Lin S, Huang J, Chang PR, Wei S, Xu Y, Zhang Q. Structure and mechanical properties of new biomass-based nanocomposite: castor oil-based polyurethane reinforced with acetylated cellulose nanocrystal. Carbohydr Polym 2013;95 (1, 5):91–99.

Lithner D. Environmental and health hazards of chemicals in plastic polymers and products. Doctoral thesis for the degree of doctor of Philosophy in Natural Science, Department of Plant and Environmental Sciences, University of Gothenburg, Gothenberg, Sweden; 2011.

Lunt J. Large-scale production, properties, and commercial applications of polylactic acid polymers. Polym Degrad Stab 1998;59:145–152.

Madival S, Auras R, Singh SP, Narayan R. Assessment of the environmental profile of PLA, PET and PS clamshell containers using LCA methodology. J Clean Prod 2009;17:1183–1194.

Manas-Zloczower, I. (2009) *Mixing and Compounding of Polymers: Theory and Practice*. Carl Hanser Verlag GmbH & CoMunich. 1185.

Momani B. Assessment of the impacts of bioplastics: Energy usage, fossil fuel usage, pollution, health effects, effects on the food supply, and economic effects compared to petroleum based plastics. Project Report Submitted to Worcester Polytechnic Institute, Worcester. 2009. Downloaded from http://www.wpi.edu/Pubs/E-project/Available/E-project-031609-205515/unrestricted/bioplastics.pdf. Accessed Sep 8, 2014.

Nath, A., Dixit, M., Bandiya, A., Chavda, S. and Desai, A. J. (2008, September). Enhanced PHB production and scale up studies using cheese whey in fed batch culture of Methylobacterium sp. ZP24. Bioresour Technol, 99 13, 5749–5755, 0960–8524.

Nonato RV, Mantelatto PE, Rossell CEV. Integrated production of biodegradable plastic, sugar and ethanol. Appl Microbiol Biotechnol 2001;57:1–5.

Okada M. Chemical syntheses of biodegradable polymers. Prog Polym Sci 2002;27:87–133.

Philp JC, Ritchie RJ, Guy K. Biobased plastics in a bioeconomy. Trends Biotechnol 2013;31 (2):65–67.

Pilz H, Brandt B, Fehringer R. *The Impact of Plastics on Life Cycle Energy Consumption and Greenhouse Gas Emissions in Europe*. Vienna: Denkstatt GmbH Hietzinger Hauptstraße 28· 1130; 2010.

Pita VJRR, Sampaio EEM, Monteiro EEC. Mechanical properties evaluation of PVC/plasticizers and PVC/thermoplastic polyurethane blends from extrusion processing. Polym Test 2002;21 (5):545–550.

Plastic News. Global thermoplastic resin capacity 2008. Plastic News 2010; (March Issue).

Reddy MM, Vivekanandhan S, Misra M, Bhatia SK, Mohanty AK. Biobased plastics and bionanocomposites: current status and future opportunities. Prog Polym Sci 2013;38 (10–11):1653–1689.

Rudel RA, Perovich LJ. Endocrine disrupting chemicals in indoor and outdoor air. Atmos Environ 2009;43 (1):170–181.

Sablani SS, Rahman MS. Food packaging interaction. In: Rahmna MS, editor. *Handbook of Food Preservation*. Boca Raton: CRC Press, Taylor & Francis; 2007. p 939–956.

Shen L, Patel M. Life cycle assessment of polysaccharide materials: a review. J Polym Environ 2008;16:154–167.

Shen L, Worrell E, Patel MK. Comparing life cycle energy and GHG emissions of bio-based PET, recycled PET, PLA, and man-made cellulosics. Biofuels Bioprod Bioref 2012;6:625–639.

Smith P, Lemstra PJ, Pijpers JPL. Tensile strength of highly oriented polyethylene. II effect of molecular weight distribution. J Polym Sci Polym Phys Ed 1982;20:2229–2241.

Solaiman DKY, Ashby RD, Foglia T, Marmer WN. Conversion of agricultural feedstock and coproducts into poly(hydroxyalkanoates). Appl Microbiol Biotechnol 2006;71 (6):783–789. 0175–7598.

Steinbüchel A. Non-biodegradable biopolymers from renewable resources: perspectives and impacts. Curr Opin Biotechnol 2005;16 (6):607–613. 0958–1669.

Thoen J, Busch R. Industrial chemicals from biomass—industrial concepts. In: Kamm B, Gruber PR, Kamm M, editors. *Biorefineries—Industrial Processes and Products: Status Quo and Future Directions*. Volume 2, Weinheim: Wiley-VCH; 2006.

Todd DB. *Plastics Compounding: Equipment and Processing*. Cincinnati: Polymer Processing Institute Books from Hanser Publishers. Hanser Gardner; 1998. p 288.

Toloken S. Asian composites growth to outpace EU, US. Plastic News 2012; (July 9 Issue).

Trucost Plc. 2013. Natural capital at risk: the top 100 externalities of business. Downloaded from www.trucost.com. Accessed Oct 8, 2014.

US Energy Information Administration. 2013. Frequently asked question. Downloaded from www.eia.gov/tools/faqs/faq.cfm?id=34&t=6. Accessed Oct 8, 2014.

Van Tuil R, Fowler P, Lawther M, Weber CJ. Properties of biobased packaging materials. In: Weber CJ, editor. *Biobased Packaging Materials for the Food Industry: Status and Perspectives*. Frederiksberg: KVL; 2000. p 136.

Vink ETH, Rábago KR, Glassner DA, Gruber PR. Applications of life cycle assessment to NatureWorks_ polylactide (PLA) production. Polym Degrad Stab 2003;80:403–419.

Vink ETH, Rábago KR, Glassner DA, Springs B, O'Connor RP, Kolstad J. The sustainability of NatureWorks_ polylactide polymers and Ingeo_polylactide fibers: an update of the future. Macromol Biosci 2004;4:551–564.

Vink ETH, Glassner DA, Kolstad JJ, Wooley RJ, O'Connor RP. The eco-profiles for current and near-future NatureWorks_ polylactide (PLA) production. Ind Biotechnol 2007;3:58–81.

Zhang L, Zhang M, Hu L, Zhou Y. Synthesis of rigid polyurethane foams with castor oil-based flame retardant polyols. Ind Crop Prod 2014;52:380–388.

5

SOCIETAL BENEFITS OF PLASTICS

When they first became commercially available, plastics were widely considered a miracle material, an exceptional example of engineering ingenuity of that generation. The promise of plastics in the future world was foreseen with surprising accuracy in the first technical volumes devoted to the material (Yarsley and Couzans, 1944). Plastics have indeed exceeded that expectation achieving the unique status of an indispensable material for modern lifestyle[1] including that in the developing world. Gaining such widespread use and consistent increase in volume production in a few decades suggests significant societal benefits to be associated with plastics (Andrady and Neal, 2009). Recently, however, serious questions have been raised as to their desirability as a material in several application areas. These concerns range from potential health hazards from additives leaching out of packaging plastics in contact with food to the occurrence of microplastics in the world's oceans posing a threat to the marine ecosystem. Despite these concerns, continuing growth of plastics and its consistent replacement of conventional materials is progressing unabated.

Advantages of using plastics in place of competing conventional materials are clearly demonstrated in at least three broad application areas: (i) energy-saving uses, (ii) uses that conserve materials, and (iii) uses that assure consumer health and safety. Value of plastic devices is apparent in medical and public health applications.

---

[1] Susan Freinkel, the author of a very readable and interesting volume, *Plastic: A Toxic Love Story*, counted the number of plastic items she came into contact in a day (Frienkel, 2011). The total was 196 items a day, twice that of nonplastic items!

Plastics and Environmental Sustainability, First Edition. Anthony L. Andrady.
© 2015 John Wiley & Sons, Inc. Published 2015 by John Wiley & Sons, Inc.

The disposable syringes, tubing, gloves and other plastic devices are low cost, disposable and can be readily sterilized. Plastic clips and sutures used for wound closure and bags for storage of blood provide a very valuable societal benefit. These benefits are consistent with the sustainability criteria discussed in Chapter 2. But societal benefits are not accrued in all applications of plastics. Also, the perceived benefits and adverse effects are of course limited to the current state of knowledge; lead paint and Freon aerosol cans were considered very desirable positive developments at the time they were introduced to the market. It was decades later that their environmental limitations were recognized and regulatory controls were introduced to limit their use.

Even with the most rigorous environmental controls in place, the use of plastics, in common with other materials, still leaves its footprint on the ecosystem. Societal benefits of plastics need to be appreciated to be able to assess if this environmental footprint is justified. Externalities are for the most part associated with the production phase of a plastic product and perhaps during its disposal. Societal benefits generally accrue during the use phase as apparent from the discussion that follows.

5.1 TRANSPORTATION APPLICATIONS OF PLASTICS

Plastics, particularly polymer-based composites (both thermoplastic and thermoset composites), have been exploited in transportation applications for decades. The use of common composites such as glass or carbon fiber reinforced unsaturated polyesters, epoxy, and poly(vinyl ester) matrices is well known. The unique advantage of composites is their strength (tensile strength 4–6 times that of steel) and stiffness, despite their lightweight.

5.1.1 Passenger Cars

In 2010, there were 1.02 billion automobiles on the roads worldwide,[2] and by 2050, this number is expected to grow to 5.0 billion. 0.8 billion of these are already in use in China where car ownership is growing at the rate of >25% a year. Internal combustion engines, however, are both wasteful and polluting. Automobiles, especially larger vehicles used in the United States, are a particularly energy-inefficient (as well as polluting) means of transport or delivery of goods. Less than 12% of the energy in the fuel is actually used for transportation, the rest being lost as heat (about a third in hot exhaust gases and about the same in cooling and braking). A single-occupant light automobile still consumes approximately 80 kWh per 100 person-km of travel, nearly double that for a passenger in a full Boeing 747 (~42 kWh 100 person-km) (MacKay, 2009.) Also, 30% of the US contribution to GHG emissions is from transportation with each (US) gallon of gas burnt emitting about 24 pounds of CO_2 into the air. About 75% of this fuel consumption relates to vehicle weight

---

[2] The statistics are from Wards Auto, the information center for the global auto industry in a 2010 report. www.wardsauto.com.

TABLE 5.1   Density, Modulus, and Strength of Materials Used in Automobiles

Material	Density (g/cm^3)	Modulus (GPa)	Tensile strength (MPa)	Embodied energy (MJ/kg)
Mild steel	7.86	200	300	32
High-strength steel	7.87	205	483	32
Aluminum	2.71	70	295	199
Magnesium	1.77	45	228	350
Polymer composite	1.57	190	810	75 GRC[a]
				150 CRC[b]

Source: Based on Cheah (2010).
[a]Glass-reinforced composite.
[b]Carbon-fiber-reinforced composite.

(McWilliams, 2011). Any decrease that can be achieved in the weight of a vehicle (without reducing its functionality) would result in very significant savings in fossil fuel use and avoided GHG emissions, provided it could be achieved without compromising passenger safety. This has driven automobile makers to increasingly substitute for metal components in automobiles with plastic or composite replacements.

The percentage of plastics in American automobiles has consistently increased over the years. But today's automobile is still over 60% iron/steel in construction, with plastics contributing only 9.3% to the weight but to about 50% of its volume (American Chemical Council, 2012). The most used plastics in passenger cars are PP (37%), PU (17%), and HDPE (11%) accounting for more than half of all plastics used in a vehicle.[3] The surface of the vehicles is protected by polymer-based coatings: usually a base electrocoat, a primer surface, and a top clear coat. It is the transparent clear coat that protects the vehicle surface from scratches (marring) and retain the gloss and color (Figure 5.1).

As seen in Table 5.1, the mechanical properties of polymer composites compare very well with those of other materials, and further substitution for at least some of the metal in an automobile is feasible. Functionally equivalent plastic components that can replace metal counterparts weigh 50–75% less. Typically, a 10% reduction in vehicle weight is estimated to reduce its fuel consumption by about 5–8%. Therefore, this weight advantage translates into very significant improved fuel efficiency, fossil fuel conservation, and avoided carbon emissions, given the size of the world fleet of vehicles.

In a recent study, the steel assist step in Chevrolet TrailBlazer/GMC Envoy was replaced by a functionally equivalent (slightly larger) but 51% lighter plastic one (PE International Inc., 2012). The impact of this substitution was studied over the vehicle lifetime using standard LCA methods. This single replacement is estimated to have saved 2.7 million gallons of fuel over the lifetimes of the full fleet of the 148,658 (2007 model year) vehicles. Fuel savings also result in avoided emission of carbon

---

[3]Information downloaded from Report by Markets and Markets Inc. www.marketsandmarkets.com/Market-Reports/automotive-plastics-market-passenger-cars-506.html

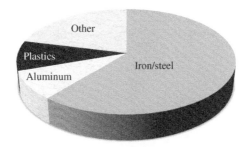

FIGURE 5.1   Fractions of different materials used in a 2011 light vehicle.

into the atmosphere. All environmental attributes did not improve as a result of the change: the acidification potential was higher for the plastic part compared to the metal step. Cost savings from the substitution of metals are readily apparent when commercial fleets of vehicles are considered. In a redesigned prototype delivery truck for a courier service,[4] the aluminum body panels were replaced with ABS plastic panels. Fuel economy gained from this change was shown to be about 40%. The savings in energy and emission accrued on switching to the lighter functionally comparable panel is obviously very significant.

National Highways Traffic Safety Administration (NHTSA) reported an interesting study (NHTSA, 2012) on material substitution from steel to polymer composite in the 2007 Chevrolet Silverado FE model. The substitution resulted in a 19% decrease in vehicle weight (from 2307 to 1874 kg) with no effect on safety in front crash tests. However, replacing metal components with composite or plastic is not a trivial process; it requires the redesign of the vehicle and therefore demands upfront capital investment as well as design time.

These savings in energy and emissions will be meaningful only if the weight reduced is not negated by the automaker adding on more features that may offset the reduction. The average curb weight of the light North American automobile has been trending up over the last two decades, increasing by 1.2% year, while the fraction of plastics/composites in the vehicles also increased during the same period (Environmental Protection Agency (EPA), 2009)! To achieve significant fuel savings, however, materials replacement must be coupled with reducing the curb weight of the vehicles (Table 5.2).

5.1.2   Air and Sea Transport

Composites have been used in aircraft construction for several decades but, until very recently, never in fuselage or wing construction. Conventional aircraft such as B747 or MD80 include less than 5% of carbon-reinforced composite. The new Boeing 787 commercial aircraft also called the Dreamliner (and Airbus A 350),

[4] The UPS Company tested its CV-23 prototype vehicles. The data is from UPS Composite Car Fact Sheet.

TABLE 5.2   Common Plastics Components Used in Automobiles

Plastic type	2001 (%)	2011 (%)	Where used
Polypropylene (PP)	22.1	23.9	Bumper, seating, dashboard, fuel system, body, under hood, trim, electrical, upholstery, and fluid containers
Polyurethanes (PU)	19.3	15.4	Seating and upholstery
Polyamides (nylon) (PA)	12.4	12.2	Seating, fuel system, electrical components, and fluid containers
Poly(vinyl chloride) (PVC)	7.9	7.9	Seating, trim, electrical components, and upholstery
(Acrylonitrile–butadiene–styrene) copolymer (ABS)	7.4	6.4	Bumper, seating, dashboard, trim, lighting
Polycarbonates (PC)	3.9	5.4	Dashboard and lighting
Polyethylene (PE)	4.4	4.8	Fuel system, electrical components, upholstery, and fluid containers
Other plastics[a]	22.8	24	

Source of data: JRC (2007).
[a]Includes composites based on unsaturated polyester, phenolics, acrylics, as well as other thermoplastic materials (American Chemical Council, 2012).

however, serves as a reliable window to the future innovations in composite technology. These aircraft were made of approximately 50% plastic composite material. Only 15% titanium, 20% aluminum and 10% steel were used in their construction. This reduction in weight allows for 30% fuel saving along with associated benefits of avoided GHG emissions. Compared to an Airbus A300-200 of similar size, the 787 is 30,000–40,000 lbs lighter and uses 20% less fuel in spite of having 45% more cargo space. The corresponding carbon emissions are also lower by approximately 20%, thanks to lightweight carbon-reinforced plastic composite construction.

As with conventional aircraft manufacture, numerous metal panels no longer have to be painstakingly applied to the fuselage using rivets. Large sections of it can instead be fabricated as single molding using composite material. Maintenance costs are also about 30% less with composite structures. These benefits, however, come at the cost of several inherent drawbacks of composites as a material of construction. Plastics burn readily, and burning poly(vinyl chloride) (PVC) and thermoset resins generally releases copious amounts of noxious toxic fumes. Adequate fire retardancy and other design features must be built-in to minimize this risk. Also, if subjected to a sudden impact in a crash landing, composites can crack and shatter rather than bend or dent (like malleable metals). Engineering designs need to ensure the integrity of the frame against catastrophic failure in such situations to ensure passenger safety. Concerns have been also raised on the crash survivability of 787's novel composite-wing fuel tanks (US General Accountability Office, 2011).

Ocean-going vessels have also benefited greatly from composite technology. In small pleasure vessels and fishing boats, fiber-reinforced unsaturated polyester (glass-reinforced polyester (GRP)) has been widely used. This allows for a light and reliable craft that better resists environmental corrosion relative to the wooden or metal crafts they replaced. However, the benefit of energy saving when using composites are best illustrated in larger vessels. The Visby-class stealth vessel used by the Swedish Navy is a 600 ton vessel that relies heavily on plastic composite construction to achieve good fuel efficiency, high speed, and low signature. Its hull is a sandwich construct where the foam core of PVC is covered with stiff surface laminates of carbon-fiber-reinforced poly(vinyl ester) composites (Burman et al., 2007). A vessel using composite hulls will naturally be lighter by approximately 50% compared to their metal counterpart and therefore less expensive to operate.

5.2 BENEFITS FROM PLASTIC PACKAGING

Packaging of products is a prerequisite for their sale in today's market and serves several important functions. Safe containment of the item (Buonocore et al., 2014), especially food, pharmaceuticals, or cosmetics, is the primary benefit of packaging. Plastics serve this function particularly well, and conversion of other types of packaging into plastic packaging is a common trend. The primary advantages of plastics in this application are lightweight, relatively low environmental footprint, bioinertness, and low cost. The functionality afforded by general-purpose packaging[5] includes the following:

1. Containment against damage, loss, and waste by spillage. Convenience in ease of handling and portioning in food packaging
2. Protection of contents against microbial contamination
3. Protection of contents against spoilage due to oxidation, hydrolysis, and intrusion by chemical species. Preservation of freshness of the packaged food items
4. Screening out of ultraviolet or infrared (IR) radiation that may degrade with the packaged item
5. Communication of information about the contents and their use to the consumer

Sustainable use of material resources requires the most economical and energy-efficient material be selected for packaging. A simple metric of environmental merit in packaging might be the embodied energy (EE) and the global warming potential (GWP) associated with the manufacture of the package, per unit volume of contents.

---

[5] Specialized packaging is used for instance in military rations packaging where additional functionality such as printed electronic circuitry, tamper-evident markers, or heating/cooling technologies might be integrated into the package.

TABLE 5.3   Estimate of EE and GWP (kg CO_2) Per 1 l Package

	Glass	Aluminum	Steel	PET	HDPE
Mass (g)[a]	325/433	20/45	45/102	25/62	38/38
Energy (MJ/l)	8.2	9.0	2.4	5.4	3.2
GWP (kg CO_2)[b]	0.37	0.05	0.18	0.34	0.06

Source: From Jackson and Bertényi (2005).
[a]Mass per package/approximate mass calculated for 1 Liter of contents.
[b]Estimated from (Hammond and Jones, 2008).

While the environmental footprint of a package cannot be adequately specified merely in terms of its EE and GWP, even a comparison based on just these two estimates can be useful. A wide range of environmental impacts (as discussed in Chapter 2) need to be taken into account to define the footprint comprehensively. However, the data for glass, plastic (poly(ethylene terephthalate) (PET) and HDPE), and metal cans (aluminum and steel) as packages for beverage (1l) are shown in Table 5.3. Aluminum is the worst performer using either criterion. The steel can[6] is the most energy efficient but is not the best in terms of GWP. Plastic packaging is less energy intensive than glass or aluminum and matches the GWP of the glass bottle. Assuming the functionality provided by each of these to be the same and depending on how the energy efficiency versus the GWP is weighted, steel cans or plastic bottles appear to be good choices. But the functionality of different materials is not comparable and other environmental attributes need to be included in a comprehensive assessment. More importantly, the impact of recycling (which is common for several of the packaging) has not been taken into account in this comparison. Even with these limitations, the results indicate plastics to be at the very least competitive with other packaging in terms of energy efficiency or in avoided GWP.

A US study in 2009 (funded by a plastics trade organization) reported a similar comparison of 12 oz aluminum (cans), glass bottles, and PET bottles but also taking recycling into account in the LCA. Recycling rates of different materials in the United States at the time of the study were used in developing the estimates summarized in Table 5.4.

The plastic bottle in this comparison emerges as being comparable in embedded energy cost (within the error of at least ±10%), but appears to be superior in terms of its carbon footprint, to aluminum cans. The outcome of the LCA exercise is sensitive

[6]Data from the ImpEE program, University of Cambridge, UK (Jackson and Bertényi, 2005).

TABLE 5.4   Embedded Energy, Solid Waste Generated, and GWP Per 10,000 Units of 12 oz Packages Manufactured

	Aluminum can	Glass bottle[a]	PET bottle
Energy (million BTU)	19.2	23.4	20.1
Solid waste	921	3931	554
GWP (lbs CO_2 equivalent)	3035	4252	2084
Recycling rate (%)	45.1	30.7	23.5

Source: Franklin Associates, Ltd. (2009b).
[a]Of the several choices available, the high-end data for glass bottles with steel caps was used here. Recycling rates generally reported for these materials in 2006–2007 were assumed in this calculation.

to recycling rates used, and the ranking can therefore change as recycling rates increase for different materials. Other studies (Humbert et al., 2009; Pasqualino and Meneses, 2011) also conclude plastics to have a lower carbon footprint compared to glass in specific packages.

A US study on retail packaging of tuna compared the merits of multilayer plastic pouches (PET/foil/nylon/PP) with those of conventional steel cans (Franklin Associates, Ltd., 2008). For 12 oz servings, the EE for steel can packaging was estimated to be 250% higher, and the associated GWP (lbs of CO_2 equivalents per package) was 460% higher, compared to the plastic pouch (Franklin Associates, Ltd., 2008). As the package sizes are made smaller, the environmental costs rise, and making a valid comparison becomes difficult. Generally, different LCA studies indicate the larger packages to have a smaller environmental footprint per unit quantity of packaged contents.

A comprehensive European study of 57 products compared plastic packaging (made of LDPE, LLDPE, HDPE, PP, PVC, PS, EPS, and PET) and alternative packaging (tin plate, steel, aluminum, glass, corrugated paperboard, paper, laminated paper-based composites, and wood) (Brandt and Pilz, 2011). Conventional packaging was found to be on the average 3.6 times the weight of plastic packaging for equivalent functional units. Their calculations considered the hypothetical situation where all the plastic packaging used in Europe is substituted for by a mix of alternative packaging materials. Prevailing levels of material recycling in Europe were assumed in the study. The main findings from the report are summarized in Figure 5.2. The energy saving is more than 50% when plastic packaging is used.

A similar US study in 2014 (Franklin Associates, Ltd., 2014) estimated that substitution of the 14.4 MMT of plastic packaging in the United States would require 64.4 MMT of alternative packaging materials. The substitution would result in 80% increase in cumulative energy demand and 130% higher GWP.

Advanced packaging concepts such as nanocomposite packaging for enhanced barrier performance, vacuum packaging, multilayer packaging films, modified atmosphere packaging (MAP) (Cooksey, 2014), aseptic packaging, and smart packaging developments all rely on plastics substrate for the most part. The next-generation packaging with shelf-life indicators, radio-frequency detectors, antimicrobial packaging, and biological sensors being designed is for the most part based on plastic materials.

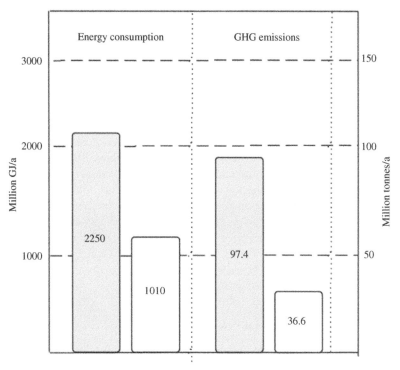

FIGURE 5.2 Effect of substituting plastic packaging materials with other packaging that provides the same functionality. Unfilled bars are for plastic packaging, and the filled bars are for a mix of other packaging. Life cycle energy consumption (scale on left) and life cycle GHG emissions (scale on right). Source: Redrawn from Brandt and Pilz (2011).

5.2.1 Waste Reduction

Food that is lost during the end of the production chain, at retail locations, and during final consumption stages, constitutes food waste (Parfitt et al., 2010). Though reliable statistics on food waste is sparse, the per capita food wastage is believed to be the highest in North America and in Western Europe, amounting to 280–300 kg/year. About 50% of edible food produced does not reach the dinner table and is wasted (Kader, 2005). Most of the food waste appears to originate from the consumption phase in Western countries; waste during processing, distribution, and storage is 2–3%, thanks to good packaging technology. Loses of 20–30% of food purchased by the consumer have been reported (Kantor et al., 1997; Mena et al., 2011; Ventour, 2008). In developing countries, most waste (50–90%) occurs in field, storage, or retail locations (Moomaw et al., 2012). The fraction of food waste in municipal solid waste streams generally decreases as the fractions of paper and board, metals, and glass packaging increase (Alter, 1989).

Properly designed packaging helps reduce some of these food losses while also reducing the environmental footprint of food production (Alter, 1989; Eide, 2002;

Marsh and Bugusu, 2007). In a study of food habits in 61 families, 20–25% of the household food waste was attributed to poor package design (Williams et al., 2012). For instance, spillage during opening packages and accessing food (especially by the elderly) was found to be an important source of waste but one that might be controlled by improved package design (Duizer et al., 2009). Ease of opening, reclosure, and features that avoid slippage during handling, can be designed into packaging to reduce waste. Package size also helps prevent waste; too much food in the package was identified as contributing to waste in a Swedish study (Williams et al., 2012). Unclear dating of freshness is another major reason for premature disposal of edible food. With advances in smart packaging (Mahalik and Nambiar, 2010), better communication of information is likely and may address this problem in the future.

Upgrading the quality of a package can sometimes reduce food waste (Williams and Williams, 2011). However, there is a trade-off between the reduction of food waste achieved and the larger environmental footprint of the upgraded package. Food waste also generates CO_2: each ton of food waste accounts for approximately 4.5 tons of the gas and also wastes embedded energy. For foods high in GWP such as meat or cheese (e.g., 25.5 kg CO_2-eq/kg bone-free meat), a substantial upgrading of packaging might be justified (Wikström et al., 2013).

5.2.2   Chemical and Microbial Protection

Packaged food is often discarded because of alterations in its odor, taste, and texture as a result of oxidation, hydrolysis, and spoilage of the contents due to degradation by bacteria. Controlling the ingress of oxygen and water vapor into food is therefore a key metric of protection afforded by food packaging. The values of permeability for O_2 and water vapor for common plastic materials are listed in Chapter 8. Plastics including blends, multilayered laminates, or filled laminates (especially with nanoscale fillers (Vermeiren et al., 2002) such as montmorillonite clay)[7] offer a wide range of barrier properties to select from (Wani et al., 2014). For instance, polymers such as poly(vinyl alcohol) (PVOH) or its copolymers in multilayer constructs provide transparent film with good barrier properties. Plastic packaging films with a thin layer of glass or ceramic surface coating deposited using physical vapor deposition provide a nearly impermeable and transparent barrier to gases and water. However, microcracking of glass during handling of the glass-coated plastic film results in some leakage of gases and water through these films.

Popularity of glass and tin cans in early packaging was due to their low cost and because they could withstand the high temperatures encountered with in-package sterilization. Low-temperature sterilization (peroxide treatment, use of short bursts of ultrahigh temperatures and gamma ray irradiation) is used in modern sterilization to minimize the alteration of taste and aroma of the food. Plastics can withstand all the common sterilization treatments without significant degradation. Where transparency is not a requirement, metallized plastic films or metal–plastic laminates offer

---

[7]With Meals Ready to Eat (MRE), the multilayer packaging used in military field rations, the packaged food is edible after 3 years of storage at 80°F (27°C).

impermeability along with the convenience of heat sealability. Natural preservatives (such as eugenol) can be incorporated into plastic packaging to help delay ripening of fruit (Serrano et al., 2008).

Unlike with paper packaging, microbial passage across a nonporous plastic package is not possible. Equal microbial protection is offered by metal cans (Ahmed Idris and Alhassan, 2010) or glass bottles. Yet, the lightweight, low cost, convenience of use, design freedom and lack of potential hazard in use, generally make plastics the more desirable packaging material.

5.3 PLASTICS IN AGRICULTURE

Modern agricultural practices as well as food processing (more than 75% of grown food is processed prior to consumption) are particularly energy-intensive processes. They also result in serious water and air pollution in addition to carbon release into the environment. With food production being such a critical need, inefficient expenditure of energy, and the large environmental footprint of the food industry, is acceptable to the society. But any technologies that extend the growing season, reduce leaching of fertilizer, encourage early harvests, and increase the yield of produce (Li et al., 1999) are aggressively pursued. Plastic mulch films and greenhouse technology provide such improvements (see Fig. 5.3).

Mulch (soil cover) is used in arid and semiarid zones to conserve water used in irrigation (Li et al., 2008). Plastic mulch is lightweight and nonporous and is a good thermal insulator, serving the role particularly well (Liu et al., 2009). The polyethylene (LDPE) and PVC films keep the soil temperature high during cold season and reduce the evaporation of soil water (Li et al., 2013) allowing economical drip

FIGURE 5.3 Plastic films used as mulch in agriculture.

irrigation to be used. The retention of soil moisture and the buildup of relatively high concentrations of CO_2 under impermeable plastic mulch films result in a microclimate conducive to crop growth (Bonachela et al., 2012). Not only do plastic mulch films produce crops earlier, they can also be optimally colored (Decoteau et al., 1998) to reflect or scatter appropriate wavelengths of light to match needs of the crop to ensure higher yield of produce.

However, the capital cost of plastic mulch is generally higher than that for conventional natural mulch, and it also has to be removed from the field annually and either reused or replaced in the next growing season. This labor-intensive and expensive practice is avoided by using environmentally photodegradable mulch film; these films left on the field after the season are broken down by photodegradation (see Chapter 6) into small fragments and are mixed in with the soil. The plastic fragments generated are too small to be discernible against the soil background.

Commercial use of greenhouse technology increases crop yield by providing an extended growing season as well by allowing closer control of the growing environment. Vegetables grown in greenhouse environments required less irrigation compared to conventional vegetable cultivation and provide a higher yield (Chang et al., 2011).

Rigid plastics such as polycarbonate (PC), fiber-reinforced polyester (GRP), impact-modified acrylics, as well as flexible films such as polyethylene (PE), PVC, and sometimes PET are used as glazing material in commercial greenhouses (Giacomelli, 1999; Govind et al., 1987). Glass by comparison is fragile, expensive to install, and heavy and may not withstand the weight of accumulated snow/ice. Even though plastics have a shorter service life compared to glass and the 4–6 ml flexible films are easily prone to tears, they are still a popular and economical choice of greenhouse material. A comparison of the optical and thermal performance of these glazing materials with glass is given in Table 5.5. The transmittance of photosynthetically active radiation (PAR)[8] in sunlight through glass and plastics is about the same, but rigid plastics transmit less of UV radiation (UV-B radiation that is considered detrimental to plant growth), while flexible plastics transmit approximately 50% of the IR radiation as well. The IR $\lambda > 3000$ nm radiation possibly improves the temperature in the greenhouse in colder climates. The transmission rates for PAR by plastic mulch can be modified by incorporating dyes/modifiers into the films (Lamnatou and Chemisana, 2013).

5.4   BUILDING INDUSTRY APPLICATIONS

Building construction (the second largest application sector after packaging) accounts for 25–30% of plastic production worldwide, and the specific uses of plastics in buildings are summarized in Figure 5.4. The figure is based on data compiled for 2005 in Europe, but the information is about the same for the United States. In the United States, however, about 30% of the exteriors of residential buildings are covered with vinyl siding. The most used plastic in building construction is PVC with about 3/4th of the production used in this sector. Products include pipes, fittings, cladding, flooring,

---

[8]PAR is the light in the wavelength range of 400–700 nm.

TABLE 5.5   Greenhouse Glazing Materials and their Characteristics

		Transmittance (%)			
Material	Life (years)	PAR[b]	UV[c]	IR	R[a]
Glass	30+	90	60–70	<3	1.1
Acrylic (twin wall)	20	86	44	<3	1.67
Polycarbonate (twin wall)	10	75	18	<3	1.67
Polyethylene (one layer)	3–4	88	80	~50	0.91

Source: Data from Both (2008) and Eugene A. Scales & Associates, Inc. (2003).
[a]The higher the R-value, the better the material is as thermal insulation.
[b]Photosynthetically active radiation.
[c]$\lambda = 300$–$400\,nm$

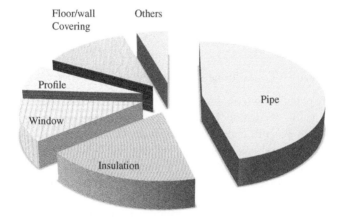

FIGURE 5.4   Main uses of plastics in building applications. Source: Drawn from data in Cousins (2012).

ductwork (rainwater goods), and window frames. These use the highest volume of plastics followed closely by foam insulation. Most building products have a long service life of several decades except for the floor/wall coverings and fitted furniture. These latter items constitute the common building-related waste in the municipal waste stream.

5.4.1   Pipes, Conduit, and Cladding

Because of their relatively lower cost, plastic pipes have for the most part replaced metal pipes in residential construction. Rigid PVC, chlorinated PVC (CPVC), PE, ABS, and polybutene (PB) plastic pipes are commonly used. GRP thermoset pipes are also used in civil engineering. They are lighter compared to available alternatives and much simpler to install, have relatively lower EE, and save energy by reducing the loss of heat from hot-water pipes (because the thermal conductivity of plastics is orders of magnitude lower than that of metals such as copper). Unlike metal pipes, they do not corrode or build up minerals over time requiring higher pumping pressures. No metal contaminants are added from them into potable water.

Rigid PVC pipes are extensively used for plumbing and waste water systems in residential housing because it is economical. A 2006 study by the building industry found PVC or CPVC pipe-based plumbing system for a 12-story residential tower costs approximately 75% less in material costs and 39% less in labor costs compared to the use of metal pipes (cast iron and copper) plumbing system (JB Engineering, 2006). The EE and GWP for the lighter plastic pipes will be much lower than that for cast iron or copper pipes in any event. Especially in developing countries, plastic pipe systems have contributed to the availability of potable water to rural populations. For potable water and hot water, CPVC, cross-linked polyethylene (XPE), or PB pipes are used. These can withstand continuous operating temperatures of approximately 80°C. Their susceptibility at high temperatures and flammability is a disadvantage compared to metal pipes. Conduits that carry electrical wiring as well as cable covering are also made out of rigid PVC.

5.4.2 Extruded PVC Cladding and Window Frames

A particularly popular low-cost choice (more than 50% of the market) of cladding for exterior façade of residential units in the United States is rigid PVC siding. Accessories such as soffit and fascia as well as rainwater systems are also usually made of the plastic and are gaining market share mainly because of low cost and good performance relative to metal. Not only are they lightweight, noncorroding, and easy to install, but they can also be fabricated in a wide range of colors. An opacifier additive, usually rutile TiO_2 filler, is used in PVC siding intended for use outdoors to control light-induced yellowing discoloration or surface chalking damage on extended exposure. Despite the stabilization afforded by pigment, the lifetime of cladding is still limited by uneven yellowing discoloration and surface degradation leading to "chalking." Recent innovations include lighter microfoam cladding and insulated PVC cladding with polystyrene foam (EPS) backing. Including recycled content in cladding improves its environmental characteristics. According to a leading manufacturer,[9] 10–13% of the foam backing, 79–96% of the PVC, and <1% of the glue in a foam-backed cladding are based on recycled material (including postindustrial as well as postconsumer contributions).

The use of extruded rigid PVC windows and doorframes is common in residential building practice in the United States. An estimated 3 MMT of PVC (or about 8% of the production) is used in window- and door-frame applications (Stichnothe and Azapagic, 2013). Comparisons of the performance of PVC windows to aluminum or wood widows taking (Asif et al., 2005; Citherlet et al., 2000) recycling

---

[9] Data is from CertainTeed Siding Products LCA Study (CertainTeed, Valley Forge, PA). Downloaded from http://www.certainteed.com/resources/vs_lca_report_cts417.pdf

TABLE 5.6   A Comparison of Leading Materials Used in Window Frames

Frame	Embodied energy (MJ/kg)[a]	Carbon emissions[a] kg eq CO_2/kg	Thermal conductivity (W/(m·K))	Mean estimated lifetime[b] (years)
Aluminum	214 (34.1)	11.2 (1.98)	205	43.6
PVC	77.2	2.41	0.19	24.1
Hardwood	7.8	0.47	0.07–0.17	39.6

[a]Hammond and Craig (2012). Value for recycled metal is shown in parenthesis.
[b]Asif et al. (2005) for each material.

into account (Stichnothe and Azapagic, 2013), are published. Generally, wood and aluminum-clad wood windows are found to have lower overall impacts compared to aluminum or PVC windows. A review of the available LCAs on windows (Salazar and Sowlati, 2008), however, concluded that lifetime environmental impact is not strongly determined by the choice of material as no one material has an advantage in all impact categories. It is the design features, such as thermal transmittance coefficient of the entire window and air tightness of the window to limit the operational energy losses, that define the environmental impact in the use phase (Tarantini et al., 2011). Use-phase impacts therefore dominate the lifecycle impacts for windows.

PVC window frames have a thermal conductivity that is at least 1–2 orders of magnitude lower than that of aluminum and about the same as that of wood windows. A PVC window with double glazing significantly lowers thermal losses compared to standard aluminum frames and helps make homes energy efficient (Table 5.6). Toughness of a PVC-extruded profile can match that of aluminum frames. But plastic windows are less durable compared to the aluminum windows (though still better than wooden frames in this regard).

5.4.3   Foam Insulation

Plastic foam insulation is typically made of rigid polyurethane (PU), polyisocyanurate, or expanded polystyrene (EPS). The thermoset foams include the sprayed in-place formulations where liquid reactants generate the foam *in situ*. All insulation plays a very important role in conserving energy used to heat/cool homes. In the United States, space heating/cooling is the second largest use of fossil fuel energy accounting for 19% of the total energy usage. Efficient low-cost insulation of buildings results in considerable savings in fossil fuel resources and therefore, also avoided GHG emissions.

Heat losses in residential buildings occur primarily via convection from the roof (~25–35%) followed by air leakage and walls (~15–25% each). Adequate insulation in roof and exterior walls is therefore a significant energy-saving investment. Table 5.7 compares thermal conductivity and environmental merits of different insulation materials (Dewick and Miozzo, 2002). The insulation efficiency is typically

TABLE 5.7   The Thermal Conductivity and Environmental Performance Rating of Common Building insulation Materials

Insulating material	Density (kg/m³)	Conductivity at 10°C[a] (W/m·K)	Environmental impact rating
Rigid PU foam	35–50	0.023	5.75
Expanded PS foam	15–30	0.033–0.038	5.25
Extruded PS foam	28–45	0.026	5.25
Phenolic foam	60	0.022	5.75
PVC foam	40–300	0.029–0.048	6.00
Glass wool	16–80	0.031–0.037	4.50
Rock wool (from slag)	23–80	0.033–0.037	4.50

Source: From Dewick and Miozzo (2002).

[a]Conductivity (λ) is temperature dependent: $\lambda = \lambda_0 + bT$, where λ_0 is the value of λ at 273 K.

measured in R values (the higher the value, the better the insulation). Values of $R \sim 4$–5 for PU and extruded polystyrene foams are higher than that for glass wool or wood ($R \sim 3.0$–3.5) as well as for plate glass ($R \sim 0.2$).

Recent studies (Franklin Associates, Ltd., 2009a, 2012) modeled energy saving from polystyrene (EPS) insulation used in residential housing. A comparison was made between foam-insulated structural panels (5 5/8 in EPS with strand board) and stick-framed conventional wall with R-19 fiberglass insulation. The payback period for the additional cost of EPS insulation in the wall was estimated to be 2.7–7.8 years for the United States and 1.4–3.9 years for Canada. If the structure lasted for 50 years (the insulation effectiveness of the EPS survives but may slowly decrease over this long a period), the average energy savings were found to be as high as 10 times the cost of the insulation. Energy saved in residential heating/cooling is also linked to very significant savings in avoided GHG emissions expressed as CO_2 equivalents.

Plastic insulation compares particularly well with competing materials (see Table 5.7). However, unlike inorganic insulation materials such as glass fiber, cork, perlite, or rock wool, all polymer foams, especially PU, will readily burn in a fire emitting smoke as well as CO and other gases (Stec and Hull, 2011; Yi et al., 2011). Smoke inhalation is well known to be the predominant cause of death in building fires. PU, being a thermoset, does not melt in a fire, whereas thermoplastic polystyrene does melt around 240°C. Flammability of insulation is controlled to some extent by using flame-retardant additives in the insulation.

The various applications of plastic materials in building construction are summarized in Table 5.8, which include both thermoplastics and thermoset materials. The number of + symbols qualitatively indicates the volume use of the particular plastic material in the specific product category. In addition to the main product categories discussed already, the table includes low-volume products such as plastics used in roofing or as moisture control films.

TABLE 5.8 Main Types of thermoplastics Used in Building Construction

	PVC	HDPE	LDPE	EPS	PU[a]	ABS[b]	PB	EPDM[c]
Pipe and conduit	+++	+++	+			+	+	
Siding, fascia, soffit	+++		+					
Rainwater systems	++							
Insulation				+++	+++			
Windows/doors	++							
Flooring	++							
Moisture control			++					
Roofing membrane[d]	++							++

[a]Polyurethane and polyisocyanurate foam.
[b](Acrylonitrile–butadiene–styrene) copolymer.
[c](Ethylene–propylene–diene monomer) polymer is an elastomer.
[d]Plasticized grade of PVC is used.

5.4.4 Wood–Plastic Composites

Berge (2009) discussed trends in sustainability of building materials and construction. There is growing interest in the construction industry to use greener materials and in certifying construction projects as being "green" or "sustainable." Using more of low-impact materials such as wood is an integral part of this effort. A material that benefits from this trend toward "green" buildings is thermoplastic composites with bio-fiber fillers, called wood–plastic composites (WPCs). The 2010 global market for WPC was 2.3 MMT and is expected[10] to grow to 4.6 MMT by 2016, given the short-term projected growth rate of 13.8%. It is used in outdoor furniture and decking as well as in the construction of bridges (see Fig. 5.5). The longest plastic bridge in the United States the 24.6 feet long and is the Onion Ditch Bridge in Logan County, Ohio, is made of recycled plastic.

The plastic fraction used in WPC is usually recycled PE, PP, PET, PVC PS, or a mix of these. Wood fiber as well as other natural lignocellulosic fibers can be used as the filler; the characteristics of fibers that are good candidates are shown in Table 5.9. Being of natural origin, a given type of fiber can have a large variation in properties (Kouini and Serier, 2012). The use of compatibilizers or pretreatment of the wood fraction is often employed to ensure good dispersion of filler in the plastic matrix.

WPC does not outperform wood in most environmental attributes. For instance, a recent (Bolin and Smith, 2011) cradle-to-grave LCA was carried out on the environmental merits of WPC based on recycled plastics and those of treated wood, for deck construction (Clemons, 2002). In every category, the treated wood (with copper oxide and dodecyl dimethyl ammonium chloride) outperformed WPC material. The energy cost of WPC was >8.5 times that of treated wood. But when compared to plastic building materials, WPC nearly always turns out to be better from an environmental standpoint, based on in several LCA studies (Corbiere-Nicollier et al., 2001; Wotzel et al., 1999).

---

[10]BCC Research Market Forecasting (2011). Global market for wood–plastic composites will pass 4.6 million metric tons by 2016, http://bccresearch.blogspot.com/2011/11/global-market-for-wood-plastic.html

FIGURE 5.5   A deck made of wood–plastic composites.

TABLE 5.9   Candidate Wood Fibers For Wood–plastic Composites

	Glass	Flax	Hemp	Jute	Ramie	Coir	Sisal
Density (g/cm³)	2.55	1.4	1.48	1.46	1.5	1.25	1.33
Tensile strength (MN/m²)	2400	800–1500	550–900	400–800	500	220	600–700
Stiffness (GPa)	73	60–80	70	10–30	44	6	38
Water absorption (%)	—	7	8	12	12–17	10	11

Source: Reproduced with permission from Pritchard (2007).

The wood component in WPC absorbs light and undergoes facile light-induced degradation. Plastic component also undergoes weathering degradation depending on the thermoplastic used. Of the common plastics, polystyrene (Lee et al., 2012) is the most susceptible to such degradation. The light stabilizers used in bulk plastics such as hindered amine light stabilizers (HALS) (Xue et al., 2012) or light absorbers (Chaochanchaikul and Narongrit, 2011) can be successfully used with WPCs as well.

5.5   ORIGINAL EQUIPMENT MANUFACTURE (OEM)

Most electronic and other equipment used at home have a considerable amount of plastics in their design. Plastics are good electrical and thermal insulators, and being lighter than conventional materials used in OEM (such as metals), results in energy saving during transport. The classes of plastics commonly used in home appliances are summarized in Table 5.10. In some of these uses, the specific plastics are virtually irreplaceable without significantly increasing the cost, and in others, volume

TABLE 5.10 Main Types of Plastics Used in Equipment and Household Goods Manufacture

	PP	PS	PPE/PS	SAN	ABS	PC/ABS	PET	PA	PVC
Printers and fax machines	X	X	X	X	X				
Telecom equipment		X			X	X			
TV housing		X	X			X	X		
Toys	X	X			X			X	X
Monitors		X			X	X			
Computer housing		X			X	X			
Dishwashers		X			X				X

Source: Based on Mudgal and Lyons (2011).
The leading material in each category is highlighted. PPE - Poly(phenylene oxide).

production of the equipment at an affordable cost is only possible because plastic materials are used in their fabrication.

A salient use of plastics is in refrigerators and freezers where closed-cell insulating foams such as PU are used. These result in significant energy savings that can be appreciated by comparing the life cycle energy costs of refrigerators that use the alternative fiberglass insulation with those using PU foam. One such study (Franklin Associates, Ltd., 2000) found appliances with PU insulation to require only 61% of the energy used by those relying on fiberglass insulation. The study was based on foam using the blowing agent HFC141b, now phased out pursuant to the Montreal Protocol. But the general conclusion of the study still holds at least qualitatively.

5.6 USING PLASTICS SUSTAINABLY

Consumption invariably results in an environmental footprint. That resulting from using food, water, and medicine, essential for existence, and others such as transportation and illumination that allows for higher quality of life, are accepted because of the high level of benefits these services provide. Plastics too provide a range of societal benefits that are very much a part of modern lifestyle. The availability of these is at the cost of approximately 8% of the annual fossil fuel use along with the associated GHG and other emissions.

The investment provides a ubiquitous building material, a range of packaging that delivers food safely from the farm to the table, transportation that is lighter (conserving non-renewable fuels), medical devices that protect human health, and a myriad of other conveniences. The nontangible benefits of these conveniences also contribute to their societal benefits. There are also applications where plastics either directly or indirectly help energy production. For instance, development of wind energy relies heavily on thermoset composites. A blade of a standard 1.5 MW windmill measures 35–40 m and weighs 6–7 tons. These are fabricated from GRP or epoxy composites. In photovoltaic (PV) modules, plastics protect the active material within PV modules, replacing glass in the application. Plastic components are used in nuclear power plants and in some hydroelectric plants as well.

Can the investment in plastics be justified on the basis of the impressive societal benefits they deliver? Defining environmental sustainability in material-specific terms (implied in the above query) is an oversimplification. No material plastics, glass, or paper can be rejected as being unsustainable as a class; each may be sustainable or not in some of its specific applications. Regardless of the class of material, it is ultimately that which delivers the required functionality and places the minimum burden on the environment that is sustainable. In most applications, plastics do qualify in this regard. Nevertheless, within each class of materials, the goal should be to innovate, improve, and recover for reuse materials and make them safer to use and increasingly environmentally sustainable.

REFERENCES

Ahmed Idris YM, Alhassan IH. Effect of packaging material on microbiological properties of Sudanese white cheese. Int J Dairy Sci 2010;5 (3):128–134.

Alter H. The origins of municipal solid waste: the relations between residues from packaging materials and food. Waste Manag Res 1989;7 (2):103–114.

American Chemical Council. 2012. Chemistry and light vehicles. Available at http://www.plastics-car.com. Accessed October 8, 2014.

Andrady AL, Neal MA. Applications and societal benefits of plastics. Philos Trans R Soc Lond B Biol Sci 2009;364:1977–1984.

Asif M, Muneer T, Kubie J. Sustainability analysis of window frames. Build Serv Eng Res Technol 2005;26 (1):71–87.

Berge B. *Ecology of Building Materials.* 2nd ed. Oxford: Architectural Press; 2009.

Bolin CA, Smith S. Life cycle assessment of ACQ-treated lumber with comparison to wood plastic composite decking. J Clean Prod 2011;19 (6–7):620–629.

Bonachela MR, Granados JCL, Hernández J, Magán JJ, Baeza EJ, Baille A. How plastic mulches affect the thermal and radiative microclimate in an unheated low-cost greenhouse. Agric Forest Meteorol 2012;152 (15):65–72.

Both AJ (2008). Greenhouse glazing. New Jersey Agricultural Extension Station. Available at http://njveg.rutgers.edu/assets/pdfs/ajb/Glazing.pdf. Accessed October 8, 2014.

Brandt B, Pilz, H. The impact of plastic packaging on life cycle energy consumption and greenhouse gas emissions in Europe. Plastic Europe and Denkstatt GmbH Hietzinger Hauptstraße 28·1130 Vienna, Austria. 2011. Available at www.denkstatt.at. Accessed October 8, 2014.

Buonocore G, Sico G, Mensitieri G. Safety of food and beverages: packaging material and auxiliary items. In: Yasmine M, editor. *Encyclopedia of Food Safety.* Waltham: Academic Press; 2014. p 384–396.

Burman M, Lingg B, Villiger S, Enlund H, Hedlund-Åström A, Hellbratt S-E. Cost and energy assessment of a high speed ship. Project LASS Report. KTH, School of Science, Sweden. 2007. Downloaded from http://www.escm.eu.org/docs/eccm13/0116.pdf. Accessed October 8, 2014.

Chang J, Wu X, Liu A, Wang Y, Xu B, Yang W, Meyerson LA, Gu B, Peng C, Ge Y. Assessment of net ecosystem services of plastic greenhouse vegetable cultivation in China. Ecol Econ 2011;70 (4):740–748.

Chaochanchaikul K, Narongrit S. Stabilizations of molecular structures and mechanical properties of PVC and wood/PVC composites by Tinuvin and TiO_2 stabilizers. Polym Eng Sci 2011;51 (7):1354–1365.

Cheah LW. Cars on a diet: the material and energy impacts of passenger vehicle weight reduction in the U.S [Ph.D. Thesis]. Cambridge, MA: Massachusetts Institute of Technology. 2010.

Citherlet S, Di Guglielmo F, Gay J-B. Window and advanced glazing systems life cycle assessment. Energy Build 2000;32 (2000):225–234.

Clemons C. Wood–plastic composites in the United States: the interfacing of two industries. For Prod J 2002;52 (6):10–18.

Cooksey K. Modified atmosphere packaging of meat, poultry and fish. In: Han JH, editor. *Innovations in Food Packaging.* 2nd ed. Food Science and Technology International Series San Diego: Academic Press; 2014. p 475–493.

Corbiere-Nicollier T, Gfeller Laban B, Lundquist L, Leterrier Y, Manson JAE, Jolliet O. Life cycle assessment of biofibres replacing glass fibres as reinforcement in plastics. Resour Conserv Recycl 2001;33:267–287.

Cousins K. Plastics in building construction. Technology & Engineering. Shropshire: Smithers Rapra Publishing. 124pp; 2012.

Mudgal MS, Lyons L. Plastic waste and environment. Report prepared for the European Commission (DG ENV). Bio-Intelligence Service, Paris, France; 2011.

Decoteau DR, Kasperbauer MJ, Daniels DD, Hunt PG. Plastic mulch color effects on reflected light and tomato plant growth. Sci Hortic 1998;34 (3–4):169–175.

Dewick P, Miozzo M. Sustainable technologies and the innovation–regulation paradox. Futures 2002;34 (9–10):823–840.

Duizer M, Robertson T, Han J. Requirements for packaging from an ageing consumer's perspective. Packag Technol Sci 2009;22 (4):187–197.

Eide MH. Life cycle assessment (LCA) of industrial milk production. Int J Life Cycle Assess 2002;7 (2):115–126.

Environmental Protection Agency (EPA). *Light-Duty Automotive Technology, Carbon Dioxide Emissions, and Fuel Economy Trends: 1975 through 2009.* Washington, DC: U.S. Environmental Protection Agency; 2009.

Eugene A. Scales & Associates, Inc. Energy conservation opportunities for greenhouse structures. St. Paul: Minnesota Department of Commerce Energy Office; 2003.

Lee C-H, Hung K-C, Chen Y-L. Effects of polymeric matrix on accelerated UV weathering properties of wood-plastic composites. Holzforschung 2012;66 (8):981–987.

Li F-M, Guo A-H, Wei H. Effects of clear plastic film mulch on yield of spring wheat. Field Crop Res 1999;63 (1):79–86.

Franklin Associates, Ltd. Plastics energy and greenhouse gas savings using refrigerator and freezer insulation as a case study. Prairie Village: Franklin Associates, Ltd.; 2000.

Franklin Associates, Ltd. *LCI Summary of Six Tuna Packaging Systems.* Prairie Village: Franklin Associates, Ltd; 2008.

Franklin Associates, Ltd. Life cycle inventory of three single-serving soft drink containers. Prairie Village: Franklin Associates, Ltd.; 2009a.

Franklin Associates, Ltd. Energy and greenhouse gas savings for EPS foam insulation applied to exterior walls of single family residential housing in the U.S. and Canada (EPSMA\ KC092060). Prairie Village: Franklin Associates, Ltd; 2009b.

Franklin Associates, Ltd. ICCA Building Technology Roadmap. Sponsored by International Association of Chemical Industries [ICCA]. Prairie Village: Franklin Associates, Ltd; 2012.

Franklin Associates, Ltd. Impact of plastics packaging on life cycle energy consumption & greenhouse gas emissions in the United States and substitution analysis. Prairie Village: Franklin Associates, Ltd; 2014.

Freinkel S. *Plastic: A Toxic Love Story*. Boston: Houghton Mifflin Harcourt; 2011. p 336.

Giacomelli GA. Greenhouse glazings: alternatives under the sun. Department of Bioresource Engineering, Cook College, Rutgers University; 1999. Available at http://AESOP. RUTGERS.EDU/~ccea/publications.html. Accessed October 8, 2014.

Govind R, Bansal NK, Goyal IC. An experimental and theoretical study of a plastic film solar greenhouse. Energy Convers Manag 1987;27 (4):395–400.

Humbert S, Rossi V, Margni M, Jolliet O, Loerincik Y. Life cycle assessment of two baby food packaging alternatives: glass jars vs. plastic pots. Int J Life Cycle Assess 2009;14 (2009):95–106.

Hammond G, Craig G. Inventory of carbon and energy. University of Bath, Sustainable Energy Research Team. 2012. Available at http://wiki.bath.ac.uk/display/ICE/Home+Page; jsessionid=2358DA949145979F0F8865783188344D. Accessed October 8, 2014.

Jackson S, Bertényi T 2005. Recycling of plastics. Energy Data from ImpEE. Available at www-g.eng.cam.ac.uk/impee/topics/RecyclePlastics/files/Recycling%20Plastic% 20v3%20PDF.pdf. Accessed October 8, 2014.

JB Engineering. Cost analysis of high-rise plumbing piping system; 2006. Report Nr 06A0706E1. Munster: JB Engineering

JRC. Assessment of the environmental advantages and drawbacks of existing and emerging polymers recovery processes. 2007.

Kader AA. Increasing food availability by reducing post harvest losses of fresh produce. Proceedings of the 5th International Postharvest Symposium; 2005. Available at http://ucce. ucdavis.edu/files/datastore/234-528.pdf. Accessed October 8, 2014.

Kantor L, Lipton K, Manchester A, Oliveira V. Estimating and addressing America's food losses. Food Rev 1997;20 (1):2–12.

Kouini B, Serier A. Properties of polypropylene/polyamide nanocomposites prepared by melt processing with a PP-g-MAH compatibilizer. Mater Des 2012;34:313–318.

Lamnàtou C, Chemisana D. Solar radiation manipulations and their role in greenhouse claddings: fluorescent solar concentrators, photoselective and other materials. Renew Sustain Energy Rev 2013;27:175–190.

Li S, Kang S, Li F, Zhang L. Evapotranspiration and crop coefficient of spring maize with plastic mulch using eddy covariance in northwest China. Agric Water Manag 2008;95 (11):1214–1222.

Li SX, Wang ZH, Li SQ, Gao YJ, Tian XH. Effect of plastic sheet mulch, wheat straw mulch, and maize growth on water loss by evaporation in dryland areas of China. Agric Water Manag 2013;116 (1):39–49.

Liu CA, Jin SL, Zhou LM, Jia Y, Li FM, XiongYC, Li XG. Effects of plastic film mulch and tillage on maize productivity and soil parameters. Eur J Agron. 2009, Nov;31(4): 241–249.

MacKay DJ. *Sustainable Energy without the Hot Air.* Cambridge, UK: UIT Cambridge Limited; 2009.

Mahalik NP, Nambiar AN. Trends in food packaging and manufacturing systems and technology. Trends Food Sci Technol 2010;21 (3):117–128.

McWilliams A. Lightweight materials in transportation. Wellesley: BCC Research. Report # AVM056B; 2011, February. Available at http://www.bccresearch.com. Accessed October 28, 2014.

Marsh K, Bugusu B. Food packaging roles, materials, and environmental issues. J Food Sci 2007;72 (3):39–55.

Mena C, Adenso-Diaz B, Yurt O. The causes of food waste in the supplier–retailer interface: evidences from the UK and Spain. Resour Conserv Recycl 2011;55 (6):648–658.

Moomaw W, Griffin T, Kurczak K., LomaxJ. The critical role of global food consumption patterns in achieving sustainable food systems and food for all. A UNEP Discussion Paper. United Nations Environment Programme, Division of Technology, Industry and Economics, Paris, France; 2012.

National Highways Traffic Safety Administration (NHTSA). Investigation of opportunities for lightweight vehicles using advanced plastics and composites. 2012. Report Nr DOT HS 811 692WTH Consulting LLC, Apex.

Parfitt J, Barthel M, Macnaughton S. Food waste within food supply chains: quantification and potential for change to 2050. Philos Trans R Soc Lond B Biol Sci 2010;365 (1554):3065–3081.

Pasqualino M, Meneses FC. The carbon footprint and energy consumption of beverage packaging selection and disposal. J Food Eng 2011;103 (2011):357–365.

PE International Inc. *Life Cycle Assessment of Polymers in an Automotive Assist Step.* Boston: PE International Inc; 2012.

Pritchard G. Plants move up the reinforced plastics agenda, Reinf Plast. 2007;51(10):28–31. Report by Markets and Markets, Inc., 2013. Available at www.marketsandmarkets.com/Market-Reports/automotive-plastics-market-passenger-cars-506.html. Accessed October 8, 2014.

Salazar J, Sowlati T. Life cycle assessment of windows for the residential market in North America. Scand J For Res 2008;23 (2):121–132.

Serrano M, Martinez-Romero D, Guillen F, Valverde JM, Zapata PJ, Castillo S, Valero D. The addition of essential oils to MAP as a tool to maintain the overall quality of fruits. Trends in Food Science & Technology. In: Coles R, Kirwan MJ, editors. *Food and Beverage Packaging Technology.* 2nd ed. Hoboken: Wiley-Blackwell; 2008.

Stec AA, Hull TR. Assessment of the fire toxicity of building insulation materials. Energy Build 2011;43 (2–3):498–506.

Stichnothe H, Azapagic A. Life cycle assessment of recycling PVC window frames. Resour Conserv Recycl 2013;71:40–47.

Tarantini M, Loprieno AD, Porta PL. A life cycle approach to Green Public Procurement of building materials and elements: a case study on windows. Energy 2011;36 (5):2473–2482.

US General Accountability Office. Status of FAA's actions to oversee the safety of composite airplanes. GAO-11-849, General Accountability Office. A report to congressional requesters; 2011.

Ventour L. *The Food We Waste.* Banbury: WRAP; 2008.

Vermeiren L, Devlieghere F, Debevere J. Effectiveness of some recent antimicrobial packaging concepts. Food Addit Contam 2002;10:77–86. In: Coles R, Kirwan MJ. *Food and Beverage Packaging Technology.* 2nd ed. Hoboken: Wiley-Blackwell; 2011. p 83.

Wani AA, Singh P, Langowski H-C. Food technologies: packaging. In: Motarjemi Y, editor. *Encyclopedia of Food Safety*. Waltham: Academic Press; 2014. p 211–218.

Wards Auto. no date. Available at http://wardsauto.com/ar/world_vehicle_population_110815. Accessed October 8, 2014.

Wikström F, Williams H, Verghese K, Clune S. The influence of packaging attributes on consumer behaviour in food-packaging LCA studies – a neglected topic. J Clean Prod 2013;73:100–108.

Williams H, Williams FW. Environmental impact of packaging and food losses in a life cycle perspective: a comparative analysis of five food items. J Clean Prod 2011;19:43–48.

Williams H, Wikström F, Otterbring T, Martin L, Anders G. Reasons for household food waste with special attention to packaging. J Clean Prod 2012;24:141–148.

Wotzel K, Wirth R, Flake R. Life cycle studies on hemp fibre reinforced components and ABS for automotive parts. Angew Makromol Chem 1999;272:121–127.

Xue P, Jia M, Wang K, Ding Y, Wang L. Effect of photostabilizers on surface color and mechanical property of wood-flour/HDPE composites after weathering. J Wuhan Univ Technol Mater Sci Ed 2012;27 (4):621–627.

Yarsley WE, Couzans EG. *Plastics*. London: Pelican Publishers; 1944.

Yi A-h, Liu J-y, Zhao X, Yang Z. Study of the combustion performance of three kinds of organic heat insulation materials. Procedia Eng 2011;11:614–619.

6

DEGRADATION OF PLASTICS IN THE ENVIRONMENT

Though the plastics used in high volume are evidently recalcitrant materials, being organic compounds, they invariably do break down or degrade on exposure to the environment. Except when exposed outdoors to solar ultraviolet radiation (UVR) where measurable rates of loss in mechanical integrity are obtained, the process is painstakingly slow in most environments. Under solar irradiation, plastics gradually loses strength and mechanical properties, finally weakening to a point where surface cracks appear and brittle fracture fragments the material. The small fragments have high specific surface area, and if they are in a biotic medium, they will undergo biologically mediated degradation, further weakening and ultimately converting the material into inorganic molecules such as CH_4, NH_3, and CO_2. This last step where the plastic material is converted into inorganic molecules is "mineralization." However, because this entire process can take decades or centuries, plastics are often perceived as being "forever."[1] The biodegradation rates of plastics in most environments are far too slow to play any meaningful role in removing plastic litter from the environment. As all plastics being invariably biodegradable, the widely used terms "biodegradable plastics" and "degradable plastics" are strictly speaking, misnomers. These terms merely mean that these particular plastics degrade or break down at a much faster, generally at an observable or measurable rate, compared to the conventional common plastics such as polyethylenes (PE).

[1] A recent nonfiction best seller *The World Without Us* by Alan Weisman has a chapter with the title "Polymers Are Forever."

Plastics and Environmental Sustainability, First Edition. Anthony L. Andrady.
© 2015 John Wiley & Sons, Inc. Published 2015 by John Wiley & Sons, Inc.

Two requirements must be met to obtain biologically-mediated mineralization of a polymer: it must be exposed to a biotic environment conducive to degradation and its chemical structure must be amenable to breakdown by the available set of microbial enzymes in that environment. Thus, even the inherently biodegradable starch in grains (barley and wheat) found in Egyptian pyramids[2] was preserved over millennia because of the dry and dark environment they were stored in.

An often-posed question is how long it would take for a given plastic to completely degrade (or mineralize) outdoors, in soil, or in the ocean.[3] Because mineralization rates depend on the type of polymer and biotic characteristics of the environment it is placed in, this is a question that cannot be answered without qualification. A valid question might be how long does low density polyethylene (LDPE) takes to mineralize completely in garden soil? Direct data is not available to answer even this question reliably for any plastic material. Estimates based on extrapolation of laboratory data on accelerated biodegradation are sometimes quoted in response to such queries because of the lack of good information. Such estimates of lifetimes of common plastics are unreliable and inaccurate as extrapolations cannot be justified given the long durations of exposure involved. However, the process is slow enough to suggest that all common plastics ever produced (and not incinerated) still survive somewhere in the environment, undergoing slow biodegradation.

6.1   DEFINING DEGRADABILITY

At the outset, it is important to define the term "degradation" in the present context; it is a chemical change that alters the properties of the material (this change is called "degradation" because useful properties of interest such as high strength or high stiffness of the plastics are compromised in the process). For instance, polyolefins such as PE exposed to solar radiation undergo photo-initiated oxidation, the principal degradation reaction outdoors. In this case, oxidation is also accompanied by the scission of long chain-like polymer molecules. It is the chain scission (rather than the incorporation of oxygen into the molecular structure) that decreases the useful mechanical properties of the material and lead to brittleness that causes fracture. As discussed in Chapter 3, the unique, desirable properties of plastics are a consequence of their long-chain architecture; any shortening of the average chain length or DP markedly reduces the useful properties of the material.

Common degradation processes are conveniently categorized according to the principal agencies that bring about the chemical change. Figure 6.1 is self-explanatory and lists the different degradation mechanisms. Of these, only light-induced degradation breaks down common plastics at rates that are readily measurable. Biodegradation is perhaps the slowest of these. Hydrolysis as a mechanism of

---

[2]Egypt's pyramids, the largest and the oldest stone structures, were built about 4500 years ago.
[3]Lifetimes of plastics outdoors or in the ocean, sometimes published in the popular literature, are speculative and not reliable.

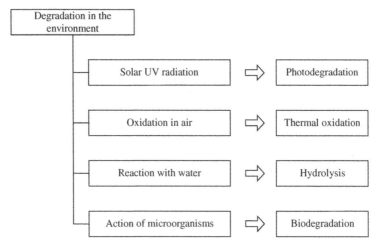

FIGURE 6.1 Principal agents of plastics degradation in the environment.

environmental degradation is available only to a few select plastics such as poly(lactic acid) (PLA) (Höglund et al., 2012; Muller et al., 1998). Common plastics such as PE, polypropylenes (PP), PS (polystyrene), and polyvinyl chloride (PVC) do not hydrolyze appreciably under ambient environmental conditions.

Photo-oxidation is the common mode of degradation for plastics exposed outdoors and is due to two processes acting in concert, photoinitiation, and thermal oxidation. In many instances, it is common to have several of these agents operate together, or sequentially, to degrade the plastic in a given environment.

Regardless of the mechanism involved, the extent of degradation increases with the duration of exposure of the material to the agent. Over very long durations, all polymers will invariably be completely degraded, both abiotically from thermo-oxidation followed by environmental biodegradation of the oxidized polymer. Complete biodegradation amounts to mineralization where the entire sample is converted into small molecules such as CO_2 and water, allowing the carbon sequestered in the plastic to join the nature's carbon cycle. At intermediate durations of exposures, a partially degraded plastic with a reduced average molecular weight **Mn** (g/mol) or sometimes with a higher average cross-link density, but always with markedly compromised useful properties, is obtained.

6.2 CHEMISTRY OF LIGHT-INDUCED DEGRADATION

Plastics used outdoors are routinely exposed to sunlight and undergo facile light-induced degradation (Hamid et al., 1995; Ranby, 1989; Sheldrick and Vogl, 2004). This is often the only significant degradation mechanism relevant to urban plastic litter and plastics used outdoors, such as in the exterior of buildings. Solar radiation at

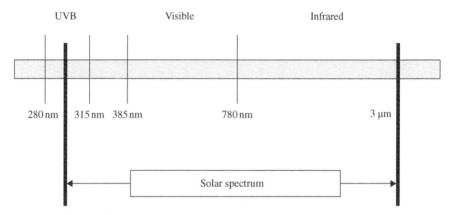

FIGURE 6.2 Regions of the solar spectrum reaching the Earth's surface.

the Earth's surface, facilitating degradation, can be divided into three broad regions as shown in Figure 6.2. It is only the visible-light wavelengths in the spectrum that can be perceived by the human eye but it is the solar UV-B radiation ($\lambda = 290–315$ nm) that is the most efficient in initiating degradation reactions (Torikai, 2000). Terrestrial solar UV spectrum contains only a small (~5%) fraction of UV-B radiation, most of it being filtered out by the stratospheric ozone layer. The spectral irradiance distribution of sunlight as well as the total solar radiation dose varies widely with the geographic location. In places such as Arizona, it can be as high as 8,000 MJ with approximately 330 MJ (total UVR) per year.

An important prerequisite for light-induced degradation is that the plastic absorb solar UVR or visible radiation[4] as only the absorbed light can degrade the plastic. Polyolefins in theory do not have chromophores that can absorb light effectively, but trace impurities in the plastic and some of the additives in the material act as good chromophores. Plastics such as PS, polycarbonate (PC), and polyethylene terephthalate (PET), have aromatic functional groups that absorb UV radiation.

Absorbed solar UV-Visible radiation can result in several different types of degradation reaction in a polymer.

1. Photolysis of the covalent bonds: The photon energy in solar radiation (especially the UVR) exceeds the bond dissociation energy for common covalent bonds in polymers. Despite this, direct photolysis of covalent bonds in polymers is rare, especially with the predominant C–C or C–H bonds in common

---

[4]This is an axiomatic principle of photochemical phenomena and is referred to as the Grotthus–Draper law.

polymers and only occurs in relatively weaker bonds such as peroxy linkages {–O–O–} as seen in the oxidation of polyolefins. Photolysis of polymer hydro-peroxide ROOH results in radical products that are able to initiate the oxidation reactions.

$$ROOH \rightarrow RO \cdot + \cdot OH$$

Photolysis is also involved in chain scission of (ethylene–carbon monoxide) copolymers irradiated with UV radiation.

$$-CH_2-CH_2-\overset{\overset{O}{\|}}{C}-CH_2-CH_2- \longrightarrow -CH_2-CH_2-\overset{\overset{O}{\|}}{\underset{\bullet}{C}} + H_2\overset{\bullet}{C}-CH_2-$$

2. Photomodification of chemical structure without any chain scission: Photoreaction may modify the main chain or the side chains of an irradiated polymer. The average chain length (or DP) does not change in the process even though **Mn** (g/mol) may be reduced. In the light-induced dehydrochlorination of PVC (D'Aquino et al., 2012) for instance, HCl is evolved without any significant main-chain scission or cross-linking, introducing conjugated unsaturation into the main chain.

$$\sim CH_2 - CHCl - CH_2 - CHCl - CH_2 - CHCl \sim \rightarrow$$
$$\sim CH_2 - CH = CH - CH = CH - CHCl \sim \cdots + \cdots 2HCl$$

It is the absorption of blue wavelengths in the white light spectrum by these polyene sequences that make degraded PVC surfaces appear yellow. Surface discoloration of PVC cladding (siding) and window frames exposed to sunlight is well known (Edge et al., 2010) and is a result of this reaction.

3. Photo-initiated oxidation accompanied by scission: With PE and PP, light-initiated oxidation is the predominant reaction, and extensive main-chain scission, sometimes accompanied by cross-linking, takes place with rapid loss of mechanical integrity. The autoxidation cycle associated with photo-initiated scission is discussed in detail in Figure 6.3.

4. Rearrangement of structure without scission: Photo-rearrangement reactions may take place, especially in pendant side chains. In bisphenol A polycarbonates, photodegradation by solar radiation occurs via two concurrent mechanisms. One is oxidative main-chain scission somewhat similar to that in polyolefins and results in a reduction of both DP and **Mn** (g/mol). The other is a structural rearrangement in the polymer chain as shown below. Exposure to wavelengths less than 300 nm (solar UV-B radiation) leads primarily to this (photo-Fries) rearrangement while oxidation of side chains occurs due to longer solar wavelengths (Pickett, 2011; Rivaton, 1995).

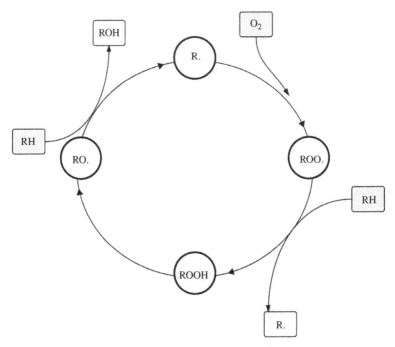

FIGURE 6.3   The cyclic autoxidation reactions for a polyolefin RH.

Photo-Fries Rearrangement

6.2.1   Light-Initiated Photo-Oxidation in PE and PP

The mechanism of light-induced degradation in PE and PP is autocatalytic oxidation of the polymer that is accompanied by main-chain scission (Singh and Sharma, 2008; Balabán et al., 1969). Solar UVR initiates this reaction by creating free radicals that propagate via a well-established cyclic sequence of reactions. Thermal oxidation of these polymers also proceeds via the same sequence of reactions but is slow at ambient temperatures. It is, however, significant during high-temperature processing of plastics, and thermal stabilizers are typically incorporated into plastics compounds to avoid degradation during processing. Light stabilizers are used in formulations intended for outdoor applications (such as siding, stadium seating, mulch films, glazing, or rope). The basic autoxidation reaction scheme involved is very well established and is schematically shown in Figure 6.3.

1. Initiation:

$$RH \rightarrow Free\,radicals, e.g., R\cdot, H\cdot$$

2. Propagation:

$$R\cdot + O_2 \rightarrow ROO\cdot$$
$$ROO\cdot + RH \rightarrow ROOH + R\cdot$$

3. Termination:

$$ROO\cdot + ROO\cdot \rightarrow ROOR + O_2$$
$$R\cdot + R\cdot \rightarrow R-R$$
$$RO\cdot + H\cdot \rightarrow ROH$$
$$R\cdot + H\cdot \rightarrow RH$$

As only hydrogen atoms in the polymer are abstracted in the reaction sequence, the polymer is designated as RH. The initiation reaction yields R· or alkyl radicals that add to molecular oxygen to form peroxy radicals (ROO·) that in turn reacts with the polymer (RH) to form hydroperoxide (ROOH) and an R· radical. As long as there is a steady supply of RH and oxygen, these two reactions can go on to convert the available –RH to –ROOH functionalities. The –ROOH moieties themselves undergo photolysis to yield RO· radicals making this an autocatalytic process. Some radicals may combine pairwise at high enough concentrations, or get otherwise deactivated, to end the cyclic process. At high extents of oxidation, the initial reaction products themselves get oxidized leading to the formation of a mix of aldehydes, carboxylic acids, ketones, and alcohols (Andrady et al., 2007). This makes the carbonyl index (the ratio of the >C=O band at $1740\,cm^{-1}$ to the "thickness" band at $2020\,cm^{-1}$ in infrared (IR) spectrum of the polymer) that increases with weathering (Andrady et al., 1993b; Roy et al., 2007), a convenient spectroscopic means of quantifying oxidation.

Chain scission accompanying these oxidative reactions, believed to occur during the propagation step of the sequence, is mostly responsible for the loss of useful properties in the polymer. As these reactions generate a high concentration of polymer radicals, some cross-linking (joining of chain radicals) can also occur concurrently with chain scission. But chain scission usually dominates over cross-linking in polyolefins exposed to solar-simulated radiation (Craig et al., 2005). Photo-oxidized polyolefins generally show reduced **Mn** (g/mol) (Andrady et al., 1993b) at relatively low levels of oxidation. A particularly sensitive and practical bulk mechanical property that reflects the extent of degradation is the tensile extensibility of the plastic (Andrady, 2007). In semi-crystalline polymers, chain scission occurs predominantly in the amorphous regions that oxygen can readily diffuse into. Therefore, the percent crystallinity of the polymer (for instance with PE) will generally increase with photo-initiated oxidation of these polymers. But the presence of additives and fillers can modify the matrix and this increase is not always observed in field exposure experiments.

6.2.2   Embrittlement and Fragmentation

In semi-crystalline polymers such as polyolefins, initial oxidation occurs in the amorphous tie molecules between crystallites and sometimes even in the inter-lamellar amorphous chains within crystallites. This allows a relatively low levels scission to cause a disproportionately large changes in mechanical properties due to brittle failure in the amorphous regions (Celina, 2013). In PP for instance, beginning stages of bulk embrittlement corresponded to only 0.01% of oxidation (Fayolle et al., 2004).

Plastics weathered outdoors or in the laboratory show surface cracking and pitting because of extensive degradation at the surface layer (Küpper et al., 2004; Qayyum and White, 1993; Yakimets et al., 2004; see Fig. 6.4). Solar UV radiation can only penetrate a thin surface layer of the plastic due to the high absorption coefficients of polymers at those wavelengths (Gulmine et al., 2003). This will likely localize degradation and lead to cracking of the surface. The particle size distribution yielded by fragmentation depends on whether bulk or surface fragmentation is dominant. Surface layer embrittlement leads to a large number of smaller fragments (Balabán et al., 1969) with one dimension equal to the thickness of the surface layer.[5]

Once the mechanical integrity of the plastic is drastically reduced, diurnal hot–cold cycles, action of animals, and friction from rainwater and wind movement can break up the surface as well as the bulk of weakened material. Litter is believed to fragment into progressively smaller pieces by this mechanism until the size range is small enough for it to blend in with the sand or soil in the background. Plastic fragments mixed in with the soil or water can then undergo slow biodegradation.

Fragmentation is an important precursor to biodegradation as it increases the surface area of plastic available to microbes to interact with. However, the kinetics of fragmentation (as opposed to that of degradation) and the evolution of particle sizes during fragmentation of plastics outdoors are essentially unknown.

A distinction needs to be made between degradation-induced fragmentation described earlier and that due to entirely physical causes. Marine borers consuming polystyrene (EPS) foam floats (Davidson, 2012) and insects or rodents attacking

---

[5] $N = 8 + (6d^2/t^2) - (12d/t)$ spherical daughters are formed where t is the thickness of peripheral layer (and therefore the diameter of daughter particle) and d is the diameter of the parent particle.

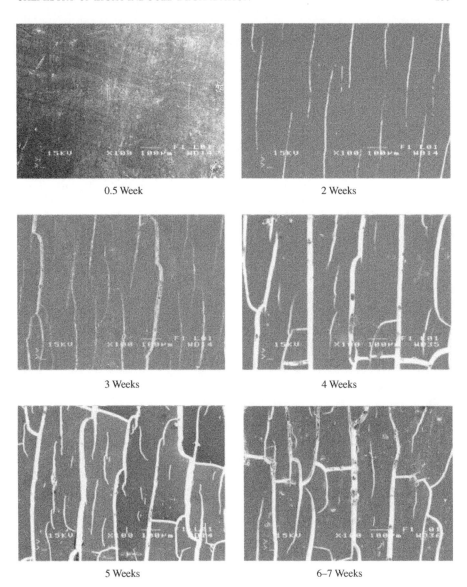

FIGURE 6.4 Development of surface cracks on PP surfaces on exposure to a filtered xenon light source ($600\,W/m^2$) at 42°C and at different durations of exposure. Source: Reproduced with permission from Yakimets et al. (2004).

plastic cable covers and gas pipes also create small plastic fragments (Chalot, 2011). But neither the average molecular weight **Mn** (g/mol) or the chemical structure of the polymer are altered in these processes. Therefore, not being true degradation, these changes are referred to as "biodeterioration." The same is true of composite plastics mixed with biodegradable materials, for instance, LDPE/starch[6] or LDPE/chitosan (Sunilkumar et al., 2012; Rutkowska et al., 2002) blends that are sometimes mislabeled as "biodegradable plastics." On environmental exposure, only the starch or chitosan fraction in these will readily biodegrade, and the resulting porous recalcitrant LDPE residue crumbles into small fragments (Jbilou et al., 2013). The **Mn** (g/mol) of the LDPE particles so produced is not reduced in the process.

6.2.3 Temperature and Humidity Effects on Degradation

In common with all chemical reactions, the rate of oxidative degradation also increases with the temperature of the substrate (Andrady et al., 2003). The magnitude of acceleration depends on the activation energy of the degradation process. The common form of the Arrhenius equation relates the rate constant k of a reaction to the temperature T (K).

$$\ln k = -\frac{\Delta E}{RT} + Z \tag{6.1}$$

where R is the gas constant (1.986 cal/mol), E is the activation energy, and Z is a constant. Therefore,

$$\ln\left(\frac{k_1}{k_2}\right) = -\frac{\Delta E}{R}\left(\frac{1}{T_1} - \frac{1}{T_2}\right) \tag{6.2}$$

The Equation 6.2 must be applied with caution as the rate must be time independent for the expression to apply. Kinetics of polymer degradation can change at $T > $ Tg (°C); therefore, values of E determined below Tg cannot be assumed at be the same at $T > $ Tg (°C). If the mechanism of degradation changes with temperature, Equation 6.2 cannot be used to estimate the value of ΔE. Typical values for ΔE for unstabilized PP was reported to be 27 kJ/mol (François-Heude et al., 2014).

Temperature effect is especially pertinent to outdoor exposure of plastics. Nearly 52% of the sun's energy is in the IR range, with most of it in the near-IR range (NIR, 700–1200 nm). NIR radiation interacts with the plastic increasing its bulk temperature[7] well above that of the surrounding air. The difference in temperatures between a surface painted with white pigment (reflects >80% sunlight) and a carbon black dispersion (reflects <5% sunlight) and exposed to solar radiation outdoors is more than 25°C.[8] Coloration of the plastic therefore significantly affects rates of degradation because of differences in heat buildup. The thermal conductivity

---

[6] A host of such blends including starch + cellulose acetate, PCL, starch + PP; and starch + copolymers are commercially available.

[7] The temperature of the plastic heated by sunlight will not be high enough to cause thermolysis or pyrolysis of the plastic. But it can be high enough to very noticeably accelerate the photooxidation process.

[8] Data from "Paint it Cool" Brochure" BASF Chemical Company. www.dispersions-pigments.basf.com/portal/streamer?fid=560474

of the backing material used (such as air, metal, and wood) in samples exposed also makes a difference in the outcome of weathering experiments because the temperature of samples depend on the backing.

6.2.4 Wavelength-Dependent Photodamage

The efficiency of photodegradation is inversely proportional to the wavelength of absorbed light; in sunlight, the solar UV-B wavelengths (290–315 nm) are therefore the most damaging radiation[9] followed by UV-A. Wavelength sensitivity information for deterioration of selected relevant properties (such as yellowing or changes in tensile strength) for common plastic resins and some of their common formulations have been reported (Andrady, 1997; Torikai and Hasegawa, 1998). These are plots of the efficiency of photodamage process (the extent of selected modes of damage per mole of photons incident on the sample) when the plastic is exposed to monochromatic light) versus the wavelength, λ (nm) (Andrady, 1997). An example of a wavelength sensitivity plot (also called an action spectrum) for light-induced yellowing of ligno-cellulose or mechanical pulp is shown in Figure 6.5. The data were generated by exposing the plastic to a beam of near-monochromatic radiation at different wavelengths followed by measuring the yellowness index with a colorimeter.

An alternative representation of the wavelength sensitivity of a polymer is the activation spectrum (which is specific to the source of light used). It is useful in assessing broad regions of the solar spectrum most effective in promoting degradation of the

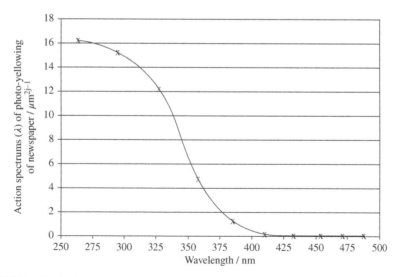

FIGURE 6.5 Action spectrum for the light-induced yellowing of mechanical pulp. Source: Reproduced with permission from Heikkilä and Kärhä (2014).

[9]Shorter wavelength (<290 nm) called UV-C is far more damaging than UV-B, but this radiation is filtered out by the stratospheric ozone layer and does not reach the Earth's surface.

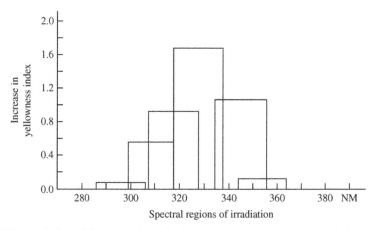

FIGURE 6.6 Effect of different solar radiation wavebands on the yellowness index of unstabilized Lexan polycarbonate film (0.70 mm) exposed to natural sunlight facing 26° South in Miami, FL. Source: Reproduced with permission from Andrady et al. (1992).

plastic material. Unlike single-wavelength (or monochromatic) sensitivity plots, these provide information on the effects of broad spectral bands of the solar spectrum. They yield information on synergistic and antagonistic effects encountered at different wavelengths. The effects of different wavebands in the solar-simulated spectrum are isolated using a series of cut-on filters.[10] Andrady (1997; Andrady et al., 1992) gives a good description of the cut-on filter technique and how it is used to generate activation spectra. Figure 6.6 shows an activation spectrum for the change in yellowness index of unstabilized Lexan PC film (0.70 mm) exposed to sunlight (Andrady et al., 1992). The 320–338 nm waveband yields the most degradation. Note that activation spectra depend on both wavelength sensitivity and the fraction of radiation in the wavelength interval of interest in the solar spectrum. As the fractional UVA in solar radiation is much higher than that of UVB, the most damaging region for plastics tend to be located at the UVA wavelengths.

As a first approximation, the change in a property D, ΔD, of a plastic material on exposure to solar radiation (or other radiation source) with a spectral irradiance distribution $H(\lambda)$ is assumed to be an additive function depending on the wavelength sensitivity function of the degradation process $E(\lambda)$.

$$\Delta D = \int H(\lambda) \cdot E(\lambda) \cdot d\lambda \tag{6.3}$$

The damage ΔD is the product of the quantum efficiency and the radiation absorbed by the sample (Martin et al., 2003) As the action spectrum $E(\lambda)$ is unitless, the activation spectrum has units (W/m²/nm). Then, at any given wavelength

$$\Delta D = H(\lambda)\{1 - \exp(-A(\lambda))\} \cdot \Phi(\lambda) \tag{6.4}$$

[10] A good review of the technique is found in the study of Torikai (2000).

TABLE 6.1   Most Damaging Range of Wavelengths in Sunlight for Common Thermoplastics

Plastic materials	Cut-off wavelength (nm)[a]	Damaging range (nm)[b]	Property of interest
LDPE/HDPE	<180	260–360	Optical density, extensibility
PP	<180	320–350	Tensile properties
		~320	Yellowing
PVC	<240	300–320	Yellowing
PS	<270	~280	Chain scission
		~330	Yellowing
PC	<280	310–340	Yellowing
Paper[c]	–	334–354	Yellowing

Source: Data from Searle et al. (2010).
[a]Wavelength at 5% absolute transmission is commonly used as the cut-on wavelength.
[b]The wavelength range as demarcated by activation spectra generated using a series of cut-on filters and solar simulated radiation.
[c]Groundwood pulp (newsprint) paper is included for comparison. Paper data from Andrady and Searle (1995).

where, $\Phi(\lambda)$ is the quantum efficiency of the degradation. The activation spectrum, or the damage that is attributed to any waveband ($\lambda_1 < \lambda < \lambda_2$), is obtained by integrating the expression above over the wavelength interval. With most plastics, however, the simple additivity implied by Equation 6.3 does not apply as there can be synergy or mutual cancellation of the effects of different wavelengths.

The solar absorption cutoff and the most effective solar wavelength range for selected modes of damage for different plastics based on reported activation spectra are summarized in Table 6.1. These data are for non-UV stabilized plastics. The presence of a UV stabilizer or some additives in the compound can change the activation spectra very significantly.

6.2.5   Testing Plastics for Photodegradability

Weatherability of plastic materials can be investigated under natural or accelerated laboratory exposure. Typically, standard ASTM (dog-bone shaped) test pieces of the plastic of interest are exposed outdoors in natural weathering exposure sites. Ideally, locations of high insolation (such as AZ and FL) are selected to ensure fast degradation. The samples exposed outdoors, usually on South-facing racks, are periodically removed and tested to determine their extent of degradation.

Tensile extensibility is a particularly sensitive measure of the extent of weathering. Alternatively, the development of carbonyl functionalities (>C=O) or other spectral changes in samples determined in Fourier Transform Infra-Red Spectroscopy, FTIR, might be used to assess degradation. In PE (Andrady et al., 1993b), the development of the carbonyl functionalities correlates well with the decrease in extensibility. Plastics such as PVC, PC, or PS (Ghaffar et al., 1976) turn yellow on photodegradation. Surface discoloration is a property of interest because it is the uneven discoloration and "chalking" of the surface (not the loss of mechanical integrity) that leads to replacement of products such as PVC siding. A solids colorimeter is used to measure tristimulus color values, and these are reported in terms of the parameters L, a, and b or as calculated values of yellowness index (YI) or whiteness index (WI).

Weatherability of plastics can also be measured in the laboratory with exposure of samples to simulated sunlight (xenon lamp with double borosilicate filters) or to UV wavelengths in sunlight using a UV 340 fluorescent sunlamp with or without water spray. The advantage of this "laboratory-accelerated" exposure is the high degree of control over spectral quality, light/dark cycles, and temperature. Other sources (such as fluorescent sunlamp UV-315) that emit a disproportionately higher amount of the UV radiation in its spectrum relative to that in terrestrial solar spectrum, allow for accelerated degradation. Lamps that emit wavelengths shorter than 290 nm, however, must be used with caution, because these wavelengths are typically not present in solar spectrum that reaches the Earth's surface.[11] At shorter wavelengths of exposure or at excessive sample temperatures not typically encountered outdoors, photoreactions that are not typical in environmental exposure to sunlight could take place. This is also the reason why the intensity of solar-simulated light should not be increased indiscriminately in accelerated weathering tests.[12]

Most plastic products, such as plastic building products, outdoor furniture, and artificial turf used outdoors, are stabilized against solar UV-induced damage to ensure full service life outdoors. A discussion of the different stabilization mechanisms and the interesting chemistry associated with stabilization are beyond the scope of this chapter. The reader is directed to other works that discuss the topic comprehensively (Wypych, 2010).

Stabilizers generally protect the polymer against solar UV damage via three strategies:

1. Absorbing incident UV radiation using organic (e.g., with benzophenones and benzotriazoles) or inorganic (e.g., rutile titanium dioxide) additives.
2. Quenching the photo-excited species formed in the polymer (e.g., with nickel dibutyl dithiocarbamate).
3. Removing the free radicals formed in the polymer (e.g., with hindered amine light stabilizers (HALS)).

The first two classes of compounds inhibit initiation reactions that form radical products, while the third mops up the radicals after they are formed.

---

[11] Some of the available standard acceleration techniques expose the sample to as much as 6 suns of radiation but at a controlled temperature.

[12] Minsker et al. (1982) have suggested that the intensity not be increased beyond three times the standard solar intensity in the geographic region. But this arbitrary value is likely to be excessive. It is safer to use the highest natural sunlight intensity (1 Sun) and lengthen the light-time as opposed to dark-time in the accelerated weathering exposure.

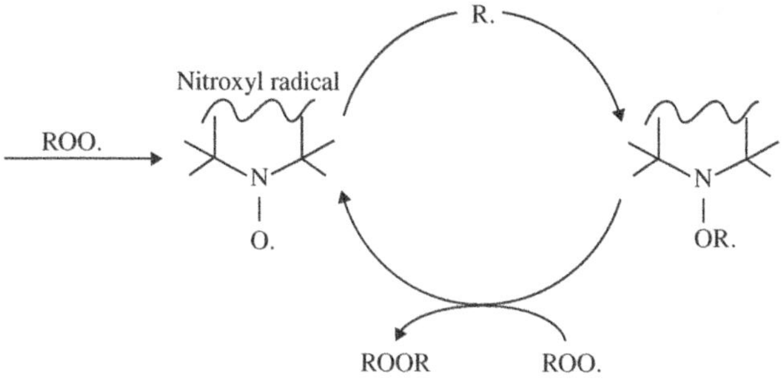

FIGURE 6.7 Simplified schematic of the mechanism of UV stabilization by HALS. P refers to polymer chain.

With rigid PVC products used in exterior building applications, approximately 9–13% of rutile titanium dioxide is often used as an opacifier pigment. The pigment absorbs incident UV radiation shielding the bulk polymer from exposure. Organic UV absorbers such as benzophenone or benzotriazoles compete with the chromophores in the plastic in absorbing radiation. For instance, benzophenones reversibly isomerize in the process.

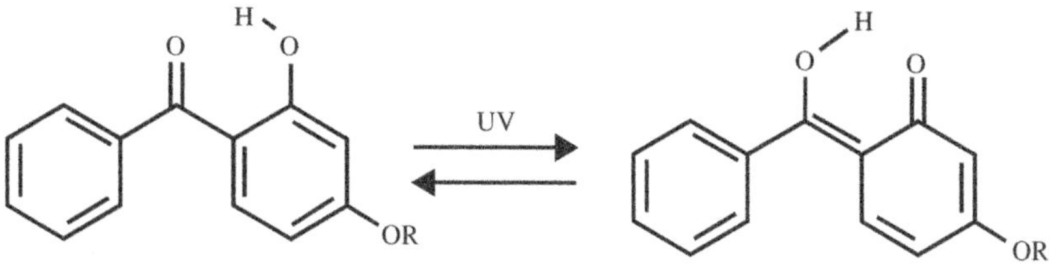

HALS are by far the most efficient class of stabilizer for polyolefins and are not consumed in the stabilization process (Gugumus, 1991; Malik et al., 1995). They work via the nitroxyl radical species that reacts with RO· radicals, deactivating them (see Fig. 6.7). The species is regenerated as shown in Figure 6.7, making it a very effective radical-mop even at very low concentrations (Gugumus, 2002a). These can be used only at levels of 0.05% by weight and still deliver excellent protection against UV damage in polyolefins. HALS cannot be used in PVC formulations as HCl from dehydrochlorination deactivates them.

Hindered phenols and quinone stabilizers also work via a similar mechanism by removing free radicals from the system. Often, two or more classes of stabilizers are used together to take advantage of synergistic enhancement in stabilization (Gugumus, 2002b; Gugumus, 2002c). For instance, combinations of two UV absorbers (an oxanilide with benzophenone or benzotriazole or a phenylhydroxytriazine) were reported to be systematically synergistic in both PE and PP. While most such combinations are merely additive, some can even be antagonistic.

6.3 ENHANCED PHOTODEGRADABLE POLYOLEFINS

Plastics can be designed to photodegrade at rates that are several times faster than expected. These "photodegradable plastics"[13] find popular use in agricultural films as they degrade and disintegrate into small pieces that blend in with the soil at the end of the growing season, saving the expense of removal of used mulch from the field for disposal. However, the long-term impact of accumulating plastic (usually PE) fragments in the soil might be unacceptable (Kitch, 2001). These are also used in the degradable six-pack rings[14] that when discarded as litter, breaks down and embrittles on exposure outdoors within 6–8 weeks.

Accelerated photodegradation is achieved either using structural modifications of the resin itself or by using catalysts or additives that accelerate the oxidation reactions (Koutny et al., 2006; Wiles and Scott, 2006). A copolymer of ethylene with approximately 1% of carbon monoxide (ECO copolymer) includes carbonyl functionalities in its main chain[15] and is an example of a photodegradable plastic (Scoponi et al., 1993). Carbonyls are potent chromophores that absorb solar UV-B radiation (Andrady et al., 1993a). On absorbing the radiant energy, the copolymer undergoes photolysis via the Norrish II reaction leading to chain scission and therefore weakening of polymer.[16] A hydrogen atom in position 3 from the carbonyl group is needed in the structure to effect this reaction as shown below. Norrish II reaction was also suggested in recent work on photodegradation of PLA (Tsuji et al., 2006).

A second approach uses transition metal pro-oxidant catalysts as an additive (Pablos et al., 2010; Roy et al., 2009). Several "photodegradable" and "oxo-biodegradable" plastic products rely on this approach. Transition metals can act as redox catalysts to accelerate degradation via catalyzed peroxide decomposition into radicals.

---

[13] The term "photodegradable plastics" is in quotes to recall that it is a misnomer. All plastics are photodegradable; these are designed to photodegrade at a faster rate in sunlight.
[14] Photodegradable six-pack rings have been available since early 1990s. At least 16 states require these to be enhanced photodegradable.
[15] A vinyl ketone comonomer would result in ketone side chains.
[16] Discussions of polymer photodegradation often include a comparison of the main-chain bond energies with the wavelength-dependent photon energies. This should not be taken to mean that photons at these wavelengths directly dissociate these bonds; most photodegradation reactions are secondary processes.

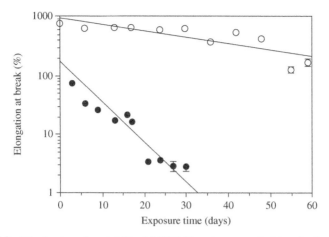

FIGURE 6.8   Weathering of unstabilized LDPE films (*open symbols*) and enhanced photo-degradable ECO copolymer (*filled symbols*) exposed outdoors in Miami, FL.

TABLE 6.2   Location-dependent Enhancement in Photodegradation Obtained Using ECO Copolymer in Place of LDPE Laminate of Same Thickness (Andrady et al., 1993a)

Location of exposure	Degradable plastic ECO $(B \times 10^3)$ days^{-1}	Control plastic LDPE $(B \times 10^3)$ days^{-1}	B^*/B
Cedar Knolls, NJ	52	9.5	5.5
Chicago, IL	87	4.3	20
Miami, FL	69	14	4.9
Seattle, WA	40	4.2	9.5
Whitman, AZ	257	25	10

$$Fe^{2+} + ROOH \rightarrow Fe^{3+} + RO \cdot + OH^-$$
$$Fe^{3+} + ROOH \rightarrow Fe^{2+} + ROO \cdot + H^+$$

Typically, transition metals, such as manganese, iron, cobalt, and nickel (but not heavy metals), are used to catalyze peroxide decomposition (Andrady et al., 1996). Their additive masterbatches are used at very low levels, and the metals are therefore present at very low concentrations of about 0.01 and 0.5% weight in the plastic product (Arnaud et al., 1994). Some of the plastics with these additives have been approved for use in food contact plastics in the United States and in Canada.

Figure 6.8 compares the change in tensile elongation at break (or extensibility) of LDPE and the ECO copolymer films at different exposure durations. For clarity, the logarithm of tensile extensibility is plotted in the figure. Steeper the gradient of the linear plots, faster is the rate of degradation. The ratio of the gradients is a measure of the degree of "enhancement" afforded by the ECO copolymer over conventional LDPE. The data in the figure for exposure in Miami, FL, in (summer) show an enhancement (Andrady et al., 1993a) factor of 5 for the photodegradable ECO material. Table 6.2

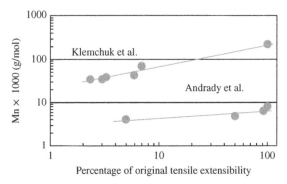

FIGURE 6.9 Two sets of data showing the relationship between number–average molecular weight and the percent retention of extensibility of degraded polyethylene. The upper set is for data on high-density polyethylene oxidized in oxygen at 100°C (Klemchuk and Horng, 1984). The lower set is for poly(ethylene-*co*-carbon monoxide) exposed outdoors at ambient temperature in air (Andrady et al., 1993a). Source: Reproduced with permission from Andrady (2011).

compares the values of gradient B (days^{-1}) in plots of changes in tensile elongation at break for copolymer of ethylene with (1%) carbon monoxide (ECO) and that for the control LDPE film of the same thickness (60 mil), at different locations in the US. The ECO laminates used are of the same grade as in photodegradable six-pack rings and were supplied by the manufacturer (ITW HiCone Company). The ratio of the values of gradient ($B*/B$) is an approximate measure of the enhancement of degradation using this approach and ranges from 5 to 20 depending on the location (Andrady et al., 1993a).

6.3.1 Effects of Photodegradation on Biodegradation

Photo-oxidative degradation results in several changes in the polymer. Primarily, the average molecular weight **Mn** (g/mol) is reduced along with the mechanical integrity of the material. Also, it increases the surface hydrophilicity encouraging faster subsequent biofilm formation (Donlan, 2002), a prelude to potential biodegradation. Finally, any brittle fragmentation reduces particle size and therefore increases the specific surface area favoring faster biodegradation. It is interesting to study the change in **Mn** (g/mol) with the decreased extensibility, particularly at embrittlement when the plastic is reduced to small fragments.

In Figure 6.9, the average tensile extensibility, ε, is plotted versus the **Mn** (g/mol) of the plastics weathered to different extents as determined by GPC for polyethylene (LDPE) and ECO copolymer. Interestingly, even the completely embrittled material (ε<5%) is still a high polymer (Bates and Sidwell, 2006), with molecular weights far too high for any appreciable biodegradation to occur. Mineralization of common plastics with **Mn** (g/mol) > 10^3 (g/mol) has not been reported in the literature. It is sometimes suggested that extensive photo-degradation reduces **Mn** (g/mol) sufficiently to facilitate faster biodegradation. These findings are not consistent with such a claim.

Complete mineralization of the photo-degraded polymer fragments has not been experimentally established (Koutny et al., 2006; Rojas and Greene, 2007; Artham et al., 2009) for any of the common plastic materials. Moderate biodegradation rates were reported in several studies (Chiellini and Corti, 2003; Yoon et al., 2012) on highly pre-degraded (photo- and thermally degraded) plastics. But these initial degradation rates are likely due to the rapid biodegradation of the small amounts of very low molecular weight fraction in the pre-degraded plastics. These have low enough molecular weights to be microbially assimilated. In other studies, PE with photodegradable additives were pre-degraded thermo-oxidatively, but still biodegraded at about the same rate as PE with no additive (Ojeda et al., 2009; Yashchuk et al., 2012).

6.4 BIODEGRADATION OF POLYMERS

The degradation of a substrate mediated by living organisms, usually microorganisms, is termed biodegradation.[17] Typically, microorganisms first attach on to the surface of a substrate, such as the plastic material, as a biofilm (Lucas et al., 2008) and secrete exo-enzymes to break down the plastic. As common polymers are water insoluble and have molecules that are far too large to diffuse into the microbial cell, heterogeneous biodegradation occurs outside the cell via the enzymes secreted by the organisms. Any soluble products of biodegradation are low enough in molecular weight to be absorbed or assimilated by the microorganisms. Cellular metabolism of the sorbed nutrients yields as CO_2 and water, liberating energy (stored as intracellular ATP) in the process.

Biodegradation can be either oxidative (aerobic) or reductive (anaerobic) depending on the environment the polymer is placed in. In the latter case, CH_4, H_2 and sometimes NH_3 are formed as by-products of the degradation. For glucose, the biodegradation reaction can be summarized in the following reactions.

Aerobic biodegradation:

$$C_6H_{12}O_6 + 6O_2 \rightarrow 6CO_2 + 6H_2O \quad \Delta G = -2870\,kJ\,/\,mol$$

Anaerobic biodegradation:

$$C_6H_{12}O_6 \rightarrow 3CO_2 + 3CH_4 \quad \Delta G = -390\,kJ\,/\,mol$$

There is net energy gain to the microorganism from biodegradation and they utilize the derived energy for their growth and reproduction. Growth incorporates some of the carbon from the biodegraded polymer into new biomass. In common with other reactions, biodegradation is also accelerated at higher temperatures and also requires a minimal level of moisture. The latter is essential for the microbial consortia to exist and is therefore a prerequisite to environmental biodegradation.[18] Figure 6.10 is a

---

[17] Biodegradation also includes non-microbially mediated degradation such as in biodegradable human implants (Brach del Prever et al., 1996).

[18] In the laboratory, enzymes isolated from living organisms can also affect biodegradation in systems that have no living cells.

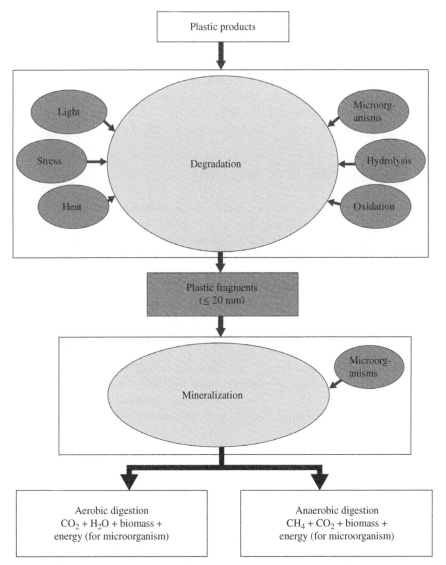

FIGURE 6.10 A schematic diagram of biodegradation of a solid polymer showing the two main stages of primary abiotic degradation to embrittlement followed by biodegradation of fragmented residue. Source: Modified and reproduced with permission from Krzan et al. (2006).

schematic diagram of the environmental degradation of plastics material. It is important to appreciate that a polymer that is readily biodegradable under aerobic exposure (such as poly(e-caprolactone) or PCL) need not be necessarily biodegradable under anaerobic conditions (Mohee et al., 2008).

6.4.1 Terminology and Definitions

The terminology relating to biodegradation used in technical and trade literature can be inconsistent and confusing. The consensus definitions of the terms by testing organizations such as ASTM, DIN, or ISO only partially address the inherent difficulties in developing nomenclature on biodegradable products. Specifying a plastic or other material as being "enhanced biodegradable" or "biodegradable" suggests, that the particular plastic can be demonstrated to environmentally biodegrade at a significantly faster rate compared to conventional plastics. For brevity and convenience, the term "biodegradable plastics" will be used here to mean "enhanced biodegradable plastics." In specifying a plastic material as being biodegradable, several key issues need to be addressed:

1. As biodegradation of a substrate depends on it, the relevant environment should also be specified when claiming a polymer as being biodegradable (e.g., "chitosan is 90% biodegradable in coastal marine sediment in 60 days" or "polycaprolactone is 100% biodegradable in *rhodochrous* and *C*. lab culture in 30 days" or "poly(lactic acid) is 90% biodegradable under composting conditions in 1.5 months"). Percentages here refer to the percent mineralization or carbon conversion of the polymer. Generically, describing a plastic as being "biodegradable" has little meaning.

2. Exposure environments are complex, being defined by the availability of oxygen, temperature, availability of humidity, consortia of biomass, pre-adaptation of the microbes, as well as other factors. A "garden soil" environment or a sewage sludge inoculum used in tests, for instance, is highly variable depending on the source and can yield variable test results (Mezzanotte et al., 2005). Healthy garden soil will have diverse consortia of bacteria and fungi ranging in 10^6–10^8 cells/g (Martin and Focht, 1977). Such a soil will also have active free enzymes liberated by lysis of living cells that may contribute to biodegradation. Table 6.3

TABLE 6.3   Main Environments in Which Plastic Litter is Found

Environment	Aerobic	Temperature	Other contributing mechanisms
Natural environments			
Surface soil	Yes	Ambient (T)	Photo-thermal oxidation
Subsurface soil	Yes	Ambient (T)	Thermal oxidation
Marine beach	Yes	$\gg T$	Photo-thermal oxidation
Marine surface water	Yes	Slightly $< T$	Photo-thermal oxidation
Marine coastal sediment	Yes	Slightly $< T$	Hydrolysis, biodegradation
Deep sea sediment	No	$\ll$ ambient (T)	Hydrolysis
Man-made environments			
Landfills	No	Slightly $< T$	Hydrolysis, biodegradation
Composting facilities	Yes	Much higher $> T$	Thermal oxidation
Anaerobic digesters	No	Slightly $> T$	Hydrolysis, biodegradation

TABLE 6.4   A Listing of ASTM Test Methods Related to Degradation of Plastics

ASTM standards	Description
D3826-98 (2002)	Standard practice for determining degradation end point in degradable polyethylene and polypropylene using a tensile test
D5071-99[a]	Standard practice for exposure of photodegradable plastics in a xenon arc apparatus
D5208-01	Standard practice for fluorescent ultraviolet (UV) exposure of photodegradable plastics
D5209-92	Standard test method for determining the aerobic biodegradation of plastic materials in the presence of municipal sewage sludge
D5210-92 (2002)	Standard test method for determining the anaerobic biodegradation of plastic materials in the presence of municipal sewage sludge
D5247-92	Standard test method for determining the anaerobic biodegradability of degradable plastics by specific microorganisms
D5271-02	Standard test method for determining the anaerobic biodegradation of plastic materials in an activated-sludge-wastewater-treatment system
D5272-92 (1999)	Standard practice for outdoor exposure testing of photodegradable plastics
D5338-98 (2003)	Standard test method for determining aerobic biodegradation of plastic materials under controlled composting conditions
D5510-94 (2001)[a]	Standard practice for heat aging of oxidatively degradable plastics
D5511-02	Standard test method for determining anaerobic biodegradation of plastic materials under high-solids anaerobic-digestion conditions
D5526-94 (2002)	Standard test method for determining anaerobic biodegradation of plastic materials under accelerated landfill conditions
D5951-96 (2002)	Standard practice for preparing residual solids obtained after biodegradability standard methods for plastics in solid waste for toxicity and compost quality testing
D5988-03[a]	Standard test method for determining aerobic biodegradation in soil of plastic materials or residual plastic materials after composting.
D6002-96 (2002)e1	Standard guide for assessing the compostability of environmentally degradable plastics
D6003-96	Standard test method for determining weight loss from plastic materials exposed to simulated municipal solid-waste (MSW) aerobic compost environment
D6340-98	Standard test methods for determining aerobic biodegradation of radiolabeled plastic materials in an aqueous or compost environment
D6400-99e1[a]	Standard specification for compostable plastics
D6691-01	Standard test method for determining aerobic biodegradation of plastic materials in the marine environment by a defined microbial consortium

TABLE 6.4 *(Continued)*

ASTM standards	Description
D6692-01	Standard test method for determining the biodegradability of radiolabeled polymeric plastic materials in seawater
D6776-02	Standard test method for determining anaerobic biodegradability of radiolabeled plastic materials in a laboratory-scale simulated landfill environment
D6852-02	Standard guide for the determination of bio-based content, resources consumption, and environmental profile of materials and products
D6866-04a[a]	Standard test methods for determining the bio-based content of natural range materials using radiocarbon and isotope ratio mass spectrometry analysis
D6868-03	Standard specification for biodegradable plastics used as coatings on paper and other compostable substrates
D6954-04	Standard guide for exposing and testing plastics that degrade in the environment by a combination of oxidation and biodegradation
D7026-04	Standard guide for sampling and reporting of results for the determination of bio-based content of materials via carbon isotope analysis

Source: Reproduced with permission from Krzan et al. (2006).
[a] Indicates that a revision is being proposed to the standard.

summarizes the common environments where plastics wastes are likely to end up in, and the mechanisms of degradation available in each case.

Adhering to a standardized test protocol such as the relevant ASTM standards in assessing degradability helps avoid some of this variability. Naming the test method used to assess biodegradability increases the credibility of the claim (e.g., chitin was found to be 80% biodegradable in seawater in 60 days when tested according to ASTM D6691-09. Though there is no guarantee that inter-laboratory data will be strictly comparable, specifying the standard at least informs how exactly the test was carried out.

The ASTM has developed 26 standards related to test methods on degradation of plastics materials. These are shown in Table 6.4.

3. How much faster should the rate of biodegradation of a plastic material be for it to be reasonably called "biodegradable?" The biodegradation must be rapid enough to be readily observable experimentally; the mineralization experiment must be completed in a reasonable timescale of weeks or months rather than years. A specific rate or the extent of biodegradation might be agreed upon for this purpose; the ASTM D5338-93 as well as European standard (EN 13432) requires a material to be at least 90% biodegraded in less than 6 months. The ASTM D5988 requires mineralization $\geq 70\%$ at $25 \pm 2°C$ within 6 months. Alternatively, the rate might be benchmarked to the biodegradation of selected natural materials such as dead dry biomass or a biopolymer such as cotton.

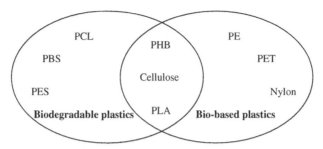

FIGURE 6.11   Diagram illustrating the potential enhanced biodegradability of only some bio-based plastics. Source: Redrawn from Tokiwa et al. (2009).

(ASTM D5338-93 allows a comparison with the rate for cellulose as the criteria for biodegradability.) However, the suitability of cellulose as a positive control is not clear (Mezzanotte et al., 2005).

4. The terms *biodegradable* plastics and *bio-based* or *bio-derived* plastics were already defined in Chapter 4. Plastics can be classified in terms of the source of raw materials into four classes: plastics based on fossil-fuel feed stocks, bio-polymers made by living organisms, modified biopolymers, and bio-based plastics derived from renewable biomass feedstock. Members of each of these classes can be either inherently "biodegradable" or "recalcitrant" when placed in an appropriate biotic environment. Biodegradability is a property or characteristic of plastics and is independent of the feedstock it is based on. This distinction was discussed in detail in Chapter 4 and is further illustrated in Figure 6.11.

6.4.2   Biodegradable Plastics

Biodegradable polymers are broadly classified into two groups: biopolymers and other (bio-derived and fossil-fuel derived) polymers. The first includes well-known examples of biopolymers such as cellulose and chitin that rapidly break down in most environments. It is their rapid biodegradation that is responsible for the removal of waste biomass (by mineralization) from environment. Biodegradable polymers are not used widely as packaging materials though the potential for their use clearly exists. The second class includes conventional polyesters such as PCL, those derived from bio-based raw material such as PLA, and bacterial polyesters such as poly(hydroxyl butyrate) (PHB). Aliphatic polyesters are susceptible to hydrolysis by lipolytic enzymes commonly secreted by microorganisms (Tokiwa et al., 1976). These enzymes are widely distributed in different environments (Mergaert and Swings, 1996). The populations of aliphatic polymer-degrading microorganisms in different ecosystems was found to be generally in the following order: PHB = PCL > PBS > PLA (Tokiwa et al., 2009). Three crucial factors that facilitate high rates of biodegradation of these polymers are the low average molecular weight, **Mn**, low degree of crystallinity (Tsuji and Miyauchi, 2001), and the lack of (or minimal) side chains in their structure (molecules with side chains are relatively more difficult to be assimilated by microorganisms (Tokiwa et al., 1976). The amorphous matrix with a looser structure is relatively easier to biodegrade because of better accessibility to enzymes compared to the crystalline domains. Also, higher the melting point Tm (°C) of the polymer, lower will be its biodegradability. Some examples of common biodegradable plastics are introduced below.

PCL $\{-OCH_2CH_2CH_2CH_2CH_2CO-\}n$) is a partially-crystalline polyester that is biodegraded by microbial lipases and esterases. The plastic is made from petrochemical feedstocks. It has too low a melting point (60°C) to be useful in any packaging applications. Higher aliphatic polyesters such as poly(butylene succinate) (PBS) $(-O(CH_2)_4OOC(CH_2)_2CO-)n$ and poly(ethylene succinate) (PES) $(-O(CH_2)_2OOC(CH_2)_2CO-)n$ are also biodegradable at a rate that depends on environmental factors (Kasuya et al., 1997). They have higher melting points of 112–114°C and 103–106°C, respectively, and the properties compare well to those of polyolefins. As succinic acid can be derived from plant sources, the polysuccinates can be potentially a bio-based polymer.

PLA $\{O(CH_3)CHCO-\}n$ is a semi-crystalline polymer that can be made from bio-derived raw materials. Its properties are comparable to PS and even approaches those of PET (Scaffaro et al., 2011). PLA-degrading microorganisms are not widely distributed in nature and soil biodegradation rates are slow (Ohkita and Lee, 2006) but PLA is rapidly compostable compared to PCL or poly-3-hydroxybutyrate-co-3-hydroxyvalerate (PHBV). PLA resin is now in small-scale commercial production and is likely to be the leading compostable resin available in the near term. It is used in disposable single-use items such as food service items, yogurt cups, water bottles, and thermoformed boxes. The claimed superior environmental benefit of using biodegradable PLA over a conventional resin such as PP in food packaging, however, is not always obvious (Hermann et al., 2011). Poly(glycolic acid) (PGA) is a biodegradable polymer that is less hydrophobic than PLA. Both copolymers and blends of PGA with PLA are used in medical applications as the metabolites they produce on degradation are nontoxic.

Poly(hydroxyalkanoates) are polyesters synthesized by microorganisms grown under special conditions. They are biopolymers similar to cellulose or chitin, but treated separately because they are produced only by stressed organisms and because of their potential as a commercial biodegradable plastic. For instance, PHB $\{-O(CH_3)CHCH_2CO-\}n$ is a natural polymer produced in high yield by many species of bacteria grown under special culture conditions. Its properties are similar to those of PP (Mooney, 2009). Up to 10% of soil microbial colonies can degrade PHB at ambient temperatures and the polymer is readily compostable as well. A popular biodegradable copolymer is PHBV with 12–72 mol% of hydroxyl valerate. In soil media seeded with compost (mixed microbial environment), 90% biodegradation can be expected in 10–22 months at 25°C, when ASTM D 5988-03 was followed (Arcos-Hernandez et al., 2012; Kim et al., 1999). Films of PHBV exposed to soil undergo ready biodegradation and show surface colonization by microbial species and pit and crevice formation on surface (Sang et al., 2002).

However, under anaerobic conditions (such as in a landfill), PHBV does not break down readily (Ishigaki et al., 2004). Figure 6.12 shows the weight loss versus

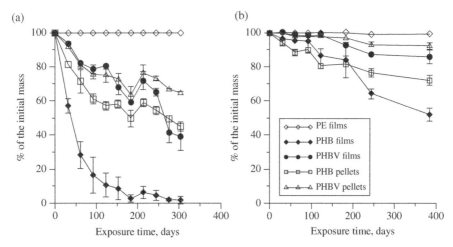

FIGURE 6.12 Weight loss curves for PHB and PHBV (films and pellets) incubated in tropical garden soil at two exposure sites in Russia: (a) Hoa Lac and (b) Dam Bai. Source: Reproduced with permission from Boyandin et al. (2013).

duration of soil exposure for PHB and PHBV film and pellet samples at two Russian exposure sites and illustrates the ease of biodegradation of the biopolymer PHBV and PHB.

6.4.3   Testing Readily Biodegradable Plastics

Testing and quantifying polymer biodegradation is difficult because simulated biotic environments are difficult to create in the laboratory in a consistent manner. For instance, "compost" as used in D5338-92, "marine sediment" in D6691-01, or "sewage sludge" as used in D5210-92 can have very different microbial profiles depending on their source, and how they were collected or cultured subsequently. Exposure of plastics to a monoculture of specific microorganisms in the laboratory, followed by visual evidence of colonization (ASTM G21-90 or G22-76) or CO_2 evolution (D6691-01) can be carried out reproducibly but has little relevance to real environments. Using "natural exposure" or field testing does not overcome this difficulty because of their high degree of variability. This limitation introduces significant variability into the standard test data on biodegradation of plastics.

Andrady (2000) has reviewed the four common approaches available for studying biodegradation of polymers. These involve experiments where one of the reactants or a product is closely monitored. These are as follows:

1. Monitoring the accumulation of biomass (Seal, 1994). ASTM (1996) G22-76, ASTM (2009) G21-96.
2. Monitoring the depletion of substrates either by simple "weight loss" (Tsuji and Suzuyoshi, 2002). ASTM D6003-96 or D5247-92). Alternatively, in

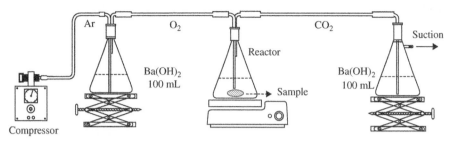

FIGURE 6.13 Respirometry experiment for measuring evolved CO_2 in biodegradation studies. Source: Reproduced with permission from Calil et al. (2006).

oxidative biodegradation, oxygen demand can be used as a measure (e.g., D5271-02).

3. Monitoring carbon dioxide (or methane in case of anaerobic breakdown) evolution (e.g., D5209-92, D5210-92, D5338-92, D6691-01, D6692-01, D6340-98).

4. Monitoring the changes in substrate properties. Changes in mechanical or other properties of degrading plastics substrate (as in D5247-92), which is an indirect measure of the degradation of the substrate.

Of these, only the third approach or respirometry allows a convenient means of establishing the degree of carbon conversion or mineralization. In simple respirometry experiments, finely divided plastic materials are mixed in with moist garden soil (~60% humidity) fortified with N and P salts and incubated along with activated sewage sludge inoculum. The outcome of the test is therefore to some extent dependent on variability of the biotic composition of the inoculum. Biodegradation of the polymer and other available soil biomass releases CO_2. Some experimental arrangement to sorb the CO_2 liberated in alkali for subsequent titrimetric determination is employed. For instance, an air stream passed over the soil sample might be bubbled into one or more sorption flasks carrying dilute alkali (see Fig. 6.13). The alkali is periodically titrated to calculate the CO_2 evolved by the biodegradation process, and fresh aliquot of alkali is substituted for it. Continuous measurement of the evolved gas by automated titrimetry (Pagga et al., 2001) constitutes significant recent improvements in the technique. Alternatively infrared spectroscopic analysis (Calmon et al., 2000) or automated gas chromatographic detection of the CO_2 might be used. A control flask with the same amount of soil but without any polymer substrate is used to establish background levels of CO_2 evolution. Subtracting the background level from the data allows the net CO_2 evolution attributable to the polymer to be determined. This is generally plotted as a percentage of carbon converted versus the duration of exposure (Andrady and Song, 1999; Narayan, 2011).

Andrady and Song (1999) reported the design for a convenient self-contained biometer flask (see Fig. 6.14) for respirometry of polymers. The polymer is incubated in the biotic medium in the upper chamber and generates CO_2 within that

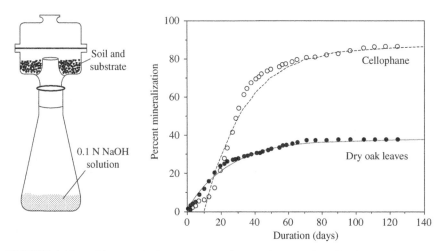

FIGURE 6.14 A biometer flask respirometer for carrying out mineralization studies. A respirometry curve for cellophane (regenerated cellulose sheet) compared to that of oak leaves. Source: Reprinted with permission from Andrady and Song (1999).

chamber. The dense gas drops down to dissolve in the alkali ($Ba(OH)_2$ solution) contained in the flask. Periodically, the flask is removed (ideally in a CO_2-free air chamber) and the alkali with sorbed CO_2 titrimetrically with a dilute acid. It is immediately replaced with a fresh flask of alkali. An identical control chamber with no polymer is used as a control sample to allow subtraction of the background levels of CO_2 evolution. The difference in CO_2 estimated for background and experimental runs will be that due to the biodegradation of polymer sample. Triplicate flasks are generally used to estimate the CO_2 levels at each of the sampling durations.

A biopolymer such as cellulose or a readily biodegradable synthetic polymer such as PCL yields good results with such a technique. The plot on the right-hand side of Figure 6.14 compares the mineralization curve for dry oak leaves (*Quercus alba*) with that for regenerated cellulose (cellophane) used as a positive control. Note that for oak leaves, the rate of biodegradation is slower and the extent of mineralization (carbon conversion) is much lower than for cellulose. The slower rate is likely due to the presence of lignin, with only the cellulose fraction being mineralized within the duration of observation. Even with regenerated cellulose that should be 100% biodegradable, carbon conversion is less than 100% as some of the carbon is used to generate new biomass as the microbial population in the flask grows.

Biodegradability of the substrate is calculated as percentage of the overall CO_2 production based on the determined carbon content of the samples. The value needs to be corrected for endogenous emissions from inoculum and medium, obtained from the control flask. The time dependence of the carbon conversion C_m is given by:

$$C_m = C_{max} \cdot a \cdot \{1 - \exp(-kc)t\} \tag{6.5}$$

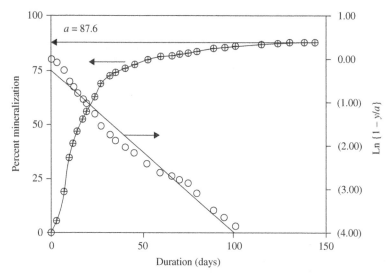

FIGURE 6.15 Gas evolution data (filled symbols) plotted as percent mineralization for the biodegradation of bleached paperboard packaging material in a respirometer. Soil media (70 wt% humidity) with sewage sludge inoculum was used. Also included is a plot of the data (open symbols) as suggested by Equation 6.2. Source: Reproduced with permission from Andrady and Song (1999).

where k is the rate constant, C_m the percent biodegradability of the substrate, a and c, constants. The value of k can be conveniently obtained by plotting the data as follows (see Fig. 6.15).

$$\mathrm{Ln}\left\{1 - \frac{C_m}{a}\right\} = -kt + kc \qquad (6.6)$$

For bleached paperboard for instance, Andrady et al. found $k \sim (0.14\text{–}0.18)\,\mathrm{days}^{-1}$ (Andrady and Song, 1999).

In practice, 100% of a plastic material will not mineralize and a small residue remains in the environment. These may include fillers, catalyst residues, and recalcitrant additives. It is important to ensure that these and their reaction products are non-toxic and do not harm soil organisms or affect plant growth.

6.5 BIODEGRADABILITY OF COMMON POLYMERS

Slow biodegradation of common polymers such as PE, PET, or PP does occur in nature (Shah et al., 2008; Tsao et al., 1993) but at rates that are several orders of magnitude lower than that observed with biopolymers such as cellulose and chitosan or biodegradable synthetic polymers such as PCL. These rates are too low to have any significant impact on litter management, and conventional plastics as are not

(a)   (b)

(c)   (d)

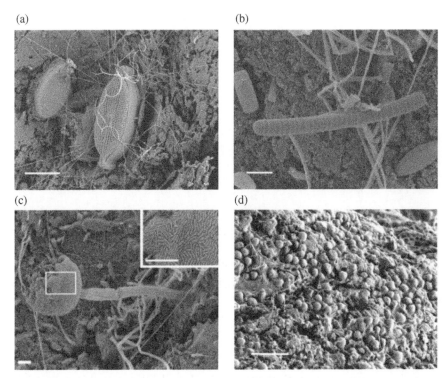

FIGURE 6.16   Electron micrographs (a–c) showing the diversity of microbial flora on polyolefin debris surfaces exposed to marine environments. Micrograph (d) shows pitting around the microbes. All *scale bars* are 10 μm. Source: Reprinted with permission from Zettler ER, Mincer TJ, Amaral-Zettler LA. Life in the "plastisphere": microbial communities on plastic marine debris. Environ Sci Technol 2013;47 (13):7137–7146. Copyright (2003) American Chemical Society.

usually thought of or referred to as being "biodegradable." The few microorganisms capable of biodegrading these are usually not abundant in natural settings and are often overwhelmed by native species.

Plastics in soil or water environments develop a microbe-rich biofilm on the surface. Polyethylene, the plastic used in highest volume, when exposed to seawater, soil, or other biotic environment is readily populated on the surface by a consortium of microorganisms (see Fig. 6.16; Zettler et al., 2013). Gilan et al. (2004) attributed the biofilm formation to slow reduction in hydrophobicity of the surface. Microscopic examination of the surface of plastics exposed to biotic environments (for instance marine sediment) for longer periods of time show surface pits and depressions where microorganisms appear to have settled in and biodegraded the surface (Bonhomme et al., 2003). The pitted and eroded area under the foot of an attached organism is believed to be biodegraded (Bonhomme et al., 2003), and the local molecular weight was shown to be lower than that of the surrounding polymer (Ohtaki et al., 1998). The rate of this degradation is very slow because the species of interest are not major constituents in natural consortia and also because of the availability of alternative carbon sources that are easier to digest and assimilate by microorganisms.

PE is known to undergo biodegradation by numerous microorganisms: a recent review (Restrepo-Flórez et al., 2014) lists 17 genera of bacteria and 11 genera of fungi that biodegrade PE. These include *Rhodococcus ruber* C-208 that secretes a laccase enzyme (Santo et al., 2013) well known to degrade lignins (Coll et al., 1993). In the study of Sivan et al. (2006), cell-free laccase enzymes catalyzed by copper were able to reduce the **Mn** (g/mol) of PE by 15–20%. Incubating the plastic in a laboratory culture of live cells can also lead to significant degradation. Reduction of **Mw** (g/mol) by up to 25% was reported (Hadad et al., 2005) in PE films incubated with the thermophilic bacterium *Brevibacillus borstelensis*. Many other microbial species biodegrade PE, (Sudhakar et al., 2008): These include fungal species (*Aspergillus* sp., *Chaetomium globosum*, *Penicillium funiculosum*, *Pullularia pullulans*), bacterial species (*Pseudomonas aeruginosa*, *Bacillus cereus*, *Coryneformes*, *Bacillus* sp., *Mycobacterium*, *Nocardia*, *Corynebacterium*, *Candida*, and *Pseudomonas*), and Actinomycetales (*Streptomycetaceae*). PVC was shown to be biodegraded by *Pseudomonas putida* (Anthony et al., 2004) and white rot fungi (Kirbas et al., 1999). Species such as *Curvularia senegalensis*, *Fusarium solani*, *Aureobasidium pullulans* Howard (2002), and *Pseudomonas chlororaphis* (Zheng et al., 2005) biodegrade the thermoset polyurethane (PU).

The mechanism of polyolefin biodegradation is not entirely clear but a likely hypothesis is that it follows paraffin biodegradation by microorganisms (Albertsson, 1978). In this case, abiotic degradation converts the polyethylene into low molecular weight fatty acids that are sorbed by the cells and undergo β oxidation.

However, as already alluded to, lab-culture data do not always translate into measurable rates in natural or field exposures of PE. Early data on incubating PE in soil show minimal weight loss even over very long periods of exposure. Reported experimental results on PEs (Table 6.5) are consistent with very slow biodegradation.

6.5.1 Additives that Enhance Degradation in Common Polymers

Oxo-biodegradable additives claim to combine both photo- and biodegradability enhancement in a single package (Chiellini et al., 2007). These are the first additives[19] to claim biodegradability in treated PET used in soda bottles. The system consists of a photoinhibitor to prevent premature degradation, a metal salt prooxidant (e.g., stearates of iron, cobalt, and manganese) and a biodegradation promoter that includes microcellulose powder (Ammala et al., 2011). Adding 1–5% by weight of additive masterbatch to polyolefin is claimed to render the material environmentally degradable. There is little doubt that oxo-biodegradables effectively catalyze photo-degradation but only a modest improvement in biodegradability is usually obtained (Yashchuk et al., 2012). After a year of natural weathering (abiotic degradation), the

---

[19] Reverte oxo-biodegradable additive masterbatch is intended for use with PE, PP, and PET. Other such additives are available from Addiflex and Symphony (Ammala et al., 2011).

TABLE 6.5   Summary of Estimates of Biodegradation of Polyethylenes in Natural Environments by Weight Loss Method

Environment	Plastic	Duration (days)	Weight loss (%)	References
Different soils and coal waste	LDPE	225	0.13–0.28	Nowak et al. (2011)
Seawater		365	1.9	Artham et al. (2009)
Soil		800	0.1	Albertsson (1980)
Soil		3650	0.2	Albertsson and Karlsson (1990)
Rhodococcus ruber		30–56	2.5–7.5	Hadad et al. (2005); Santo et al. (2013); Sivan et al. (2006)
Seawater	HDPE	365	1.6	Artham et al. (2009)
Soil		800	0.4	Albertsson (1980)

Source: Summarized from Restrepo-Flórez et al. (2014).

oxo-biodegradable films of PE incubated in compost at 58°C showed 12.4% mineralization over a period of 3 months; biodegradation at 25°C was only 5.4% (Ojeda et al., 2009). Chiellini and Corti show a mineralization of up to 50% for pre-degraded oxo-biodegradable PE in soil (over a period of 550 days) but lower rates have been reported as well (Rojas and Greene, 2007). A study funded by the European Commission (Feuilloley et al., 2005) found the pre-photodegraded PE with the additive to only biodegrade approximately 15% after 1 year in soil.[20] These materials are not compostable according to ASTM 6400 or EN 13432 (Thomas et al., 2010).

6.5.2   Degradable Plastics and Sustainable Development

It is interesting to assess the contribution of using enhanced-degradable plastics in place of conventional varieties to sustainable development.

1. The use of enhanced photo- or biodegradable plastics has a potential negative impact on recyclability of plastic waste. The potential of degradable plastics contaminating a recovered plastics waste stream intended for recycling, poses a serious problem (EUPR, 2009; Kitch, 2001). Sorting cannot remove degradable plastics effectively from the recycling stream. This is especially true of photodegradable plastics; most biodegradable plastics can be sorted with automated NIR systems (but these do not work well with plastic film). The presence of even 10% degradable material (with residual pro-oxidant) adversely affected the

---

[20] Advertised claims of 100% biodegradability of PVC, PET, and PE are misleading and have not been supported with repeatable robust data. The US National Advertising Division (Better Business Bureau) recently recommended that a leading manufacturer stop using such claims.

quality of recycled HDPE (Samsudin et al., 2013). A recent study[21] by European Plastic Convertors (EuPC) concluded that as little as 2 wt.% of oxo-degradable material can affect recyclate quality adversely. Some biodegradable plastics or additives are hydrophilic, can release sorbed water as steam, or thermally degrade or even caramelize at polyolefin reprocessing temperatures, discoloring or weakening the recycled plastic product.

2. The use of photodegradable or oxo-biodegradable plastics can effectively reduce the threat of entanglement (e.g., use in six-pack yokes) of animals in litter such as six-pack rings, plastic strapping bands, or netting and also help control aesthetic degradation by litter. Several states legally require their six-pack rings to be photodegradable. This advantage of preventing entanglement and perhaps reducing visible litter, however, has to be weighed against the emerging concern on microplastics in the environment (Andrady, 2011). Weathering degradation of plastics into fragments in a very short timescale (compared to regular plastics) presents a potential ecological hazard. The implications of the marine organisms ingesting these microplastic fragments (see Chapter 10) are not well understood at this time. But the adverse ecological impacts of plastics debris now shift from the larger fauna to smaller, less visible, genera including zooplanktons that are vital to the marine food pyramid.

Urban litter is aesthetically unacceptable and enhanced photodegradable plastics do address the problem but by fragmenting larger waste into minute fragments that are not easily discernible against the background. Littering, however, is primarily a behavioral problem that will worsen with the ongoing trend toward overpopulated urban centers. Degradable plastics cannot be expected to address this impending problem in its entirety. There is even the possibility that "degradability" label may in fact be misinterpreted leading to more littering as the waste is expected to degrade!

Compostable plastics in products such as compost bags or food-contact packaging are commercially available. Plastics, however, are not good candidates for composting[22] and do not yield compost (humic residues like biomass does). They are in fact removed from municipal waste streams intended for composting. Composting releases sequestered carbon in the plastic back into the atmosphere, contributing to global warming. The releases of this carbon back into the carbon cycle is of little merit and is not a persuasive argument in favor of composting or biodegrading, as the carbon in plastics was not removed from the operational carbon cycle to start with. It was sequestered for millions of years as fossil fuels.

Compostable plastics may have a niche role to play where its use contributes to better compost quality by minimizing the plastic residue in the product

---

[21] The 2013 study was conducted by TCKT (Transfer Center for Polymer Technology) for EuPc (Brussels). Based on the findings, EuPc urges better sorting and collection of waste degradable plastics. Available at www.plasticsconverters.eu/uploads/FINAL%20Impact%20of%20Degradable%20 Plastic%20Carrier%20Bags%20on%20mechanical%20recycling.pdf

[22] Composting makes sense when there is a market demand for compost as a product. In such situations, there is enough readily biodegradable biomass in the MSW to generate compost.

(Goldstein and Block, 2000). This can be true, for instance, with plastic food packaging that is difficult or expensive to remove from the food residue and is disposed of in a compost stream.

All plastic waste including litter is a material of value in terms of the embedded energy and the nonrenewable materials that comprise it. It is clearly a resource or a raw material that must ideally be recycled and its material and/or energy resources extracted for reuse. Degradability and compostability do not facilitate this key sustainability objective.

REFERENCES

Albertsson AC. Biodegradation of synthetic polymers. 2. Limited microbial conversation of C-14 in polyethylene to (CO-2)-C-14 by some soil fungi. J Appl Polym Sci 1978;22:3419–3433.

Albertsson AC. The shape of the biodegradation curve for low and high density polyethenes in prolonged series of experiments. Eur Polym J 1980;16:623–630.

Albertsson AC, Karlsson S. The influence of biotic and abiotic environments on the degradation of polyethylene. Prog Polym Sci 1990;15:177–192.

Ammala A, Bateman S, Dean K, Petinakis E, Sangwan P, Wong S, Yuan Q, Yu L, Patrick C, Leong KH. An overview of degradable and biodegradable polyolefins. Prog Polym Sci 2011;36 (8):1015–1049.

Andrady AL. Wavelength sensitivity in polymer photodegradation. Adv Polym Sci 1997;128:47–94.

Andrady AL. In: Hamid SH, editor. Handbook of Polymer Degradation. 2nd ed. New York: Marcel Dekker; 2000. p 441.

Andrady AL. Ultraviolet radiation and polymers. In: Mark JE, editor. Physical Properties of Polymers Handbook. 2nd ed. Boca Raton: Springer; 2007. p 857–867.

Andrady AL. Microplastics in the marine environment. Mar Pollut Bull 2011;8:1596–1605.

Andrady AL, Searle ND. Photoyellowing of mechanical pulps II: activation spectra for light-induced yellowing of newsprint paper by polychromatic radiation. TAPPI J 1995;78 (5):131.

Andrady AL, Song YE. Aerobic mineralization of paperboard materials used in packaging applications. J Appl Polym Sci 1999;74:1773–1779.

Andrady AL, Searle ND, Crewdson LFE. Wavelength sensitivity of unstabilized and UV stabilized polycarbonate to solar simulated radiation. Polym Degrad Stab 1992;35 (3):235–247.

Andrady AL, Pegram JE, Nakatsuka S. Studies on controlled lifetime plastics. 1. The geographic variability in out-door lifetimes of enhanced photodegradable polyethylene. J Environ Polym Degrad 1993a;1 (1):31.

Andrady AL, Pegram JE, Tropsha Y. Changes in carbonyl index and average molecular weight on embrittlement of enhanced photo-degradable polyethylene. J Environ Degrad Polym 1993b;1:171–179.

Andrady AL, Pegram JE, Searle ND. Wavelength sensitivity of enhanced photodegradable polyethylenes, ECO, and LDPE/MX. J Appl Polym Sci 1996;62:1457–1463.

Andrady AL, Hamid HS, Torikai A. Effects of climate change and UV-B on materials. J Photochem Photobiol Sci 2003;2:68–72.

Andrady AL, Hamid HS, Torikai A. Effects of stratospheric ozone depletion and climate change on materials damage. Photochem Photobiol Sci 2007;6:203–207.

Anthony SD, Meizhong L, Christopher EB, Robin LB, David LF. Involvement of linear plasmids in aerobic biodegradation of vinyl chloride. Appl Environ Microbiol 2004;70:6092–6097.

Arcos-Hernandez MV, Laycock B, Pratt S, Donose BC, Nikolić MAL, Luckman P, Werker A, Lant PA. Biodegradation in a soil environment of activated sludge derived polyhydroxyalkanoate (PHBV). Polym Degrad Stab 2012;97 (11):2301–2312.

Arnaud R, Dabin P, Lemaire J, Al-Malaika S, Chohan S, Coker M, Scott G, Fauve A, Maaroufi A. Photooxidation and biodegradation of commercial photodegradable polyethylenes. Polym Degrad Stab 1994;46 (2):211–224.

Artham T, Sudharkar R, Venkatesan C, Nair M, Murty KVGK, Doble M. Biofouling and stability of synthetic polymers in sea water. International Biodeterioration & Biodegradation 2009;63 (7):884–890.

Artham T, Mukesh D. Biodegradation of physicochemically treated polycarbonate by fungi. Biomacromolecules 2010;11 (1):20–28.

Balabán L, Majer J, Vesely K. Photooxidative degradation of polypropylene. J Polym Sci Part C Polym Symp 1969;22:1059–1071.

Bates I, Sidwell J. The comparative GPC analysis of aged and unaged polyethylene film samples. Shawbury: Smithers Rapra, RAPRA Report CTR 46303; 2006.

Bonhomme S, Cuer A, Delort AM, Lemaire J, Sancelme M, Scott C. Environmental biodegradation of polyethylene. Polym Degrad Stab 2003;81:441–452.

Boyandin AN, Svetlana VP, Karpov VA, Ivonin VN, Đỗ NL, Nguyễn TH, Lê TMH, Filichev NL, Levin AL, Filipenko ML, Volova TG, Gitelson II. Microbial degradation of polyhydroxyalkanoates in tropical soils. Int Biodeterior Biodegrad 2013;83:77–84.

Brach del Prever E, Crova M, Costa L, Dallera A, Camino G, Gallinaro P. Unacceptable biodegradation of polyethylene in vivo. Biomaterials 1996;17 (9):873–878.

Calil MR, Gaboardi F, Guedes CGF, Rosa DS. Comparison of the biodegradation of poly(ε-caprolactone), cellulose acetate and their blends by the Sturm test and selected cultured fungi. Polym Test 2006;25 (5):597–604.

Calmon A, Dusserre-Bresson L, Bellon-Maurel V, Feuilloley P, Silvestre F. An automated test for measuring polymer biodegradation. Chemosphere 2000;41 (5):645–651.

Carmona HA, Wittel FK, Kun F, Herrmann HJ. Fragmentation processes in impact of spheres. Phys Rev Lett 2004;93:035504.

Celina M. Review of polymer oxidation and its relationship with materials performance and lifetime prediction. Polym Degrad Stab 2013;98:2419–2429.

Chalot R. Rodents and termite repellents. Compd World 2011; (July/August):43.

Chiellini E, Corti A. Simple method suitable to test the ultimate biodegradability of environmentally degradable polymers. Macromol Symp 2003;197:381–395.

Chiellini E, Corti A, D'Antone S. Oxo-biodegradable full carbon backbone polymers—biodegradation behaviour of thermally oxidized polyethylene in an aqueous medium. Polym Degrad Stab 2007;92 (7):1378–1383.

Coll PM, Fernandes-Abalos JM, Villanueva JR, Santamaria R, Perez P. Purification and characterization of a phenoloxidase (laccase) from the lignin-degrading basidiomycete PM1. Appl Environ Microbiol 1993;59:2607–2613.

Craig IH, White JR, Shyichuk AV, Syrotynska I. Photo-induced scission and crosslinking in LDPE, LLDPE, and HDPE. Polym Eng Sci 2005;3:579–587.

D'Aquino CA, Balmant W, Ribeiro RLL, Munaro M, Vargas JVC, Amico SC. A simplified mathematical model to predict PVC photodegradation in photobioreactors. Polym Test 2012;31 (5):638–644.

Davidson TM. Boring crustaceans damage polystyrene floats under docks polluting marine waters with microplastics. Mar Pollut Bull 2012;64:1821–1828.

Donlan R. Biofilms: microbial life on surfaces. Emerg Infect Dis 2002;8:881–890.

Edge M, Liauw CM, Allen NS, Herrero R. Surface pinking in titanium dioxide/lead stabiliser filled PVC profiles. Polym Degrad Stab 2010;95 (10):2022–2040.

European Plastics Recyclers (EUPR). *OXO Degradables Incompatibility with Plastics Recycling*. Brussels: EUPR; 2009. Available at http://www.fkur.com/fileadmin/user_upload/MediaInfo/Heisse_Eisen/oxo_degradables_incompatible_with_recyling.pdf. Accessed October 10, 2014.

Fayolle B, Audouin L, Verdu J. A critical molar mass separating the ductile and brittle regimes as revealed by thermal oxidation in polypropylene. Polymer 2004;45:4323.

Feuilloley P, César G, Benguigui L, Grohens Y, Pillin I, Bewa H, Lefaux S, Jamal M. Degradation of polyethylene designed for agricultural purposes. J Polym Environ 2005;13:349–355.

François-Heude A, Richaud E, Desnoux E, Colin X. Influence of temperature, UV-light wavelength and intensity on polypropylene photothermal oxidation. Polym Degrad Stab 2014;100:10–20.

Ghaffar A, Scott GA, Scott G. The chemical and physical changes occurring during UV degradation of high impact polystyrene. Eur Polym J 1976;11:271–275.

Gilan I, Hadar Y, Sivan A. Colonization, biofilm formation and biodegradation of polyethylene by a strain of *Rhodococcus ruber*. Appl Microbiol Biotechnol 2004;65 (1):97–104.

Goldstein, N., Block, D. Sorting out the plastic, Biocycle J Compost Organics Recycl, 41, 8, 2000.

Gugumus F. Mechanisms and kinetics of photostabilization of polyolefins with N-methylated HALS. Polym Degrad Stab 1991;34 (1–3):205–241.

Gugumus F. Possibilities and limits of synergism with light stabilizers in polyolefins. 1. HALS in polyolefins. Polym Degrad Stab 2002a;75:295–309.

Gugumus F. Possibilities and limits of synergism with light stabilizers in polyolefins. 2. UV absorbers in polyolefins. Polym Degrad Stab 2002b;75:309–320.

Gugumus F. Possibilities and limits of synergism with light stabilizers in polyolefins. 1. HALS in polyolefins. Polym Degrad Stab 2002c;75 (2):295–308.

Gulmine JV, Janissek PR, Heise HM, Akcelrud L. Degradation profile of polyethylene after artificial accelerated weathering. Polym Degrad Stab 2003;79 (3):385–397.

Hadad D, Geresh S, Sivan A. Biodegradation of polyethylene by the thermophilic bacterium *Brevibacillus borstelensis*. J Appl Microbiol 2005;98:1093–1100.

Hamid SH, Amin MB, Maadhah AG. Weathering degradation of polyethylene. In: Hamid SH, Amin MB, Maadhah AG, editors. *Handbook of Polymer Degradation*. New York: Marcel Dekker; 1995.

Heikkilä A, Kärhä P. Photoyellowing revisited: determination of an action spectrum of newspaper. Polym Degrad Stab 2014;99:190–195.

Hermann BG, Debeer L, De Wilde B, Blok K, Patel MK. To compost or not to compost: carbon and energy footprints of biodegradable materials' waste treatment. Polym Degrad Stab 2011;96 (6):1159–1171.

Höglund A, Odelius K, Albertsson AC. Crucial differences in the hydrolytic degradation between industrial polylactide and laboratory-scale poly(l-lactide). Appl Mater Interfaces 2012;4 (5):278.

Howard GT. Biodegradation of polyurethane: a review. Int Biodeterior Biodegrad 2002;40:245–252.

Ishigaki T, Sugano W, Nakanishi A, Tateda M, Ike MM, Fujita M. The degradability of biodegradable plastics in aerobic and anaerobic waste landfill model reactors. Chemosphere 2004;54 (3):225–233.

Jbilou F, Joly C, Galland S, Belard L, Desjardin V, Bayard R, Dole P, Degraeve P. Biodegradation study of plasticised corn flour/poly(butylene succinate-co-butylene adipate) blends. Polym Test 2013;32 (8):1565–1575.

Kasuya K, Takagi K, Ishiwatari S, Yoshida Y, Doi Y. Biodegradabilities of various aliphatic polyesters in natural waters. Polym Degrad Stab 1997;59:327–332.

Kim MN, Lee A, Lee K, Chin I, Yoon J. Biodegradability of poly(3-hydroxybutyrate) blended with poly(ethylene-co-vinyl acetate. Eur Polym J 1999;35 (6):1153–1158.

Kirbas Z, Keskin N, Guner A. Biodegradation of polyvinylchloride (PVC) by white rot fungi. Bull Environ Contam Toxicol 1999;63:335–342.

Kitch D. Biocycle international—global overview—biodegradable polymers and organics recycling—an international perspective provides some insights into markets, policies, opportunities, constraints and trends that impact use of bioplastics. Biocycle 2001;42 (2):74.

Klemchuk PP, Horng P. Perspectives on the stabilization of hydrocarbon polymers against thermo-oxidative degradation. Polym Degrad Stab 1984;7 (3):131–151.

Koutny M, Lemaire J, Delort A-M. Biodegradation of polyethylene films with prooxidant additives. Chemosphere 2006;64:1243–1252.

Krzan A, Hemjinda S, Miertus S, Corti A, Chiellini E. Standardization and certification in the area of environmentally degradable plastics. Polym Degrad Stab 2006;91 (12):2819–2833.

Küpper K, Gulmine JV, Janissek PR, Heise HM. Attenuated total reflection infrared spectroscopy for micro-domain analysis of polyethylene samples after accelerated ageing within weathering chambers. Vib Spectrosc 2004;34 (1):63–72.

Lucas N, Bienaime C, Belloy C, Queneudec M, Silvestre F, Nava-Saucedo J. Polymer biodegradation: mechanisms and estimation techniques—a review. Chemosphere 2008;73 (4):429–442.

Malík J, Tuan DQ, Špirk E. Lifetime prediction for HALS-stabilized LDPE and PP. Polym Degrad Stab 1995;47 (1):1–8.

Martin JP, Focht DD. In: Elliot LF, Stevenson FJ, editors. Soils for Management of Organic Wastes and Waste Waters. Madison: CSSA; 1977. p 115.

Martin JW, Chin JW, Nguyen T. Reciprocity law experiments in polymeric photodegradation: a critical review. Prog Org Coat 2003;47:292–311.

Mergaert J, Swings J. Biodiversity of microorganisms that degrade bacterial and synthetic polyesters. J Ind Microbiol 1996;17:463–469.

Mezzanotte V, Bertani R, Innocenti FD, Tosin M. Influence of inocula on the results of biodegradation tests. Polym Degrad Stab 2005;87 (1):51–56.

Minsker KS, Kolesov SV, Zaikov GE. *Degradation and Stabilization of Vinylchloride-Based Polymers*. New York: Pergamon Press; 1982.

Mohee R, Unmar GD, Mudhoo A, Khadoo P. Biodegradability of biodegradable/degradable plastic materials under aerobic and anaerobic conditions. Waste Manag 2008;28 (9):1624–1629.

Mooney BP. The second green revolution? Production of plant-based biodegradable plastics. Biochem J 2009;418 (2):219–232.

Muller RJ, Witt U, Rantze E, Deckwer WD. Architecture of biodegradable copolyesters containing aromatic constituents. Polym Degrad Stab 1998;59:203–208.

Narayan R. Carbon footprint of bioplastics using biocarbon content analysis and life-cycle assessment. MRS Bull 2011;369:716–721.

Nowak B, Pająk J, Drozd-Bratkowicz M, Rymarz G. Microorganisms participating in the biodegradation of modified polyethylene films in different soils under laboratory conditions. Int Biodeterior Biodegrad 2011;65:757–767.

Ohkita T, Lee SH. Thermal degradation and biodegradability of poly(lactic acid)/corn starch biocomposites. J App Polym Sci 2006;100:3009–3017.

Ohtaki A, Sato N, Nakasaki K. Biodegradation of poly-ε-caprolactone under controlled composting conditions. Polym Degrad Stab 1998;61:499–505.

Ojeda TFM, Dalmolin E, Forte MMC, Jacques RJS, Bento FM, Camargo FAO. Abiotic and biotic degradation of oxo-biodegradable polyethylenes. Polym Degrad Stab 2009;94 (6):965–970.

Pablos JL, Abrusci C, Marín I, López-Marín J, Catalina F, Espí E, Corrales T. Photodegradation of polyethylenes: comparative effect of Fe and Ca-stearates as pro-oxidant additives. Polym Degrad Stab 2010;95 (10):2057–2064.

Pagga U, Schefer A, Muller RJ, Pantke VM. Determination of the aerobic biodegradability of polymeric material in aquatic batch tests. Chemosphere 2001;42:319–331.

Pickett JE. Influence of photo-Fries reaction products on the photodegradation of bisphenol-A polycarbonate. Polym Degrad Stab 2011;96 (12):2253–2265.

Qayyum MM, White JR. Effect of stabilizers on failure mechanisms in weathered polypropylene. Polym Degrad Stab 1993;41:163–172.

Ranby B. Photodegradation and photo-oxidation of synthetic polymers. J Anal Appl Pyrolysis 1989;15:237–247.

Restrepo-Flórez J, Bassi A, Thompson MR. Microbial degradation and deterioration of polyethylene—a review. Int Biodeterior Biodegrad 2014;88:83–90.

Rivaton A. Recent advances in bisphenol-A polycarbonate photodegradation. Polym Degrad Stab 1995;49:163–179.

Rojas E, Greene J. Performance evaluation of environmentally degradable plastic packaging and disposable food service ware, Final Report. Sacramento: Integrated Waste Management Board, State of California, 2007. p 1–70.

Roy PK, Surekha P, Rajagopal C, Chatterjee SN, Choudhary V. Studies on the photo-oxidative degradation of LDPE films in the presence of oxidised polyethylene. Polym Degrad Stab 2007;92 (6):1151–1160.

Roy PK, Surekha P, Raman R, Rajagopal C. Investigating the role of metal oxidation state on the degradation behaviour of LDPE. Polym Degrad Stab 2009;94 (7):1033–1039.

Rutkowska M, Heimowska A, Krasowska K, Janik H. Biodegradability of polyethylene starch blends in sea water. Pol J Environ Stud 2002;11 (3):267–274.

Samsudin MSF, Wahab MAA, Ahamid Z. Effect of recycled HDPE with pro-oxidant on the photodegradation of HDPE film. Prog Rubber Plast Recycl Technol 2013;29 (2):69–79.

Sang BI, Hori K, Tanji Y, Unno H. Fungal contribution to in situ biodegradation of poly (3-hydroxybutyrate-co-3-hydroxyvalerate) film in soil. Appl Microbiol Biotechnol 2002;58:241–247.

Santo M, Weitsman R, Sivan A. The role of the copper-binding enzyme—laccase—in the biodegradation of polyethylene by the actinomycete *Rhodococcus ruber*. Int Biodeterior Biodegrad 2013;84:204–210.

Scaffaro R, Morreale M, Mirabella F, La Mantia FP. Preparation and recycling of plasticized PLA. Macromol Mater Eng 2011;296 (2):141–150.

Scoponi M, Pradella F, Carassiti V. Photodegradable polyolefins. Photo-oxidation mechanisms of innovative polyolefin copolymers containing double bonds. Coord Chem Rev 1993;125:219–230.

Seal KJ. Test methods and standards for biodegradable plastics. In: Griffin GJL, editor. Chemistry and Technology of Biodegradable Polymers. London: Blackie Academic & Professional; 1994. p 116.

Searle ND, McGreer M, Zielnik A. Weathering of polymeric materials. Encyclopedia of Polymer Science and Technology. New York: John Wiley and Sons; 2010.

Shah AA, Hasan F, Hameed A, Ahmed S. Biological degradation of plastics: a comprehensive review. Biotechnol Adv 2008;26 (3):246–265.

Sheldrick GE, Vogl O. Induced photodegradation of styrene polymers: a survey. J Polym Eng Sci 2004;16 (2):65–73.

Singh B, Sharma N. Mechanistic implications of plastic degradation. Polym Degrad Stab 2008;93 (3):561–584.

Sivan A, Szanto M, Pavlov V. Biofilm development of the polyethylene-degrading bacterium *Rhodococcus ruber*. Appl Microbiol Biotechnol 2006;72:346–352.

Sudhakar M, Doble M, Sriyutha Murthy P, Venkatesan R. Marine microbe-mediated biodegradation of low- and high-density polyethylenes. Int Biodeterior Biodegrad 2008;61 (3):203–213.

Sunilkumar M, Francis T, Thachil ET, Sujith A. Low density polyethylene–chitosan composites: a study based on biodegradation. Chem Eng J 2012;204–206:114–124.

Thomas N, Clarke J, McLauchlin A, Patrick S. Assessing the environmental impacts of oxo-degradable plastics across their life cycle. EV0422. Report to the Department for Environment, Food and Rural Affairs. Loughborough University, Loughborough, England 2010.

Tokiwa Y, Ando T, Suzuki T. Degradation of polycaprolactone by a fungus. J Ferment Technol 1976;54:603–608.

Tokiwa Y, Calabia BP, Ugwu CU, Aiba S. Biodegradability of plastics. Int J Mol Sci. 2009; Sep; 10(9):3722–3742.

Torikai A. Wavelength sensitivity of the photodegradation of polymers. In: Hamid SH, editor. Handbook of Polymer Degradation. 2nd edn. New York: Marcel Dekker; 2000. p 573–604.

Torikai A, Hasegawa A. Wavelength effect on the accelerated photodegradation of polymethylmethacrylate. Polym Degrad Stab 1998;61 (2):361–364.

Tsao R, Anderson TA, Coats JR. The influence of macro invertebrates on primary degradation of starch-containing polyethylene films. J Environ Polym Degrad 1993;1:301–306.

Tsuji H, Miyauchi S. Poly(L-lactide): 6 Effects of crystallinity on enzymatic hydrolysis of poly(L-lactide) without free amorphous region. Polym Degrad Stab 2001;71:415–424.

Tsuji T, Suzuyoshi K. Environmental degradation of biodegradable polyesters poly(ε-caprolactone), poly[(R)-3-hydroxybutyrate], and poly(L-lactide) films in controlled static seawater. Polym Degrad Stab 2002;75 (2):347–355.

Tsuji H, Echizen Y, Nishimura Y. Photodegradation of biodegradable polyesters: a comprehensive study on poly(L-lactide) and poly(ε-caprolactone). Polym Degrad Stab 2006;91:1128–1137.

Wiles DM, Scott G. Polyolefins with controlled environmental degradability. Polym Degrad Stab 2006;91:1581–1592.

Wypych G. Handbook of UV Degradation and Stabilization. Toronto: ChemTec Publishing; 2010.

Yakimets I, Lai D, Guigon M. Effect of photooxidation cracks on behavior of thick polypropylene samples. Polym Degrad Stab 2004;86:59–67.

Yashchuk O, Portillo FS, Hermida EB. Degradation of polyethylene film samples containing oxo-degradable additives. Proc Mater Sci 2012;1:439–445.

Yoon MG, Jeon JH, Kim MN. Biodegradation of polyethylene by a soil bacterium and AlkB cloned recombinant cell. J Bioremed Biodegrad 2012;3:145.

Zettler ER, Mincer TJ, Amaral-Zettler LA. Life in the "plastisphere": microbial communities on plastic marine debris. Environ Sci Technol 2013;47 (13):7137–7146.

Zheng Y, Yanful EK, Bassi AS. A review of plastic waste biodegradation. Crit Rev Biotechnol 2005;25:243–250.

7

ENDOCRINE DISRUPTOR CHEMICALS

The human endocrine system consists of a set of ductless glands such as the thyroid, adrenal glands, pituitary body, and gonads that secrete different hormones to regulate physiology. Hormones secreted by these are able to effectively regulate a host of body functions including growth and reproduction. Each hormone is very specific as to its function and acts on a single target cell or tissue. These affect the physiology by interacting with receptors on the cells (including serotonin, dopamine, norepinephrine receptors in addition to the hormone receptors) and controlling the intracellular physiology and function (Pickering and Sumpter, 2003). Natural hormones, including estrogens, androgens, progesterone, thyroid hormones, as well as hypothalamic and pituitary hormones, are present in very low concentrations in the body. But they provide a critical role coordinating growth, metabolism, and reproductive processes in the body.

Endocrine-disrupting compounds (EDCs)[1] are substances that can impair the functioning of the endocrine system leading to adverse health impacts. Nearly a thousand of man-made chemicals released into the environment are recognized as EDCs (Vandenberg et al., 2009). Some of these, structurally similar to hormones, interact with relevant cell receptors to confuse the body's physiology. They either mimic or antagonize the effects of the 50 or so hormones. Others prevent the hormone–receptor interaction or interfere with hormone synthesis itself, disrupting

[1] The notion of EDCs is itself a recent one, and the term "endocrine disruptor" was first used only in 1991 at the Wingspread Conference Center in Racine, Wisconsin.

Plastics and Environmental Sustainability, First Edition. Anthony L. Andrady.
© 2015 John Wiley & Sons, Inc. Published 2015 by John Wiley & Sons, Inc.

endocrine function. EDCs typically have molecular weights less than 600 Da and like hormones are effective at very low concentrations. Most have an aromatic ring especially with phenol functionalities in their structure and are sufficiently hydrophobic to interact with the biological receptors.

The presence of EDCs in the body affects functions such as metabolic and reproductive processes including embryonic development, gonadal formation, sex differentiation, growth, and digestion. They are plausibly linked to diseases including prostate cancer, breast cancer, attention deficit/hyperactivity disorder (ADHD), asthma, and reproductive problems (Myers et al., 2009; Swan et al., 2000). Though not toxins in the conventional sense, EDCs lead to serious adverse health outcomes in both animals[2] and humans. Conditions linked to exposure to EDCs, such as reproductive problems, incidence of certain cancers, asthma, obesity, diabetes, behavioral or learning disorders, and ADHD, are on the increase worldwide. Of this disease burden, 24–33% has been attributed to environmental contributions (Smith et al., 1999).

But EDCs were never regulated until recently. This lack of scientific and regulatory recognition of the hazard posed by EDCs was mainly a result of the conventional approaches used to evaluate the toxicity of these chemicals. With EDCs, it is the developmental (*in utero* and first years of life) stages that are particularly sensitive to health effects. While a significant fraction of human diseases are attributed to environmental factors (Prüss-Üstün and Corvalán, 2006), quantifying this fraction accurately is complicated. However, a clear trend in increase in human reproductive health problems, with rising production of EDCs, has been reported. Human studies show a strong association between exposure to EDCs and the incidence of adverse health conditions (Baillie-Hamilton, 2002; Bergman et al., 2012). Conclusive etiology or cause–effect relationships are deduced mainly from animal exposure studies.

The need for precautionary action to restrict their use as a class of chemicals was agreed to at the Prague Declaration on Endocrine Disruption (June, 2005) by a group of over 100 scientists.[3] Medical professional bodies in Europe and the United States have also urged a similar stance. Research focus is finally shifting (US Environmental Protection Agency, 2012) from investigating effects of adult exposure to examining developmental exposure and related disease outcomes.

Studying the health effects of EDCs is complicated by several factors. Like hormones, the EDCs also act at very low concentrations but have a marked effect in perturbing physiological responses. These perturbations are not always immediately apparent but can still be long term (in some instances, generational), and the long latent periods involved often confound research studies. As the exposure in nature is generally to multiple EDCs, their synergistic effects (Focazio et al., 2008; Hayes et al., 2006) need to be also taken into account. Some EDCs are in fact active only

---

[2] Thinning of eggshells in certain species of birds, the decrease in Baltic Seal population, and adverse impacts on fish populations as a result of exposure to EDCs have been reported.
[3] In their statement, "In view of the magnitude of the potential risks associated with endocrine disrupters, we strongly believe that scientific uncertainty should not delay precautionary action on reducing exposures to and the risks from endocrine disrupters."

in combination, not individually (Hass et al., 2007). To complicate studies even further, subjects are not uniform in their response; particularly susceptible or resistant subgroups (e.g., children, subjects with genetic polymorphisms) may exist.

7.1 ENDOCRINE DISRUPTOR CHEMICALS USED IN PLASTICS INDUSTRY

Over 800 chemicals are known or suspected to have endocrine disruptor (ED) effects (Bergman et al., 2012). It is the use of pesticides, phytoestrogens, food chemicals, and pharmaceuticals that are mainly responsible for the presence of EDCs in air, water, and soil. Chemicals used in plastics are not the major source of EDCs in the environment but can still be a significant source of human exposure *via* packaging and consumer goods. For instance, the incidence of infertility in women working in the plastics industry was observed to be significantly higher than for the general population in at least one study (Tyl et al., 2004).

The main EDCs encountered in plastics industry include the following:

1. Bisphenol A (BPA) {2,2-bis(4-hydroxyphenyl)propane} in polycarbonates (PC), some epoxy resins, and polysulfones (annual world production ~4.4 MMT)
2. Phthalates, used as plasticizers in PVC (annual world production ~6.4 MMT)
3. Polybrominated diphenyl ethers (PBDEs) used as flame retardants in plastics (annual world production ~0.07 MMT)
4. Alkylphenol ethoxylates (APE) used in phenolic resin and as an additive (annual world production ~0.07 MMT)
5. Some heavy metal residues from cadmium and arsenic polymerization catalysts.

Table 7.1 based on a recent UNEP report summarizes the credible research on links between exposure to these classes of EDCs and the diversity of implied disease conditions.

7.2 BPA {2,2-BIS(4-HYDROXYPHENYL)PROPANE}

Over 65% of the global production of BPA is used to make PC and another 30% for epoxy resin production (Plastics Europe, 2010). Based on the 2010 data, however, only 7% of the PC was used in packaging and medical device applications. Most are instead used in construction as glazing or in media (CD and DVD applications) and in electrical/electronic applications. These non-packaging uses pose a lower risk. About a third of this global production is used in the United States. BPA is perhaps the most controversial of plastic-related EDCs present in the environment and is attributed to at least the following sources (Geens et al., 2011).

TABLE 7.1　A Summary of EDCs of Concern Relevant to Plastics and their Adverse Impacts on Human Health and on Animal Life

	PCB, PCDD, and PCDF[a]	BPA	Phthalates	PBDE	Metal[b]
Male reproductive health	X		X	X	X
Female reproductive	X	X	X	X	X
Fecundity	X				
Polycystic ovary syndrome	X	X			
Fertility issues		X	X		X
Endometriosis	X	X	X		
Uterine fibroids	X		X	X	
Prostate cancer	X	X			X
Breast cancer	X				X
Testicular cancer	X		X	X	
Thyroid cancer	X	X			
Developmental neurotoxicity	X	X	X	X	X
Metabolic syndrome	X	X	X	X	
Invertebrates		X			
Fish	X	X	X	X	
Amphibians	X	X	X	X	
Reptiles	X	X		X	
Birds	X			X	X
Mammals	X			X	X

Source: Based on Bergman et al. (2012).
[a] PCB, polychlorinated biphenyls; PCDD, dioxins; PCDFs, furans.
[b] Examples: cadmium and arsenic.

1. Polycarbonate products
   BPA PC resin is used to make baby bottles, water bottles, and food storage containers (Mercea, 2009). Residual monomer BPA in the resin is believed to leach out from the plastic package or containers into food contents. The PC baby bottles, for instance, contain 4–141 ppm (median 10.5 ppm) of residual BPA (Leo et al., 2006). In addition, hydrolysis (Aschberger et al., 2010) of PC at high temperatures may also generate BPA monomer from the plastic material.

2. The epoxy resin lining of food and beverage cans
   Metal cans that contain food or beverages employ a thin film of an epoxy polymer on its inside surface to control corrosion of the metal and to limit food contact with metal. Epoxy is ideally suited for this application, and the type of epoxy used is predominantly (over 98%) BPA-based. Both aluminum and steel beverage cans are typically sprayed with a water-based epoxy–acrylic system, often with an amino resin (or phenolic resin) to cross-link the lining (Oldring and Nehring, 2007).

The epoxy is made by reacting BPA with epichlorohydrin in the presence of a base. As with PC, the BPA derived from the epoxy lining can also leach into the consumable contents in the can. With about 17% of the US diet coming from caned food, this is a serious concern.

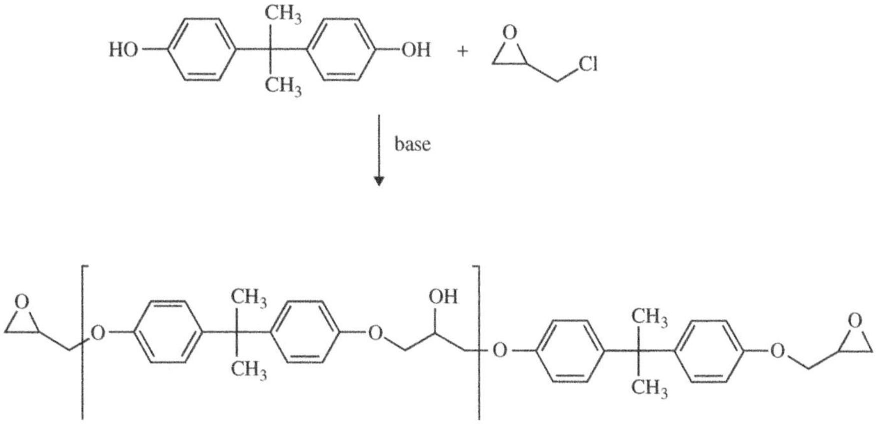

3. Dental fillings

   Some dental composite materials (Kingman et al., 2012) are based on bisphe-nol A-glycidyl methacrylate (bis-GMA) that may also contain other monomer modifiers such as bisphenol A dimethacrylate (bis-DMA) (Fung et al., 2000). BPA itself is not directly used in the composite. It is the bis-DMA and bis-GMA (the prepolymer used can have as much as 70% monomer) in the composite that can be converted into BPA and leached into the saliva of patients (within 1–3 h after dental sealant is applied) (Fleisch et al., 2010). The levels reported in saliva are typically low; the highest reported is 931 µg/l.

4. Thermally printed receipts

   Some leuco-dye-based thermal printing papers use BPA as a color developer (Terasaki et al., 2007); these are used in point-of-sale receipts, prescription labels, and some lottery tickets. Though the dye paper comes into contact with individuals only for short durations, the BPA load they deliver is 250–1000 times higher than that from plastic bottles or from can lining. BPA from this source can be accidentally ingested or transdermally acquired (Zalko et al., 2011) by handling the paper. As might be expected, store cashiers were shown to have relatively higher levels of BPA in their urine (Braun et al., 2011b). Including thermal paper in recycling, streams can unexpectedly transfer the BPA to other paper products such as toilet paper (Gehring et al., 2004)!

With so many uses for the chemical, BPA is found ubiquitously in the environment as well as in human body fluids and tissue (Murakami et al., 2007; vom Saal and Hughes, 2005). Not surprisingly, 92.6% of adults in the United States have BPA in their urine at a level averaging 2.4 µg/l in females and 2.9 µg/l in males (Calafat et al.,

TABLE 7.2   Human Body Burden of BPA (only Studies with Sensitivity < ~0.1 Ng/g Reported)

Tissue/Fluid	Technique	ng/ml	References
Healthy human serum	HPLC/coulometry	0.2–20	Vandenberg et al. (2007)
Maternal serum	GC/MS	4.4 ± 0.64	Schonfelder et al. (2002)
Fetal cord serum		2.9 ± 0.04	
Colostrum		1–7	Kuruto-Niwa et al. (2007)
Breast milk	HPLC/derivatization	0.28–0.97	Sun et al. (2004)
		0.65–0.70	Otaka et al. (2003)
Saliva	GC/MS	0.30 ± 0.043	Joskow et al. (2006)
Urine	GC/MS	1.5–5.6	Trasande et al. (2012)
		1.16[a]	Bushnik et al. (2010)
		1.79[a]	Melzer et al. (2010)

Source: Reproduced with permission from Geens et al. (2011).
[a]Median value (interquartile range).

2008). Lang et al.'s (2008) study has correlated the BPA concentration in the urine of 1445 adults with a range of their medical conditions (prevalence of diabetes, heart disease, and liver toxicity). Particularly worrisome is the presence of BPA in follicular fluid, amniotic fluid (Ikezuki et al., 2002), and cord blood (Schonfelder et al., 2002), suggestive of fetal exposure.

Table 7.2 illustrates the body burdens of BPA from combined routes of exposure, reported in recent years.

7.2.1   Exposure to BPA

The main route of BPA into the body is via ingestion, and the role of containers, water bottles, and baby bottles has been widely discussed (see Table 7.3). For instance, PC (baby bottles) in contact with hot water or milk leach BPA monomer into the liquid at a rate determined by the temperature and duration of exposure (Le et al., 2008). The possibility of additional BPA monomer being generated in the product via hydrolysis at high temperature, especially under alkaline conditions, has been pointed out (Howdeshell et al., 2003). In recent studies on PC baby bottles, 5–10 ng/ml (or ppb) of BPA was found to leach into the liquid contents of the bottle (Brede et al., 2003; Gibson, 2007). Heating or storage of highly acidic or alkaline food in PC food storage containers also increases potential leaching and associated health risk to both children and adults. Microwave heating for short durations does not degrade PC (Ehlert et al., 2008), but heating PC containers in contact with food should still be avoided. When tested with food simulant, some of the baby bottles leached out BPA even after 169 washes (Brede et al., 2003), suggesting possible generation of the monomer by hydrolysis (Howdeshell et al., 2003). Weathered plastic, as expected, tends to produce BPA at an increased rate (Takao et al., 1999), but increased leaching from PC bottles into the contents has not been observed.

TABLE 7.3   Examples of BPA Extraction by Different Liquids in Contact with Baby Bottles

Extractant	Leaching levels (ng/ml)	Temperature (°C)	Time (min)	References
Water	0.59 and 0.75	95	30	Sun et al. (2000)
	0.2	100	0.5	D'Antuono et al. (2001)
	0.23	100	60	
Food simulant	8.4 ± 1.2 (after 51 washes) 6.7 ± 1.2 (after 169 washes)	100	60	Brede et al. (2003)
Water	0.8	37	1440	Sajiki and Yonekubo (2003)[a]
River water	4.8			
Seawater	11			
Water	0.52–2.58	100	30	Lateef (2011)

[a]This study was on polycarbonate plastic tubing.

A second important source of BPA is the consumption of canned food where BPA may leach out from the epoxy liners into the contents. Canned foods contain low levels of BPA. A recent analysis of 43 canned beverages in Belgium (Geens et al., 2010) found 1.01 ng/ml (range 0.02–8.10 ng/ml) of BPA; beverages from non-plastic packaging showed less than 0.02 ng/ml of BPA. A 2007 study found similar results: Sajiki et al. (2007) found on the average 0–842 ng/g of BPA in canned foods, while in similar foods packaged in glass and paper, the values were 0–14 and 0–1 ng/g, respectively. A more recent study (NWSM, 2010) of 50 canned food items (including soup, soda, vegetables, and fruit) found 92% of the contents to contain detectable levels of BPA with a mean value of 77.4 ppb. Concentrations found are quite variable as leaching depends on the duration of exposure, storage conditions (especially temperature), as well as the pH of the contents.

Based on the known levels of BPA in food, water, and the ambient environment, the daily intake of the EDC in units of µg/kg body weight/day might be estimated. Western adult exposure via canned food is approximately 1.5 units (European Food Safety Authority, EFSA, 2006) much lower than the infant exposure expected via feeding from PC bottles (4–13 units). BPA intake from this route accounted for 10–40% of the total daily intake (van Goetz et al., 2010), and consuming canned soup for 5 days increased the urinary BPA of adult subjects by 1000% (Carwile et al., 2011). The Cao et al. (2011) study estimated the age-dependent dietary intake of BPA to vary between 0.082 and 0.23 for infants and 0.052 and 0.081 for adults, both in units of µg/kg body weight/day. Significantly, in that population, nearly 75% of the dietary intake of BPA appears to be derived from canned soups, meats, and vegetable (corn) consumption. The recent

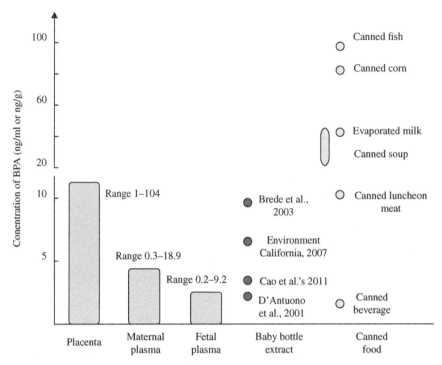

FIGURE 7.1 Approximate mean BPA concentrations in baby bottles and canned food or beverages compared to that in plasma and the placenta (Schönfelder et al., 2002). Sources: Canned food data is based on Cao et al.'s (2011) study. Canned beverage data is from Geens et al.'s (2010) study. Baby bottle data are variable because of the different temperatures, leach times, and simulants used.

estimates on intake of BPA summarized by Geens et al. (2011) show a lower value of 33–47 ng/kg body weight/day on average in the United States. Figure 7.1 shows the range of reported values.

7.2.2 Effects of Exposure to BPA

Adverse health effects of exposure to BPA have been reported at a range of doses including a substantial literature on effects at doses well below the reference dose of 50 μg/kg body weight/day deemed "safe" by the USEPA. Most of the data, however, are based on animal model studies. A detailed discussion of these effects and the possible mechanisms of toxicity involved are beyond the scope of this chapter. However, the major effects are summarized in Table 7.4. The reader is directed to references, recent books (Vaughn, 2010) and reports (Bergman et al., 2013; Wargo, 2008) on the topic for detailed information.

The available human exposure studies clearly shows the perturbation of reproductive function by BPA (Li et al., 2010a; Meeker and Ferguson, 2012) and prenatal

TABLE 7.4   A Summary of Biological Effects of Exposure to BPA at Low Doses

System affected	Adverse effect reported	References
1. Male reproductive system	• Reduced sperm count • Serum testesterone levels • Increased ano-genital distance[a] • Prostate enlargement[a]	vom Saal et al. (1998) Kawai et al. (2003); Gupta et al. (2000)[a]
2. Female reproductive system	• Abnormal ovary development • Early sexual maturation • Endometrial diseases • Adult uterine diseases[a]	Markey et al. (2005) Honma et al. (2002) Masumo et al. (2002) Newbold et al. (2007a)[a] Susiarjo et al. (2007)[a]
3. Prostate disease and breast cancer	• Prostatic lesions • Linked to prostate cancer • Breast cancer[a]	Shuk-Mei et al. (2006) Wetherill et al. (2002) Murray et al. (2006)[a]; Vandenberg et al. (2008)
4. Diabetes and weight gain	• Insulin resistance • Increased adipose and cholesterol	Ropero et al. (2008) Miyawaki et al. (2007) Takai et al. (2000)[a]
5. Behavioral changes	• Aggressive behavior • Hyperactivity • Behavior problems[a]	Kawai et al. (2002); Ishido et al. (2004); Farabollini et al. (1999)[a]; Negishi et al. (2004)[a]

Effects based on animal model studies. Low dose is one that is below the safe limit of 50 µg/kg body weight/day.
[a]Effects related to fetal or perinatal exposure.

BPA exposure to increased hyperactivity and aggression in 2 year-old girls (Braun et al., 2009, 2011b). Other reports link exposure to non-reproductive effects such as coronary heart disease, asthma, and diabetes. Transgenerational and delayed physiological effects add an additional level of complexity to the problem. Perinatal exposure to BPA can result in adverse effects that manifest much later in life. Amniotic fluid analysis during the second trimester can detect human fetal exposure to low doses of BPA (0.31–0.43 ng/ml in 20 specimens) (Edlow et al., 2012). Also, at least in mammalian animal models, exposure can even result in reproductive effects in offspring of the F3 generation (Manikkam et al., 2013; Susiarjo et al., 2007), a possibility not taken into account in conventional assessment of safe dosages.

Potential exposure to BPA from PC products can start at conception and continue through fetal stages (via maternal exposure) to infant and into the childhood years. Wong et al. (2005) estimated the worst-case dietary intake of BPA by newborn infants to be 24 µg/kg body weight/day, dropping to 15 µg/kg body weight/day at 3 months. While the estrogenic effects of BPA are well known, the full impact of fetal exposure at low concentrations remains unknown.

7.2.3   Dose–Response Relationships of BPA

Most of the controversy surrounding ED effects of BPA has to do with the shape of the unusual dose–response curve observed in laboratory and clinical studies. Conventional toxicological wisdom dictates biological response (toxicity) to be an increasing linear function of the dose administered.[4] However, with some EDCs including BPA, the dose–response function is often nonlinear over the low-dose range of interest (Lemos et al., 2009; Vandenberg et al., 2009; Wozniak et al., 2005). Nonlinear dose–response curves (NLDR curves) and life-stage-specific impacts are quite common with some hormones as well. In fact, with hormones, it is the low-dose regime that yields the maximum physiological effects (Welshons et al., 2003); the serum hormone levels (estradiol, testosterone, cortisol, and thyroid hormones) range only within 0–300 pg/ml. The low-dose activity of EDCs is therefore anticipated. For instance, low-dose effects of BPA (such as errors in chromosomal sorting) (Hunt et al., 2003) are not necessarily observed at the higher dosing levels (Welshons et al., 2006). It is the complex multimodal action of EDCs (as well as hormones) that results in the NLDR relationships. These include potential cytotoxicity, receptor downregulation, receptor competition, negative feedback loops, and tissue interaction linearity (Myers et al., 2009). The controversy on EDCs is also due to the lack of good mechanisms to explain NLDR curves at the present time.

Some NLDR curves can be U-shaped, with high responses at both low and high doses (Hugo et al., 2008) or inverted U-shaped where the greatest response is at intermediate concentrations (Jenkins et al., 2011; Newbold et al., 2007b; Welshons et al., 2006). Figure 7.2 shows two examples of nonlinear physiological dose–response curves for BPA. Unless the tests are carried out at the relevant range of concentrations with long enough durations of observation, conclusions from experimental studies will be seriously deficient. No extrapolation or interpolation of data can be relied upon to predict responses for EDCs.

7.2.4   Safe Levels of BPA

The official "safe dose" used by the FDA for BPA is the USEPA reference dose of 50 µg/kg body weight/day, based on a high-dose study on a generation of adult mice and rats (Vogel, 2009). As was the practice in the 1980s, the lowest dose at which adverse effects were observable, multiplied by safety factor of 10^3, was used to arrive at the "safe" level. Numerous later studies (Akingbemi et al., 2004; Al-Hiyasat et al., 2002; Bindhumol et al., 2003; vom Saal et al., 1998), however, show adverse responses well below this "safe" dose of 50 µg/kg body weight/day. In 2007, an expert panel of 38 scientists convened by the NIEHS, meeting in Chapel Hill, NC, reviewed the available data and arrived at the consensus view that the current levels of BPA in humans exceed those that result in adverse health impacts in animals (Expert Panel on BPA, 2007). Over 100 peer-reviewed studies show adverse effects

---

[4] Sixteenth-century observation by Paracelsus that toxicologists paraphrase as "the dose makes the poison."

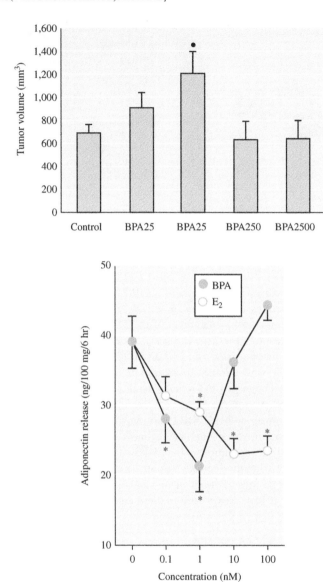

FIGURE 7.2 Examples of non-monotonic dose–response curves. Above: Effect of tumor volume in mice on the BPA levels in drinking water shows an inverted-U response. Numbers on the horizontal axis refer to μg BPA/l of drinking water available to the mice. These correspond to 0–500 pg of BPA/kg body weight. Below: Suppression of adiponectin release from human breast adipose explants by BPA and estradiol (E_2). (Hugo et al., 2008). Source: Jenkins et al. (2011).

FIGURE 7.3   Baby bottles and can liners may leach polycarbonate into food.

in animal models (vom Saal and Hughes, 2005) at levels that are sometimes orders of magnitude lower than the reference dose. In mice, for instance, ingestion of only 2 ng/g body weight of BPA resulted in changes in the preputial glands and epididymis (vom Saal et al., 1998): this dosage was estimated to be lower than that ingested in the first hour after application of a dental sealant! The National Toxicology Program (NTP) in 2008 also expressed concerns on low-dose effects of BPA.

BPA was in use prior to the enactment of the TSCA[5] and was grandfathered for use with no further testing. The growing body of data, however, has helped change the regulatory environment; in 2010, the FDA slightly changed its position on BPA agreeing with the NTP view while still maintaining BPA to be a safe chemical at the levels found in food. The use of PC in baby bottles (and baby cups) has since been banned in Canada in 2010, in the European Union, in China in 2011: the US FDA adopted a similar position in 2012.[6] BPA continues to be used in other applications but is being voluntarily phased out in can liners by a few of the manufacturers (Fig. 7.3). Analogous compounds such as BPS, BPF, and BPB, might be used in its place in several of the applications. However, the potential adverse effects of these substitute compounds (Bittner et al., 2014) have also not been studied in detail and long-term effect of exposure to these also remains essentially unknown.

7.2.5   Contrary Viewpoint on BPA

There are also some expert opinions to the contrary (Hengstler et al., 2011; Teeguarden and Hanson-Drury, 2013; Teeguarden et al., 2011) that maintain the concerns on human health hazards of BPA to be exaggerated as some of these are based on animal studies.[7] Critics also point out that some of the key studies were not repeatable (Sharpe, 2009) and the levels of exposures are minimal (Calafat et al., 2008). Also,

---

[5] The Toxic Substance Control Act of 1976 (TSCA) first gave the USEPA the authority to regulate chemicals from beginning to end of life.
[6] The FDA moved to ban BPA from baby bottle applications at the request of the American Chemistry Council and *after* its use in baby bottles has been phased out voluntarily by the industry.
[7] A website by the American Chemical Council, (factsaboutbpa.org) for instance argues the contrary position.

some of the BPA studies do not show the low-dose effects (Tinwell et al., 2002) or intergenerational effects (Tyl et al., 2002) claimed by others. The controversy still continues; in 2013, Teeguarden et al. (2013) reported that based on a sample of 150 adults exposed to high dietary levels of BPA, the serum levels of the chemical were too low to be even detected, while 84–97% of the dosage was in the urine. Since BPA is readily conjugated in the liver, unconjugated levels in the blood tend to be very low (Teeguarden et al., 2011) and implies high clearance rates of BPA from the body.

Government agencies in the United States, Canada, and Europe are in general agree that using PC in food-contact applications is safe as the levels of BPA that leach out into the food are well below that believed to pose a credible health risk. The WHO/FAO (2010) and the European Food Safety Authority (2011), German Society of Toxicology and Japan's Research Institute of Science for Safety and Sustainability also agree that polycarbonates and epoxy liners as they are commonly used today, in applications other than baby bottles, are generally safe in food-contact uses.

7.2.6 Environmental Sustainability and BPA

There is sufficient credible evidence that BPA leaches out of PC products and epoxy liners into food/beverages in contact with plastic. The critical question is whether potential adverse health impacts are encountered at the low levels of BPA in real exposure environments (as opposed to those in laboratory studies). There may not be enough research information to unequivocally answer the question.

However, environmental sustainability considerations address a slightly different question: Is there enough robust data on negative health impacts to warrant precautionary action to phase out the use of BPA? The intensity of the controversy in the science community alone answers this question in the affirmative. Precautionary avoidance of the use of BPA, at the very least in food-contact uses, pending resolution of the issue scientifically is justified. This is particularly reasonable as comparable or superior alternative materials for these applications do exist.

BPA-free plastics that can replace PC in bottles (PET and PE are widely used already) or containers with no discernible loss in functionality are available. While epoxy is an ideal, proven candidate for can liners, other plastics can be used as a replacement. A changeover to BPA-free cans has already been made in Japan back in the 1990s (Talsness et al., 2009), and BPA-free cans are already available in the marketplace.[8] Both PET and baked-on oleoresin liners can be used to replace epoxy can liners for most (except for highly acidic food) packaging uses. BPA-free dental appliances and composite fillings are also already in the marketplace. A critical requirement, however, is that these replacements are also be proven safer than the BPA they replaced. A part of the orientation toward sustainability is to impose the same or even more stringent standards on potential toxicity of the replacement for the BPA-based material. Given the resources of the industry, replacing epoxy liners, leuco dyes that use BPA, PC and polysulfones in medical applications, and PVC products with BPA, should be achievable.

---

[8]Companies such as Eden Organic, Native Forest, and Trader Joe's advertise canned food in BPA-free cans.

7.3   PHTHALATE PLASTICIZERS

Worldwide, over 6 million tons of phthalates are produced annually. They are color-less liquids having the consistency of oil, sparingly soluble in water, but very soluble in plastic matrices (as well as in adipose body tissue.) At least 95% of the production ends up in plastics and other consumer products. They are also used in cosmetic formulations such as nail polish (imparts crack resistance) and in hair spray. Its best-known use is in plastics as a plasticizer, mostly in PVC, to soften the hard, brittle plastic. It is the same plastic, PVC, that is used in siding and water pipes that is also used in soft flexible automobile seat covers and in shower curtains. In the latter cases, the PVC is intimately compounded with up to 50–60% of an organic phthalate to plasticize the resin. Notebook covers, children's toys, rainwear, and backpack compo-nents used by children as well as some flexible transparent medical tubing and blood bags are also made from plasticized PVC material. Phthalates being only dis-solved in the polymer (not covalently bound) can leach out into fluids in contact with the vinyl under some conditions.

Plasticizer molecules as a rule are highly compatible with plastics and, like all good solvents, easily disperse in between the long chain-like polymer molecules. This essentially increases the free volume between chains, reducing interchain attrac-tion and allowing increased mobility in the polymer chain segments. With reduced interchain attraction, chains can move about easier resulting in a softer polymer. A plastic such as PVC becomes progressively softer (and its glass transition tempera-ture T_g (°C) decreases), and its modulus and hardness also decrease as increasing weight fractions of plasticizer are mixed into the plastic (Fig. 7.4).

The generic chemical structures of six common phthalates are shown in Figure 7.5 and listed in Table 7.5. Of these, di(2-ethylhexyl) phthalate (DEHP) is the most commonly used and perhaps the most extensively studied member of the class (Matsumoto et al., 2002).

Once released into the environment, phthalates degrade rapidly; the half-life of DEHP (the most widely used phthalate) in water is only 2–3 weeks. But in air, or when bound to soil, phthalates can be stable for longer periods of time. The reported levels of DEHP, the most common phthalate, in air, water, and soil (see Table 7.6) illustrate this. As with other EDCs, phthalates are also ubiquitous in the environment and are present in the human body, breast milk, blood, and urine. In one study, over 75% of the US population were found to have phthalate metab-olites in the urine (Stahlhut et al., 2007). This is hardly a surprise as the available data suggests life-long human exposure to occur from *in utero* through death of individuals.

The relatively low human body load of phthalates is a result of short residence times (elimination half-life 8–10 h) in the body as phthalates are not bioaccumula-tive.[9] Better quantification of exposure might be achieved via the monitoring of primary metabolite from hydrolysis of the phthalate (Kamrin, 2009).

---

[9]Bioaccumulation occurs when the rate of removal (excretion) of a chemical from an organism is slower than the rate of intake due to exposure to the chemical.

Modulus (GPa)
Plasticizer (%)
SIEBIMM
Indentation
0
1
2
3
4
0
10
20
30
40

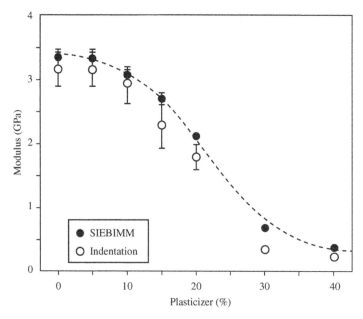

FIGURE 7.4 Modulus versus plasticizer (dioctyl phthalate) concentration for PS films. Two different techniques, indentation and strain-induced elastomer buckling instability for mechanical measurements (SIEBIMM), were used to estimate the modulus of the material. The latter technique is SIEBIMM, an optical technique for assessing the modulus of thin films of material. Source: Reproduced with permission from Stafford et al. (2004).

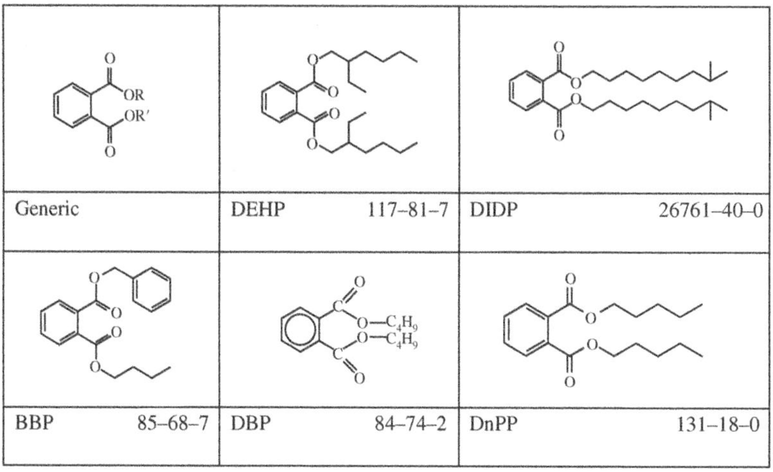

Generic
DEHP 117–81–7
DIDP 26761–40–0
BBP 85–68–7
DBP 84–74–2
DnPP 131–18–0

FIGURE 7.5 Chemical structures of some common phthalates with their CAS numbers in parenthesis. DEHP, di(2-ethylhexyl) phthalate; DIDP, diisodecyl phthalate; BBP, butyl benzyl phthalate; DBP, dibutyl phthalate; DnPP, di-*n*-pentyl phthalate.

TABLE 7.5 Common Phthalate Plasticizers and their Characteristics

Plasticizer	Code	C-chain length	log K	Solubility (mg/l)	Care[a]	Mut[b]	Ref. dosage 1 μg/kg/day	TDI 1 μg/kg/day
Di(2-ethylhexyl) phthalate	DEHP	6	7.5	0.003	C/B	+/−	20	37
Diisononyl phthalate	DINP	8–9	8.8	0.0006	N	−	120	150
Dibutyl phthalate	DBP	4	4.6	11.2	D	+/−	100	10
Diisodecyl phthalate	DIDP	9–10	8.8	0.0002	N	−	N	150
Di-n-octyl phthalate	DNOP	8	8.1	5.0×10^{-4}	N	N	N	N
Butyl benzyl phthalate	BBP	4.6	4.8	2.8×10^{-3}	C	−	200	200

[a]Carcinogenicity: Group B, possibly carcinogenic; Group C, not classified; and Group D, probably not carcinogenic to humans. N, not classified. The classification based on the USEPA and IARC.

[b]Mutagenicity RfDS and TDI values are from Bang et al. (2012).

TABLE 7.6 Typical Concentrations of DEHP in Water and Air

Medium	Units	Mean concentration	Median and range
Drinking water	μg/l	0.55	0.55 (0.16–170)
Wastewater	μg/l	27	8.3 (0.01–4400)
Rainwater	μg/l	0.17	0.17 (0.004–0.68)
Surface water	μg/l	0.21	0.05 (<0.002–137)
Outdoor air	ng/m³	5.0	2.3 (<0.4–65)
Indoor air	ng/m³	109	55 (20–240)

Source: Adapted from Clark et al. (2003).

The chronic oral reference dose of 0.02 mg/kg/day of DEHP derived by the USEPA is based on the lowest observable doses in experimental studies carried out several decades ago. As with BPA, at the time, studies on low-dose health effects of phthalates were not available. Phthalate levels in Table 7.6 are well below the oral reference level but in common with all EDCs may have low-level effects.

7.3.1 Exposure to Phthalates

The primary route of exposure to phthalates is via ingestion (Wormuth et al., 2006), while inhalation (Huang et al., 2011) and dermal contact (Koo and Lee, 2005; Rudel et al., 2003) may also play a role. Phthalates in food packaging leached into food is the primary source of ingested phthalates. High levels of DBP (5860 mg/kg) and DEHP (3680 mg/kg) in food packaging have been reported (Versar, Inc. and Syracuse Research Corporation, 2010). Partitioning of phthalates into food is particularly high with products such as cheese, butter, oils, and milk. Versar, Inc. and Syracuse Research Corporation (2010) summarized the highest reported phthalate levels (for common phthalates) in different product categories: the mean concentration in beverages (0.04–80), food (1.4–173), and food packaging (0.28–5280) expressed in mg/kg were reported. The highest levels were associated with milk, oily foods, and butter. Levels in potable water are by comparison quite low (<5 mg/l). House dust carries a much higher load by surface adsorption (10^2–10^4 mg/kg). Nail care products and some cosmetics can have levels as high as 10^6 mg/kg.

Exposure of infants and children to phthalates is a particularly serious concern. A 1994 Canadian study estimated the daily intake of DEHP for adults to be about 5.8 μg/kg body weight and found the sources to be predominantly food and indoor air. In infants and very young children, the intake is much higher (~9–19 μg/kg body weight), while the primary source still remains food, as illustrated in Figure 7.6 (Shea, 2003). Others have reported similar findings (Martinez-Arguelles et al., 2013). Human fetal exposure to phthalates occurs *in utero* or *via* maternal exposure to phthalates (Martinez-Arguelles et al., 2013) and continues on after birth through breastfeeding (Calafat et al., 2004) and commercial infant formulae (Frederiksen et al., 2007). Possible ingestion of phthalates while mouthing toys by infants is a

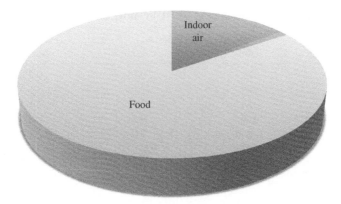

FIGURE 7.6   Intake of DEHP by source for an adult. Ingestion with food is by far the most important mechanism of exposure. Source: Reproduced with permission from Shea (2003).

particularly serious concern (Schettler, 2006). Phthalates[10] are used in formulations for children's or infant's toys; review of multiple-country data shows high phthalate levels in toys; for instance, up to 44% by weight for DEHP in PVC has been reported (Bouma and Schakel, 2002). The ban on the use of certain phthalates in children's toys by the Consumer Products Safety Commission goes a considerable way towards addressing this concern.

Infants undergoing medical procedures may have additional exposure via PVC tubing (e.g., IV tubing) (ATSDR, 2002), dialysis equipment (Center for the Evaluation of Risks to Human Reproduction, 2006), or blood bags. Adult exposure can also similarly occur through plastic medical devices[11] such as blood bags, PVC tubing (Koch et al., 2006) or personal care products. Off-gassing from surfaces such as vinyl floor tiles and wallpaper (Jaakkola and Knight, 2008) can also be a significant source of exposure. Where exposure to multiple sources of phthalates are involved, the effect is generally cumulative (Howdeshell et al., 2008).

In general, inhalation is not an important route of exposure to phthalates. But volatilized phthalates in cabin air of automobiles can contribute significantly to intake via inhalation exposure. A recent study on 23 vehicles found DBP and DEHP at concentrations ranging from 1.96 to $3.66\,\mu g/m^3$ in the cabin air (Geiss et al., 2009). Cars parked outdoors can reach high interior temperatures and correspondingly higher phthalate levels in the air. Under worst-case conditions, the concentrations of phthalates can reach concentrations of $DEHP = 2 \times 10^6\,mg/m^3$ and $BBP = 7700\,mg/m^3$ in the vehicle interior (Fujii et al., 2003)! These numbers are several times higher than even the acceptable 'safe' inhaled levels though passengers are exposed to these levels only for short durations. High levels of phthalates in paints, modeling clay, and adhesives

[10] In a soft flexible product, around 50% of the compound can be phthalate. The highest level of phthalate reported in a PVC toy is 73% of DINP, reported by the California Department of Toxic Substances Control in a 2008 study.
[11] Infants undergoing treatment in ICUs are estimated to receive phthalate doses that are 2–3 orders of magnitude higher than that for the general population (Calafat and Needham, 2004).

(~50,000 mg/kg) used as building products as well as other products such as air fresheners (15,000–500,000 mg/kg) (Versar, Inc. and Syracuse Research Corporation, 2010) might be released into indoor environments (Fujii et al., 2003).

More recent works estimate the adult daily intake to range from 4.9 to 18 μg/kg body weight. Regardless of age group, 50–98% of this DEHP intake is believed to be from food (Wormuth et al., 2006). Phthalates in indoor air are primarily derived from outgassing of vinyl floor coverings, wall coverings, and other building materials.

7.3.2 Toxicity of Phthalates

Of the common phthalates, BBP, DEHP, and DBP elicit the highest toxicity to terrestrial organisms, fish, and aquatic invertebrates (European Commission, 2008). It is the ED characteristics of phthalates (rather than their low acute toxicity or mutagenicity) that are of major concern. In common with other EDCs, the dose–response curves for biological effects of phthalates can also be non-monotonic (Andrade et al., 2006). Therefore, as with BPA, conventional toxicological screening can often be inadequate in establishing reference doses at which adverse outcomes are possible. Phthalates with linear-ester side chains with 4–6 carbons were particularly associated with the phthalate syndrome (changes in reproductive function and fetal development) in animal models. Ohtani et al. (2000) found adverse effects of DBP on the differentiation of male gonads in *Rana rugosa* tadpoles.

Fetal or infant exposure to phthalates was shown to be associated with reproductive or developmental health outcomes in males (e.g., correlation with shortened anogenital distance (Swan, 2006) or premature breast development (Colon et al., 2000)). Positive associations between phthalate metabolites in urine and symptoms of ADHD (Kim et al., 2009) as well as phthalate exposure and asthma or allergy (Bornehag et al., 2004) among school-age children have been reported.

The major health impacts of phthalates supported by recent studies and showing an association significant at $p > 0.05$ were summarized by Meeker et al. (2009b) as well as by Wargo (2008) and are shown in Table 7.7. A detailed discussion of the adverse impacts or mechanisms of endocrine disruption are beyond the scope of this chapter. Several excellent reviews (Committee on the Health Risks of Phthalates, National Research Council, 2001; Versar, Inc. and Syracuse Research Corporation, 2010) discuss the various biological impacts of exposure to phthalate.

7.3.3 Environmental Sustainability and Phthalates

Unlike with BPA, however, the government agencies in Europe, Japan, and the United States are in agreement on the need to restrict selected phthalates in some products. The US legislation (under the Consumer Product Safety Improvement Act of 2008) regulates two classes of phthalates: the first group consisting of DEHP, DBP, and BBP cannot individually exceed a level of 0.1 wt. % in toys and childcare products handled by children (younger than 12 years); the second group consisting of DINP, DIDP, and DNOP cannot individually exceed a level of 0.1 wt. % in toys or in childcare products that might be placed in a child's mouth. The European Commission

TABLE 7.7 A Summary of Effects of Human Exposure To Phthalates

System affected	Adverse effect reported	References
1. Male reproductive system	• Shorter ano-genital distance* • Smaller genitalia and incompletely descended testicles* • Semen quality deterioration	Swan et al. (2006) Lottrup et al. (2006) Duty et al. (2003)
2. Female reproductive system	• Early signs of puberty • Links to endometriosis and uterine fibromatosis • Pregnancy complications and premature deliveries	Colon et al. (2000) Latini et al. (2003) Lovekamp-Swan (2003)
3. Thyroid effects	• Lower levels of thyroid hormones	Meeker et al. (2007)
4. Asthma	• Correlation with allergic asthma	Bornehag et al. (2004)

Regulations are more stringent in that they regulate the combined level of all phthalates in a group at the same levels in toys (European Commission, 2005). These ban the use of BBP, DBP, and DEHP in all children's toys and also DIDP, DINP, and DNOP from those products that kids may place in their mouth.

A review of data on toxicity of phthalates suggests the human health risks to be generally low (Kamrin, 2009), but, there is enough uncertainty on the long-term health impacts of phthalates to justify the adoption of a precautionary stance. Of the 151 chemicals identified by the European Chemical Agency as substances of very high concern, eight are phthalates. The USEPA is working toward including the eight common phthalates to the list under TSCA Section 5(b)(4) and to regulate these under TSCA Section 6(a). The regulatory pressure on phthalates is an incentive toward developing non-phthalate plasticizers. Also, alternative plastics that can replace a large part of the soft PVC product market with the added advantages of lower metal stabilizers and reduced chlorine load in the post-use waste, are available. For instance, PVC enteral feeding tubes (EFT) and accessories in ICU procedures can be replaced by the more durable silicone or polyurethane tubes (Van Vliet et al., 2011).

Alternatives for conventional phthalate plasticizers are becoming commercially available (Schmidt, 2008; LCSP, 2011; Krauskopf, 2003); acetyl tri-*n*-butyl citrate (ATBC) are claimed to be dropped in substitute for DEPH: some of these are listed in Table 7.8.

7.4 POLYBROMINATED DIPHENYL ETHERS (PBDEs)

These compounds are flame retardants used in products such as flexible polyurethane foams (used in upholstery), plastic appliance housing, wire and cable insulation, and coatings at weight fractions that can range from 10 to 30%. About 67,000

TABLE 7.8 Some Examples of Nonphthalate Plasticizers For PVC

Plasticizer	Symbol	Chronic exposure NOAEL[a] mg/kg body weight/day	BCF[a] l/kg	Oral toxicity[a]
Acetyl tri-*n*-butyl citrate	ATBC[b]	1000	250	None
Di(2-ethylhexyl) adipate	DEHA[c]	948/1104	27	Negative effects
1,2-Cyclohexanedicarboxylic acid, dinonyl ester	DINCH[d]	40–200*	189	Negative cancer bioassay
Trioctyltrimellitate	TOTM	Not available	1–3	Negative reproductive toxicology effects
Di(2-ethylhexyl) terephthalate	DEHT	325-102/418-666	10^6	None

Phthalate substitutes can also leach out of plastics, and their toxicity to human consumers has not as yet been comprehensively studied. Once the functional equivalence of a substitute plasticizer is established, exhaustive studies need to be undertaken to unambiguously validate their safety to health and environment.
[a] Source for chronic exposure NOAEL, bioconcentration factor (BCF), and oral toxicity information is from Versar, Inc. and Syracuse Research Corporation (2010).
[b] Approved by the FDA for use as a food additive and food-contact substance.
[c] Approved by the FDA as an indirect food additive as a component of adhesives.
[d] Approval from the European Food Safety Authority (EFSA, 2005), the Japan Hygienic PVC Association (JHPA), and the German Institute for Risk Assessment (German BfR) for use as a food-contact substance. Waiting for FDA approval.
[e] Manufacturer-reported data.

tons of PBDEs were produced worldwide in 2011. Of the thermoplastic resins, PE, PP, EVA, HIPS, and ABS resins generally use PBDEs. Automotive, building construction, and textile upholstery sectors each use about a quarter of the production, and electrical products account for about another 13%. Fire protection afforded by flame retardants[12] has very significant benefits in terms of avoided property damage and lives saved in fires. The generic chemical structure of the class of compound is shown below (note structural similarity to polychlorinated biphenyls or PCBs). Structurally, there are 209 possible congeners of PBDE.

Halogenated flame retardants thermally decompose during a fire, yielding high concentrations of bromine radicals. The highly reactive bromine radicals combine with the volatilized flammable vapor from burning plastics and extract hydrogen from these, forming HBr. Removal of H and OH radicals from the hot igniting vapors controls the heat and the fire.

Naturally, the fully brominated decabromo- congener molecule (referred to as the BDE-209 congener) is the most effective and most widely used in this application.

Commercial products of PBDEs are available as mixtures of several congeners though named after the dominant component for convenience (Allen and McClean, 2007). For instance, the commercial "penta-BDE" is a mixture of tetra-, penta-, and hexacongeners. It was the most used flame retardant in polyurethane foam. Penta- and octa-BDE mixtures are no longer manufactured in (or imported into) the United States since 2004 and internationally from 2010. Only the deca-BDE is still manufactured and used in the US. The commercial product contains 97–98% deca-BDE, along with 0.3–3.0% other PBDEs, mainly nonabromodiphenyl ether. The water solubility and partition coefficients for these are listed in Table 7.9.

Though no longer manufactured, the lower-brominated congeners are widely spread in the environment because of their historic use (Gaylor et al., 2012; Renner, 2000). PBDEs have been reported from biological tissue, water, and sediment samples collected from even the most remote environments such as the Antarctic (Hale et al., 2008). It is widely present in biota and in human blood, fat tissue, and breast milk (Schecter et al., 2005). In most environmental compartments as well as in human milk, the level of BDEs has been on the increase in the past decade.

PBDEs are persistent and toxic compounds that affect the liver and, thyroid, and cause reproductive/developmental, and neurological damage. Once released to the environment, PBDEs are photolytically debrominated by exposure to sunlight and by the action of anaerobic bacteria. In HIPS blends (as in equipment casings), the deca-BDE

[12]PBDEs constitute only about 18–21% of the total flame retardants used globally. Inorganic flame retardants such as $Al(OH)_3$, $Mg(OH)_2$, Sb_2O_3, and other metal compounds are the most used class of flame retardants.

TABLE 7.9   The Solubility and Log $K_{o/w}$ for Common Classes of BDEs

Congener	Degree of Br substitution	log $K_{o/w}$	Water solubility, mg/l
Deca-BDE	10	10	0.02–0.03
Octa-BDE	8	5.50–8.90	<0.01
Hexa-BDE	6	6.86–7.92	4.08×10^{-03}
Penta-BDE	5	6.64–6.97	9.00×10^{-07}
Tetra-BDE	4	5.87–6.16	0.07

exposure to sunlight can result in the formation of polybrominated dibenzofurans (PBDFs) (Kajiwara et al., 2008) that are also toxic compounds. The deca-BDE has a short residence time in animals, and ingested material is mostly excreted within 24 h.

PBDEs are usually not used in food-contact applications ruling out their leaching into packaged food; yet, food and beverage are important routes of intake, as PBDEs are present in both water and air. However, as with phthalates, exposure via hard plastic toys can result in high values of intake approximating 10^{-4} mg/kg body weight in children (Chen et al., 2009). Inhalation of house dust loaded with the chemical (Allen et al., 2008) is the major intake mechanism. BDEs released from products or carpets may be adsorbed onto airborne particles of dust (Takigami et al., 2008) where the levels can be as high as 2.9 µg/g-dust. The total exposure from all sources was estimated to be (Health Canada, 2010) 0.05–0.08 for infants and children, 0.013–0.036 for youth, and 0.008–0.009 for adults (µg/kg body weight/day). These compare with the critical effects level of 2.22 (mg/kg bw/day) for PBDE.

7.4.1   Toxicity of PBDEs

Both animal model and *in vitro* studies suggest adverse developmental, reproductive, and neurotoxic as well as endocrine disruptive effects associated with exposure to PBDEs. Evidences for EDC activity of PBDEs are based primarily on animal model studies (Van der Ven et al., 2008). Mice and rat studies on deca-DBE at a dose level of about 5 mg/kg bw/day or above resulted in B-cell activation and a reduced number of natural killer (NK) cells in the offspring (Teshima et al., 2008). The limited human toxicity data reported are summarized in Table 7.10.

An interesting recent study (Gaylor et al., 2012) showed house crickets (*Acheta domesticus*) provided with free access to polyurethane foam with 8.7 wt. % of penta-BDE as well as uncontaminated food for 28 days accumulated substantial body burdens of 13.4 mg/kg of EDC. Even with food and water available *ad libitum*, the crickets still ingested PBDE-contaminated foam and showed no apparent distress despite the high body burdens. However, effects on the larval stages and the role of these as vectors for PBDE transfer have not been investigated.

The San Antonio Statement on Brominated and Chlorinated Flame Retardants (DiGangi et al., 2010), agreed upon at the *30th International Symposium on Halogenated Persistent Organic Pollutants* (DiGangi et al., 2010), in San Antonio, Texas, summarizes the concerns of expert scientists on PBDEs. About 145 scientists

TABLE 7.10   A Summary of Effects of Human Exposure To PBDEs

System affected	Adverse effect reported	References
1. Thyroid	• Disruption of thyroid hormone, homeostasis • Reduced levels of thyroxin in blood	Darnerud et al. (2001) Hallgren and Darnerud (1998) Turyk et al. (2008) Meeker et al. (2009b)
2. Male reproductive system	• Reduced testis size and reduced semen quality • Increased risk of cryptorchidism	Akutsu et al. (2008) Main et al. (2007)
3. Female reproductive system	• Unsuccessful or complicated pregnancies • Decreased fertility	Morreale de Escobar et al. (2000) Harley et al. (2010)
4. Cancer (human)	• Increased risk of testicular cancer (2nd generation)	Meeker et al. (2007)
5. Hepatic effects	• Liver enlargement, sometimes with degenerative changes	Zhou et al. (2001) Zhou et al. (2002)

from 22 countries developed a reasoned plea from the scientific community to consider impacts of the use of halogenated flame retardants both at present and for the future.

7.4.2   Environmental Sustainability and PBDE

The regulatory process for PBDEs appears to be on track with significant steps already been taken to phase out their use, as demanded by environmental sustainability. The decabromo- congener is expected to be voluntarily phased out in 2014. As with phthalates and BPA, the next step would be the identification of safe and viable alternatives. Some of these are already being developed. Again, ensuring the safety of available alternatives needs to be the critical focus in material substitution. For instance, hexabromocyclododecane (HBCD) is widely used (23,000 MT in 2011) as an unregulated flame retardant in polystyrene and EPS foam products including some food-contact products (at 0.7–2.5% w/w) (Alaee et al., 2003). It is bioaccumulative and, based on animal studies, is a highly toxic additive that is listed under the Stockholm Convention on Persistence Organic Pollutants to be banned for future production and use (Rani et al., 2014).

Halogen-free flame retardants that can substitute for deca-BDE are presently available. A 2009 report (Morgan and Wilkie, 2014) identified several phosphate-based flame retardants well suited for electronic equipment applications. Adopting these has to be approached with caution as organophosphorus compounds can also be potentially carcinogenic, genotoxic, or neurotoxic. As these are also used in other applications including as plasticizers, their environmental concentrations are already high (µg/g range in dust (Van den Eede, 2011)).

7.5 ALKYLPHENOLS AND THEIR ETHOXYLATES (APE)

Worldwide production of APEs is about 390,000 tons, and about 80% of it is used to produce octyl- or nonylphenol ethoxylate (NPE) nonionic surfactants. About 10% of it is used in manufacturing phosphite antioxidants (tris(4-nonyl-phenyl) phosphite used in rubber and plastics compounds. The stabilizers and plasticizers based on NP are also used in plastics intended for food-contact applications. They are not highly persistent chemicals and once discharged into the environment, APEs rapidly hydrolyze into alkylphenols that are subsequently slowly biodegraded by microorganisms. The USEPA criteria concentrations for freshwater NP exposure is CMC = 28 acute-μg/l and CCC = 6.6 chronic-μg/l:aquatic community might be exposed to >CMC concentration only briefly and <CCC concentration indefinitely.[13] The corresponding numbers for saltwater are 7.0 μg/l 1.7 μg/l.

Nonylphenol   Nonylphenol ethoxylate

Nonylphenols (NP) are used as stabilizers or plasticizers in plastics and have been reported to migrate from packaging into food (Guenther et al., 2002). The levels in lipid-rich foods such as butter or sausage were found to carry 13–14 μg/kg of NP. Not surprisingly, it is also found in human adipose tissue (Lopez-Espinosa et al., 2009) but at a lower level of 57 ng/g. Given the high volume use of NPEs in detergents, migration from plastics or food-contact packaging is likely to be relatively a minor contributor to the body burden of this EDC.

Generally, APEs are much less potent EDCs compared to those previously discussed. *In vitro* binding affinity of APEs to the receptors is much lower (4 orders of magnitude smaller) compared to estradiol. No data from human studies are available on ED effects of APEs.

7.6 EDCs AND PET BOTTLES

Water and carbonated beverage bottles are made predominantly from poly(ethylene terephthalate) (PET). The polymer is made by condensation reaction of ethylene glycol with either terephthalic acid or its dimethyl ester. The process in practice includes three steps: prepolymer formation, melt condensation to increase viscosity, and solid-state polymerization at 180–230°C to yield a resin with an average molecular weight that is high enough for use as bottle resins. Antimony trioxide is used as a catalyst in polymerization (Duh, 2002).

---

[13]CMC is the Criterion Maximum Concentration and CCC is the Criterion Continuous Concentration. These indicate concentrations at which aquatic organisms show no adverse effects under chronic and continuous exposure, respectively.

Neither BPA nor phthalate is used in manufacturing, compounding, or processing of PET into bottles. These cannot be generated as a by-product during manufacture or by the degradation of PET during use. Therefore, PET is regarded a relatively "safe" choice for packaging beverages. Consistent with the expectation, clean PET bottles generally do not show any EDCs or cytotoxic or genotoxic compounds leaching into water contained in them (Bach et al., 2013; Pinto and Reali, 2009). Leaching of trace levels of antimony catalysts into water has been reported (Keresztes, 2009).

Terephthalate group   Ethylene group

Structure of poly(ethylene terephthalate) PET

Recent reports, however, claim PET-bottled water and beverages to show, albeit very mild, but definitive ED activity (Sax, 2010). The effect, however, is inconsistent (Ceretti et al., 2010; Guart et al., 2011) and not seen in all bottles tested in a given study (Wagner and Oehlmann, 2009, 2011) and in bottles purchased at retail stores (Pinto and Reali, 2009). The estimated exposure to EDCs via this route, however, is very low (a pg to a ng estradiol per day (Kereszteset al., 2009)).

In a 2009 study, mud snails (*Potamopyrgus antipodarum*) growing in washed-out PET bottles for 56 days showed significant ($p < 0.0001$) increase in embryo production (sign of ED activity equivalent to approximately 25–75 ng/l of ethinyl estradiol) compared to those grown on glass (Wagner and Oehlmann, 2009). Other studies (Montuori et al., 2008; Pinto and Reali, 2009) show similar activity in yeast bioassay but at a lower level (Farhoodi et al., 2008). Some studies suggest the phthalates or other EDCs to have leached out of the PET (Bošnir et al., 2007; Casajuana and Lacorte, 2003). But phthalate contamination of some bottled water has been reported (Criado et al., 2005; Farhoodi et al., 2008; Montuori et al., 2008). Values reported are 0.002–11 µg/l in bottles stored for over 70 days for DMP, DEP, and DPP and 134 µg/l over 70 days for DEHP, at 25–30°C (Bach et al., 2013), compared to approximately 6000 µg/l into water from plasticized PVC containers (Mori, 1979).

Several possible explanations for the observation have been suggested: these include contamination from plant and equipment (Higuchi et al., 2004), water itself (Leivadara et al., 2008) or the filtration and purification system (Li et al., 2010), poor storage conditions of water (Pinto and Reali, 2009), and cross-contamination during recycling. However, Bicardi's (2003) study where mineral water prior to bottling was obtained from plant and bottled in PET and glass is interesting. No phthalates were observed up to 8 months' storage in either sample, but water in PET bottles for periods longer than 9 months did show phthalates. Casajuna and Lacorte (2003) also did show a similar result but with phthalates being detected in water only after 10 weeks of storage. This suggests the possibility of environmental phthalates in indoor air (Bornehag et al., 2004), rainwater (Guidotti et al., 2000), and interior of vehicles, partitioning into the PET bottle wall and being transported into the water. The observations also support the possibility of phthalates volatilized in ambient air partitioning into the PET bottle and being transported across the bottle wall into the contents.

Exposure of PET bottles filled with water to sunlight over extended durations is reported to result in accumulating degradation products on the outer layer of the bottle. These include PET monomer and dimer. However, over extended exposure period of up to 126 days, no degradation products were detected in the bottles (Wegelin et al., 2001).

Residual antimony (Sb) from oxide (catalyst used in PET manufacture, present in resin at 100–300 mg/kg resin) is an estrogenic substance (Sax, 2010) that can leach out of PET bottle into water (Shotyk and Krachler, 2007; Westerhoff et al., 2008b). The level of it in bottled water is reported to be approximately 0.156–0.343 µg/l of water (Shotyk et al., 2006; Welle and Franz, 2011). In common with phthalates, Sb levels in water also increases on storage or incubation at high temperature (Keresztes et al., 2009; Wagner and Oehlmann, 2009). The leachate concentrations achieved, however, are well below the maximum contaminant level of 6 µg Sb/l set by the USEPA, and what is leached out was mainly in the less toxic Sb(V) form as opposed to Sb(III) form (Sánchez-Martínez et al., 2013). Though the Sb levels are low, developing antimony-free PET manufacturing processes (other metals can be used as catalysts for the polycondensation synthesis) should be investigated. In any event, the reported ED effects of bottled water cannot be explained entirely on the basis of antimony (Sb) levels.

In the case of PET bottles, the reported evidence is not persuasive enough to support the notion that any EDCs originate in the bottle and leach out into water or other contents. A precautionary stance of avoiding the use of PET for food or beverage is not justified on the basis of available information. An effort needs to be taken by the industry to unequivocally and quantitatively express and attribute the some-times-observed ED activity in PET bottles to specific modes of contamination.

Comments: The discussion in this chapter suggests some cautions in using household plastics in contact with food. Avoid using food, water, or other beverage packaged in PC containers (or products with recycling symbol #7) especially where they have been stored for a period of time. Avoid consuming beverages in plastic bottles (PC bottles in particular) or food in plastic containers, left in warm interior of automobiles over long periods of time, especially if routinely exposed to sunlight. The caution is particularly important for pregnant mothers in view of recent findings linking BPA exposure to increased risk of miscarriage (Lathi et al., 2014).

Do not heat food in original metal cans they were packaged in and do not reuse the cans subsequently for other food-contact uses. Do not re-use plastic food packages, containers, bottles, or bags to heat food in microwave. The "microwaveable" packaging is not necessarily meant for repeated use. If plastic film is used to cover food being microwaved, it should not touch the food while in the oven. Remove frozen food from original packaging before microwave defrosting. Plastic film should never be used in an air oven. Plasticized plastic wraps (such as PVDC cling wrap used in early product[14]) should not be left in contact with food, especially food rich in fat (Coltro et al., 2014) for extended durations especially at higher temperatures. Most cling films today are LDPE.

---

[14] In 2004, the "Saran" brand changed to using LDPE due to environmental concerns with polyvinylidene chloride (PVDC) despite the excellent barrier properties of the latter, not matched by the LDPE film. Coltro et al., in a 2014 study, still found some commercial cling films containing phthalates.

REFERENCES

Akingbemi BT, Soitas CM, Koulova AI, Kleinfelter GR, Hardy MP. Inhibition of testicular steroidogenesis by the xenoestrogen bisphenol A is associated with reduced pituitary luteinizing hormone secretion and decreased steroidogenic enzyme gene expression in rat Leydig cells. Endocrinology 2004;145:592–603.

Akutsu K, Takatori S, Nozawa S, Yoshiike M, Nakazawa H, Hayakawa K, Makino T, Iwamoto T. Polybrominated diphenyl ethers in human serum and sperm quality. Bull Environ Contam Toxicol 2008;80 (4):345–350.

Alaee M, Arias P, Sjodin A, Bergman A. An overview of commercially used brominated flame-retardants, their applications, their use patterns in different countries/regions and possible modes of release. Environ Int 2003;29:683–689.

Al-Hiyasat AS, Darmani H, Elbetieha AM. Effects of bisphenol A on adult male mouse fertility. Eur J Oral Sci 2002;110:163–167.

Allen JG, McClean MD. Personal exposure to polybrominated diphenyl ethers (PBDEs) in residential indoor air. Environ Sci Technol 2007;41 (13):4574–4579.

Allen JG, McClean MD, Stapleton HM, Webster TF. Critical factors in assessing exposure to PBDEs via house dust. Environ Int 2008;34 (8):1085–1091.

Andrade AJ, Grande SW, Talsness CE, Grote K, Chahoud I. A dose-response study following in utero and lactational exposure to di(2-ethylhexyl)-phthalate (DEHP): non-monotonic dose-response and low dose effects on rat brain aromatase activity. Toxicology. Oct 29, 2006;227(3):185–192.

Aschberger K, Castello P, Hoekstra E, Karakitsios S, Munn S, Pakalin S, Sarigiannis D. *Bisphenol A and Baby Bottles: Challenges and Perspectives*. European Commission—Joint Research Centre—Institute for Health and Consumer Protection: Luxembourg; 2010.

Bach C, Dauchy X, Severin I, Munoz J, Serge E, Chagnon M. Effect of temperature on the release of intentionally and non-intentionally added substances from polyethylene terephthalate (PET) bottles into water. Chemical analysis and potential toxicity. Food Chem 2013;139 (1–4):672–680.

Baillie-Hamilton PF. Chemical toxins: a hypothesis to explain the global obesity epidemic. J Altern Complement Med 2002;8:185–192.

Bang DY, Kyung M, Kim MJ, Jung BY, Cho MC, Choi SM, Kim YW, Lim SK, Lim DS, Won AJ, Kwack SJ, Lee Y, Kim HS, Lee BM. Human risk assessment of endocrine-disrupting chemicals derived from plastic food containers. *Compr Rev Food Sci Food Saf* 2012;11 (5):453–470.

Bergman Å, Heindel JJ, Jobling S, Kidd KA, Zoeller RT. 2012. The state-of-the-science of endocrine disrupting chemicals—2012. Geneva: UNEP/WHO Downloaded from http://www.who.int/ceh/publications/endocrine/en/index.html. Accessed October 7, 2014.

Bergman Å, Heindel JJ, Jobling S, Kidd KA, Thomas Zoeller R, editors. State of the Science of Endocrine Disrupting Chemicals—2012. Geneva: United Nations Environment Programme and the World Health Organization; 2013.

Bindhumol V, Chitra KC, Mathur PP. Bisphenol A induces reactive oxygen species generation in the liver of male rats. Toxicology 2003;188:117–124.

Biscardi D, Monarca S, De Fusco R, Senatore F, Poli P, Buschini A, Rossi C, Zani C. Evaluation of the migration of mutagens/carcinogens from PET bottles into mineral water by Tradescantia/micronuclei test, Comet assay on leukocytes and GC/MS. Sci Tot Environ 2003;302 (1–3):101–108.

Bittner G, Yang CZ, Stoner MA. Estrogenic chemicals often leach from BPA-free plastic products that are replacements for BPA-containing polycarbonate products. Environ Health 2014;13:41.

Bornehag C. G, Sundell, J, Weschler C.J, Sigsgaard T; Lundgren B; Hasselgren M; Hägerhed-Engman L. (2004, October). The association between asthma and allergic symptoms in children and phthalates in house dust: a nested case-control study. Environ Health Perspect. 112(14):1393–1397.

Bošnir J, Puntarić D, Galić A, Škes I, Dijanić T, Klarić M, Grgic M, Curkovic M, Smit Z. Migration of phthalates from plastic containers into soft drinks and mineral water. Food Technol Biotechnol 2007;45 (1):91–95.

Bouma K, Schakel DJ. Migration of phthalates from PVC toys into saliva stimulant by dynamic extraction. Food Addit Contam 2002;19 (6):602–610.

Braun JM, Yolton K, Homrung R, Ye X, Dietrich KN, Calafat AM, Lanphear BP. Prenatal BPA exposure and early childhood behavior. Environ Health Perspect 2009;117 (12):1945–1952.

Braun JM, Kalkbrenner AE, Calafat AM, Bernert JT, Ye X, Silva MJ, Barr DB, Sathyanarayana S, Lanphear BP. Variability and predictors of urinary bisphenol A concentrations during pregnancy. Environ Health Perspect 2011a;119 (1):131–137.

Braun JM, Kalkbrenner AE, Calafat AM, Yolton K, Ye X, Dietrich KN, Lanphear BP. Impact of early life BPA exposure on behavior and executive function in children. Pediatrics 2011b;128:873–882.

Brede C, Fjeldal P, Skjevrak I, Herikstad H. Increased migration levels of bisphenol A from polycarbonate baby bottles after dishwashing, boiling, and brushing. Food Addit Contam 2003;20:684–689.

Bushnik, T., Haines, D., Levallois, P., Levesque, J., Van Oostdam, J., Viau, C. (2010, Sept). Lead and bisphenol A concentrations in the Canadian population. Health Rep, 21(3), 7–18.

Calafat A, Needham L. Exposure to di-(2-ethylhexyl) phthalate among premature neonates in a neonatal intensive care unit. Pediatrics 2004;113 (5):429–434.

Calafat AM, Slakman AR, Silva MJ, Herbert AR, Needham LL. Automated solid phase extraction and quantitative analysis of human milk for 13 phthalate metabolites. J Chromatogr B Analyt Technol Biomed Life Sci 2004;805:49–56.

Calafat AM, Ye X, Wong LY, Reidy JA, Needham LL. Exposure of the U.S. population to bisphenol A and 4-tertiary-octylphenol: 2003–2004. Environ Health Perspect 2008;116 (1):39–44.

Cao X-L, Perez-Locas C, Dufresne G, Clement G, Popovic S, Beraldin F, Dabeka RW, Feeley M. Concentrations of bisphenol A in the composite food samples from the 2008 Canadian total diet study in Quebec City and dietary intake estimates. Food Addit Contam Part A Chem Anal Control Expo Risk Assess 2011;28 (6):791–798.

Carwile JL, Ye X, Calefat AM, Michels KB. Canned soup consumption and urinary bisphenol A: a randomized crossover trial. JAMA 2011;306:2218–2220.

Casajuana N, Lacorte S. Presence and release of phthalic esters and other endocrine disrupting compounds in drinking water. Chromatographia 2003;57 (9–10):649–655.

Center for the Evaluation of Risks to Human Reproduction. 2006. NTP-CERHR monograph on the potential human reproductive and developmental effects of di(2-ethylhexyl) phthalate (DEHP). Downloaded from http://ntp.niehs.nih.gov/ntp/ohat/phthalates/dehp/dehp-monograph.pdf. NIH Publication No. 06-4476. Accessed October 7, 2014.

Ceretti E, Zani C, Zerbini I, Guzzella L, Scaglia V, Berna V, Donato F, Monarca S, Feretti D. Comparative assessment of genotoxicity of mineral water packed in polyethylene terephthalate (PET) and glass bottles. Water Res 2010;44 (5):1462–1470.

Chen SJ, Ma YJ, Wang J, Chen D, Luo XJ, Mai BX. Brominated flame retardants in children's toys: concentration, composition, and children's exposure and risk assessment. Environ Sci Technol 2009;43:4200–4206.

Clark K, Cousins IT, Mackay D, Yamada K. (2003). Observed concentrations in the environment. In *Phthalate Esters. The Handbook of Environmental Chemistry*, 3(Q), 125–177.

Cobellis L, Latini G, De Felice C. High plasma concentrations of di-(2-ethylhexyl)-phthalate in women with endometriosis. Hum Reprod 2003;18 (7):1512–1515.

Colon I, Caro D, Bourdony CJ, Rosario O. Identification of phthalate esters in the serum of young Puerto Rican girls with premature breast development. Environ Health Perspect 2000;108 (9):895–900.

Coltro L, Pitta JB, da Costa PA, Perez F, de Araújo MAVA, Rodrigues R. Migration of conventional and new plasticizers from PVC films into food simulants: a comparative study. Food Control 2014;44:118–129.

Committee on the Health Risks of Phthalates, National Research Council. *Phthalates and Cumulative Risk Assessment: The Task Ahead*. Washington, DC: National Academies Press; 2001.

Criado MV, Fernandez Pinto VE, Badessari A, Cabral D. Conditions that regulate the growth of moulds inoculated into bottled mineral water. International Journal of Food Microbiology 2005;99 (3):343–349.

D'Antuono A, Dall'Orto VC, Lo Balbo A, Sobral S, Rezzano I. Determination of bisphenol A in food-stimulating liquids using LCED with a chemically modified electrode. J Agric Food Chem 2001;49 (3):1098–1101.

Darnerud PO, Eriksen GS, Johannesson T, Larsen PB, Viluksela M. Polybrominated diphenyl ethers: occurrence, dietary exposure, and toxicology. Environ Health Perspect 2001;109 (Suppl.1):49–68.

DiGangi J, Blum A, Bergman Å, de Wit CA, Lucas, D, Mortimer D, Schecter A, Scheringer M, Shaw SD, Webster TF. (2010, Dec). San Antonio statement on Brominated and Chlorinated flame retardants, Environ Health Perspect., 118(12), A516–A518.

Duh B. Effect of antimony catalyst on solid-state polycondensation of poly (ethylene terephthalate). Polymer 2002;43:3147–3154.

Duty SM, Silva MJ, Barr DB, Brock JW, Ryan L, Chen Z, Herrick RF, Christiani DC, Hauser R. Phthalate exposure and human semen parameters. Epidemiology 2003;14:269–277.

Edlow AG, Chen M, Smith NA, Lu C, McElrath, TF. (2012, Aug). Fetal bisphenol A exposure: concentration of conjugated and unconjugated bisphenol A in amniotic fluid in the second and third trimesters. Reprod Toxicol, 34, 1. 1–7.

EFSA. Bis(2-ethylhexyl)phthalate (DEHP) for use in food contact materials. EFSA J 2005;243:1–20.

Ehlert KA, Beumer CWE, Groot MCE. Migration of bisphenol A into water from polycarbonate baby bottles during microwave heating. Food Addit Contam Part A Chem Anal Control Expo Risk Assess 2008;25:904–910.

Environmental Working Group BPA in Store Receipts. (2010, Jul). BPA coats cash register receipts. Downloaded from http://www.ewg.org/kid-safe-chemicals-act-blog/2010/07/a-little-bpa-along-with-your-change/. Accessed October 7, 2014.

European Commission, Opinion of the scientific committee on food on bisphenol A. 2002 Downloaded from http://ec.europa.eu/food/fs/sc/scf/out128_en.pdf. Accessed October 7, 2014.

European Commission. European union risk assessment report bis(2-ethylhexyl) phthalate (DEHP), Luxembourg: Office for Official Publications of the European Communities; 2008. CAS-No. 117-81-7. Vol. 80; EUR 23384EN.

European Commission—IP/05/838 05/07/ 2005. Permanent ban of phthalates: commission hails long-term safety for children's toys. Downloaded from http://europa.eu/rapid/press-release_IP-05-838_en.htm. Accessed October 7, 2014.

European Food Safety Authority (EFSA). Opinion of the scientific panel on food additives, flavourings, processing aids and materials in contact with food on a request from the commission related to 2,2-BIS(4-hydroxyphenyl)propane (bisphenol A) question number. J EFSA 2006;428:1–75.

Expert Panel on BPA. Chapel Hill bisphenol A expert panel consensus statement: integration of mechanisms, effects in animals and potential to impact human health at current levels of exposure. Reprod Toxicol 2007;24 (2):131–138.

Fakirov S. *Handbook of Thermoplastic Polyesters*. Weinheim : Wiley-VCH; 2002.

Farabollini F, Porrini S, Dessi-Fulgherit F. Perinatal human exposure to the estrogenic pollutant bisphenol A affects behavior in male and female rats. Pharmacol Biochem Behav 1999;64 (4):687–694.

Farhoodi M, Emam-Djomeh Z, Ehsani MR, Oromiehie A. Effect of environmental conditions on the migration of di(2-ethylhexyl)phthalate from PET bottles into yogurt drinks: influence of time, temperature, and food simulant. Arab J Sci Eng 2008;33 (2):279–287.

Filella M, Belzile N, Lett MC. Antimony in the environment: a review focused on natural waters. III. Microbiota relevant interactions. Earth Sci Rev 2007;80:195–217.

Fisher JS, Macpherson S, Marchetti N, Sharpe RM. Human "testicular dysgenesis syndrome": a possible model using in-utero exposure of the rat to dibutyl phthalate. Hum Reprod 2003;18:1383–1394.

Fleisch AF, Sheffield PE, Chinn C, Edelstein BL, Landrigan PJ. Bisphenol A and related compounds in dental materials. Pediatrics 2010;126 (4):760–768.

Flint S, Markle T, Thompson S, Wallace E. Bisphenol A exposure, effects, and policy: a wildlife perspective. J Environ Manage 2012;104:19–34.

Focazio MJ, Kolpin DW, Barnes KK, Furlong ET, Meyer MT, Zaugg SD, Barber LB, Thurman EM. A national reconnaissance of pharmaceuticals and other organic wastewater contaminants in the United States eII. Untreated drinking water sources. Sci Total Environ 2008;402:2–3.

Frederiksen H, Skakkebaek NE, Andersson AM. Metabolism of phthalates in humans. Mol Nutr Food Res 2007;51:899–911.

Fujii M, Shinohara N, Lim A, Otake T, Kumagai K, Yanagisawa Y. A study on emission of phthalate esters from plastic materials using a passive flux sampler. Atmos Environ 2003;37:5495–5504.

Fung EYK, Ewoldsen NO, St Germain HA, Marx DB, Miaw CL, Siew C, Chou HN, Gruninger SE, Meyer DM. Pharmacokinetics of bisphenol A released from a rental sealant. J. Am Dent Assoc 2000;131 (1):51–58.

Gaylor MO, Harvey E, Hale RC. House crickets can accumulate polybrominated diphenyl ethers (PBDEs) directly from polyurethane foam common in consumer products. Chemosphere 2012;86:500–505.

Geens T, Apelbaum TZ, Goeyens L, Neels H, Covaci A. Intake of bisphenol A from canned beverages and foods on the Belgian market. Food Addit Contam Part A Chem Anal Control Expo Risk Assess 2010;27 (11):1627–1637.

Geens T, Goeyens L, Covaci A. Are potential sources for human exposure to bisphenol-A overlooked? Int J Hyg Environ Health 2011;214 (5):339–347.

Gehring M, Vogel D, Tennhardt L, Weltin D, Bilitewski B. (2004). Bisphenol A contamination of wastepaper, cellulose and recycled paper products, Waste Manag Environ, Ii, 293–300.

Geiss O, Tirendi S, Barrero-Moreno J, Kotzias D. Investigation of volatile organic compounds and phthalates present in the cabin air of used private cars. Environ Int 2009;35 (8):1188–1195.

Gibson RL. (2007). Toxic baby bottles. Report published by the Environment California Research and Policy Center. Los Angeles, CA.

Guart A, Bono-Blay F, Borrell A, Lacorte S. Migration of plasticizers phthalates, bisphenol A and alkylphenols from plastic containers and evaluation of risk. Food Addit Contam 2011;28 (5):1–10.

Guenther K, Heinke V, Thiele B, Kleist E, Prast H, Raecker T. Endocrine disrupting nonylphenols are ubiquitous in food. Environ Sci Technol 2002;36:1676–1680.

Guidotti M, Giovinazzo R, Cedrone O, Vitali M. Determination of organic micropollutants in rain water for laboratory screening of air quality in urban environment. Environ Int 2000;26 (1–2):23–28.

Gupta C. (2000) Reproductive malformation of the male offspring following maternal exposure to estrogenic chemicals. Proc Soc Exp Biol Med. Jun; 224(2):61–68.

Hale RC, Kim SL, Harvey E, La Guardia MJ, Mainor TM, Bush EO, Jacobs EM. Antarctic research bases: local sources of polybrominated diphenyl ether (PBDE) flame retardants. Environ Sci Technol 2008;42:1452–1457.

Hallgren S, Darnerud PO. Effects of polybrominated diphenyl ethers (PBDEs), polychlorinated biphenyls (PCBs), and chlorinated paraffins (CPs) on thyroid hormone levels and enzyme activities in rats. Organohalogen Compd 1998;35:391–394.

Harley KG, Marks AR, Chevrier J, Bradman A, Sjodin A, Eskenazi B. PBDE concentrations in women's serum and fecundability. Environ Health Perspect 2010;118 (5):699–704.

Hass U, Scholze M, Christiansen S, Dalgaard M, Vinggaard AM, Axelstad M, Metzdorff SB, Kortenkamp A. Combined exposure to anti-androgens exacerbates disruption of sexual differentiation in the rat. Environ Health Perspect 2007;115 (Suppl 1):122–128.

Hayes TB, Case P, Chui S, Chung D, Haeffel EC, Haston K, Lee M, Mai VP, Marjuoa Y, Parker J, Tsui M. (2006, Apr). Pesticide mixtures, endocrine disruption, and amphibian declines: are we underestimating the impact?. Environ. Health Perspect., 114 (S–1), 40–50.

Health Canada (2012) Final human health state of the science report on decabromodiphenyl ether (decaBDE) Canada Gazette, Part I: 146(48) and download from http://gazette.gc.ca/rp-pr/p1/2012/2012-12-01/html/notice-avis-eng.html#d101. Accessed October 7, 2014.

Hengstler J.G, Foth H, Gebel T, Kramer P.J, Lilienblum W, Schweinfurth H, Völkel W, Wollin K.M, Gundert-Remy U. (2011, April). Critical evaluation of key evidence on the human health hazards of exposure to bisphenol A. Crit Rev Toxicol, 41(4), 263–291. 10.3109/10408444.2011.558487.

Higuchi A, Yoon BO, Kaneko T, Hara M, Maekawa M, Nohmi T. Separation of endocrine disruptors from aqueous solutions by pervaporation: dioctylphthalate and butylated hydroxytoluene in mineral water. J Appl Polym Sci 2004;94 (4):1737–1742.

Honma S, Suzuki A, Buchanan DL, Katsu Y, Watanabe H, Iguchi T. Low dose effect of in utero human exposure to bisphenol A and diethylstilbestrol on female mouse reproduction. Reprod Toxicol 2002;16:117–122.

Howdeshell KL, Hotchkiss AK, Thayer KA, Vandenbergh JG, vom Saal FS. (1999). Exposure to bisphenol A advances puberty. Nature 401:763–764.

Howdeshell KL, Peterman PH, Judy BM, Taylor JA, Orazio CE, Ruhlen RL, vom Saal FS, Welshons WV. Bisphenol A is released from used polycarbonate animal cages into water at room temperature. Environ Health Perspect 2003;111:9.

Howdeshell KL, Wilson VS, Furr J, Lambright CR, Rider CV, Blystone CR, Hotchkiss AK, Gray LE Jr. A mixture of five phthalate esters inhibits fetal testicular testosterone production in the Sprague Dawley rat in a cumulative, dose additive manner. Toxicol Sci 2008;105:153–165.

Huang LP, Lee CC, Hsu PC, Shih TS. The association between semen quality in workers and the concentration of di(2-ethylhexyl) phthalate in polyvinyl chloride pellet plant air. Fertil Steril 2011;96:90–94.

Hugo ER, Brandebourg TD, Woo JG, Loftus J, Alexander JW, Ben-Jonathan N. Bisphenol A at environmentally relevant doses inhibits adiponectin release from human adipose tissue explants and adipocytes. Environ Health Perspect 2008;116:1642–1647.

Hunt PA, Koehler KE, Susiarjo M, Hodges CA, Ilagan A, Voigt RC, Thomas S, Thomas BF, Hassold TJ. Bisphenol A exposure causes meiotic aneuploidy in the female mouse. Curr Biol 2003;13:546–553.

Ikezuki Y, Tsutsumi O, Takai Y, Kamei Y, Taketani Y. Determination of bisphenol A concentrations in human biological fluids reveals significant early prenatal exposure. Hum Reprod 2002;17:2839–2841.

Inoue K, Kato K, Yoshimura Y, Makino T, Nakazawa H. Determination of bisphenol A in human serum by high-performance liquid chromatography with multi-electrode electrochemical detection. J Chromatogr B 2000;749:17–23.

Inoue K, Yamaguchi A, Wada M, Yoshimura Y, Makino T, Nakazawa H. Quantitative detection of bisphenol A and bisphenol A diglycidyl ether metabolites in human plasma by liquid chromatography–electrospray mass spectrometry. J Chromatogr B 2001;765:121–126.

Ishido M, Masuo Y, Kunimoto M, Oka S, Morita M. Bisphenol A causes hyperactivity in the rat concomitantly with impairment of tyrosine hydroxylase immunoreactivity. J Neurosci Res 2004;76 (3):423–433.

Jaakkola JK, Knight TL. The role of exposure to phthalates from PVC products in the development of asthma and allergies: a systematic review and meta-analysis. Environ Health Perspect 2008;116 (7):845–853.

Jenkins S, Wang J, Eltoum I, Desmond R, Lamartiniere CA, (2011). Chronic oral exposure to bisphenol A results in a non-monotonic dose response in mammary carcinogenesis and metastasis in MMTV-erbB2 mice. Environ Health Perspect, 119, 164–1609.

Joskow R, Barr DB, Barr JR, Calafat AM, Needham LL, Rubin C. Exposure to bisphenol A from bis-glycidyl dimethacrylate-based dental sealants. J Am Dent Assoc 2006;137:353–362.

Kajiwara N, Noma Y, Takigami H. Photolysis studies of technical decabromodiphenyl ether (DecaBDE) and ethane (DeBDethane) in plastics under natural sunlight. Environ Sci Technol 2008;42 (12):4404–4409.

Kamrin MA. Phthalate risks, phthalate regulation, and public health: a review. J Toxicol Environ Health B Crit Rev 2009;12 (2):157–174.

Kawai K, Takehiro N, Nishikata H, Aou S, Takii M, Kubo C. Aggressive behavior and serum testosterone concentration during the maturation process of male mice: the effects of fetal exposure to bisphenol A. Environ Health Perspect 2003;11:175–178.

Keresztes S, Tatár E, Mihucz VG, Virág I, Majdik C, Záray G. Leaching of antimony from polyethylene terephthalate (PET) bottles into mineral water. Sci Total Environ 2009;407 (16):4731–4735.

Kim BN, Cho SC, Kim Y, Shin MS, Yoo HJ, Kim JW, Yang YH, Kim HW, Bhang SY, Hong YC. Phthalates exposure and attention-deficit/hyperactivity disorder in school-age children. Biol Psychiatry 2009;66 (10):958–963.

Kingman A, Hyman J, Masten S.A, Jayaram B, Smith C, Eichmiller F, Arnold MC, Wong PA, Schaeffer JM, Solanki S, Dunn WJ. (2012 Dec). Bisphenol A and other compounds in human saliva and urine associated with the placement of composite restorations. J Am Dent Assoc 1, 2012.

Koch HM, Preuss R, Angerer J. Di(2-ethylhexyl)phthalate (DEHP): human metabolism and internal exposure—an update and latest results. Int J Androl 2006;29:155–165.

Koo HJ, Lee BM. Human monitoring of phthalates and risk assessment. J Toxicol Environ Health A 2005;68:1379–1392.

Krauskopf LG. How about alternatives to phthalate plasticizers? J Vinyl Addit Technol 2003;9:159–171.

Kuruto-Niwa R, Tateoka Y, Usuki Y, Nozawa R. Measurement of bisphenol A concentrations in human colostrum. Chemosphere 2007;66:1160–1164.

Lang IA, Galloway TS, Scarlett A, Henley WE, Depledge M, Wallace RB, Melzer D. Association of urinary bisphenol A concentration with medical disorders and laboratory abnormalities in adults. JAMA 2008;300 (11):1303–1310. DOI: 10.1001/jama.300.11.1303. Epub 2008 Sep 16.

Lateef SS. Analysis of bisphenol A leaching from baby feeding bottles. Agilent Applications Solution. Bangalore: Agilent Technologies, Inc.; 2011.

Lathi RB, Liebert CA, Brookfield KF, Taylor JA, vom Saal, FS, Fujimoto VY. and Baker VL, (2014) Conjugated bisphenol A (BPA) in maternal serum in relation to miscarriage risk. Fertil Steril. 102(1): 123–128.

Latini G, De Felice C, Presta G, Del Vecchio A, Paris I, Ruggieri F, Mazzeo P. In utero exposure to di-(2-ethylhexyl)phthalate and duration of human pregnancy. Environ Health Perspect 2003;111 (14):1783–1785.

LCSP. Lowell center for sustainable production. Phthalates and Their Alternatives: Health and Environmental Concerns. University of Massachusetts Lowell; 2011. Available at http://www.sustainableproduction.org/downloads/PhthalateAlternatives-January2011.pdf. Accessed October 28, 2014.

Le HH, Carlson EM, Chua JP, Belcher SM. Bisphenol A is released from polycarbonate drinking bottles and mimics the neurotoxic actions of estrogen in developing cerebellar neurons. Toxicol Lett 2008;176:149–156.

Leivadara SV, Nikolaou AD, Lekkas TD. (2008) Determination of organic compounds in bottled waters. Food Chem, 108 (1) 277–286.

Lemos MFL, van Gestel CAM, Soares AMVM. Endocrine disruption in a terrestrial isopod under exposure to bisphenol A and vinclozolin. J Soil Sediments 2009;9:492–500.

Leo LW, Wong KO, Seah HL. Occurrence of residual bisphenol A and its migration levels in polycarbonate baby milk bottles. Singap J Prim Ind 2006;32:106–108.

Li X, Ying GG, Su HC, Yang XB, Wang L. (2010a) Simultaneous determination and assessment of 4-nonylphenol, bisphenol A and triclosan in tap water, bottled water and baby bottles, Environ Int, 36 (6), 557–562.

Li DK, Zhou Z, Miao M, He Y, Qing D, Wu T, Wang J, Weng X, Ferber J, Herrinton LJ, Zhu Q, Gao E, Yuan W. Relationship between urine BPA level and declining male sexual function). J Androl 2010b;31:500–506.

Li DK, Zhou Z, Qing D, He Y, Wu T, Miao M, Wang J, Weng X, Ferber JR, Herrinton LJ, Zhu Q, Gao E, Checkoway H, Yuan W. Occupational exposure to bisphenol-A (BPA) and the risk of self-reported male sexual dysfunction. Hum Reprod 2010a;25:519–527.

Lopez-Espinosa MJ, Freire C, Arrebola JP, Navea N, Taoufiki J, Fernandez MF, Ballesteros O, Prada R, Olea N. Nonylphenol and octylphenol in adipose tissue of women in Southern Spain. Chemosphere 2009;76 (6):847–852.

Lottrup G, Andersson AM, Leffers H, Mortensen GK, Toppari J, Skakkebaek NE, Main KM. Possible impact of phthalates on infant reproductive health. Int J Androl 2006;29 (1):172–180.

Lovekamp-Swan T, Davis BJ. Mechanisms of phthalate ester toxicity in the female reproductive system. Environ Health Perspect 2003;111 (2):139–145.

Main KM, Kiviranta H, Virtanen HE, Sundqvist E, Tuomisto JT, Tuomisto J, Vartiainen T, Skakkebaek NE, Toppari J. Flame retardants in placenta and breast milk and cryptorchidism in newborn boys. Environ Health Perspect 2007;115 (10):1519–1526.

Manikkam M, Tracey R, Guerrero-Bosagna C, Skinner MK. Plastics derived endocrine disruptors (BPA, DEHP and DBP) induce epigenetic trans generational inheritance of obesity, Reproductive disease and sperm epimutations. PLoS One 2013;8 (1):e55387.

Markey CM, Wadia PR, Rubin BS, Sonnenschein C, Soto AM. Long-term effects of fetal human exposure to low doses of the xenoestrogen bisphenol-A in the female mouse genital tract. Biol Reprod 2005;72 (6):1344–1351.

Markham DA, Waechter Jr., JM, Wimber M, Rao N, Connolly P, Chen Chuang J, Hentges S, Shiotsuka RN, Dimond S, Chappelle, AH. (2010) Development of a method for the determination of Bisphenol A at trace concentrations in human blood and urine and elucidation of factors influencing method accuracy and sensitivity. J Anal Toxicol, 34(6): 293–303.

Martinez-Arguelles DB, Campioli, E, Culty, M, Zirkin BR, Papadopoulos, V. (Jan 17, 2013) Fetal origin of endocrine dysfunction in the adult: the phthalate model. J Steroid Biochem Mol Biol, 137: 5–17. Available online.

Masumo H, Kidani T, Sekiya K, Sakayama K, Shiosaka T, Yamamoto H, Honda K. Bisphenol A in combination with insulin can accelerate the conversion of 3T3L1 fibroblasts to adipocytes. J. Lipid Res. 2002;43:676–684.

Matsumoto J, Yokota H, Yuasa A. Developmental increases in rat hepatic microsomal UDP-glucuronosyltransferase activities toward xenoestrogens and decreases during pregnancy. Environ Health Perspect 2002;110:193–196.

Meeker JD, Ferguson KK. Phthalates: Human exposure and related health effects. In: Schecter A, editor. *Dioxins and Health Including Other Persistent Organic Pollutants and Endocrine Disruptors. 3rd Edition* ed. Hoboken: John Wiley & Sons; 2012.

Meeker JD, Calafat AM, Hauser R. (2007). Di-(2-ethylhexyl) phthalate metabolites may alter thyroid hormone levels in men. Environ Health Perspect, Jul;115(7):1029–1034.

Meeker JD, Johnson PI, Camann D, Hauser R. Polybrominated diphenyl ether (PBDE) concentrations in house dust are related to hormone levels in men. Sci Total Environ 2009a;407 (10):3425–3429.

Meeker JD, Sathyanarayana S, Swan SH. Phthalates and other additives in plastics: human exposure and associated health outcomes. Phil Trans R Soc B 2009b;364:2097–2113.

Meeker JD, Ehrlich S, Toth TL, Wright DL, Calafat AM, Ye X, Hauser R. Semen quality and sperm DNA changes in relation to urinary BPA among men. Reprod Toxicol 2010;30:532–539.

Melzer D, Rice NE, Lewis C, Henley WE, Galloway TS. PLoS (Open Access Journal) association of urinary bisphenol A concentration with heart disease: evidence from NHANES 2003/06. PLoS One 2010;5 (1):e8673.

Mercea P. Physicochemical processes involved in migration of bisphenol A from polycarbonate. J Appl Polym Sci 2009;112 (2):579–593.

Miyawaki J, Sakayama K, Kato H (2007 Oct). Perinatal and postnatal exposure to bisphenol A increases adipose tissue mass and serum cholesterol level in mice. J Atheroscler Thromb.;14(5):245–252. Epub 2007, Oct 12.

Montuori P, Jover E, Morgantini M, Bayona JM, Triassi M. Assessing human exposure to phthalic acid and phthalate esters from mineral water stored in polyethylene terephthalate and glass bottles. Food Addit Contam Part A Chem Anal Control Expo Risk Assess 2008;25 (4):511–518.

Morgan AB, Wilkie CA. *The Non-halogenated Flame Retardant Handbook*. pp 400John Wiley & Sons: New York; 2014.

Mori S. Contamination of water and organic solvents stored in plastic bottles with phthalate ester plasticizers. Anal Chim Acta 1979;108:325–332.

Morreale de Escobar G, Obregon MJ, Escobar del Rey F. Is neuropsychological development related to maternal hypothyroidism or to maternal hypothyroxinemia? J Clin Endocrinol Metabol 2000;85 (11):3975–3987.

Murakami K, Ohashi A, Hori H, Hibiya M, Shoji Y, Kunisaki M, Akita M, Yagi A, Sugiyama K, Shimozato S, Ito K, Takahashi H, Takahashi K, Yamamoto K, Kasugai M, Kawamura N, Nakai S, Hasegawa M, Tomita M, Nabeshima K, Hiki Y, Sugiyama S. Accumulation of bisphenol A in hemodialysis patients. Blood Purif 2007;25 (3):290–294.

Murray TJ, Maffini MV, Ucci AA, Sonneschein Soto AM. Induction of mammary gland ductal hyperplasias and carcinoma in situ following fetal bisphenol A exposure. Reprod Toxicol 2006;146 (9):4138–4147.

Myers JP, Zoeller RT, vom Saal FS. A Clash of old and new scientific concepts in toxicity, with important implications for public health. Environ Health Perspect 2009;117 (11):1652–1655.

National Workgroup for Safe Markets (NWSM). 2010 An investigation into Bisphenol A in canned foods. http://nosilverlining.org/downloads/NoSilverLining-Report.pdf. Accessed October 7, 2014.

Negishi T, Kawasaki K, Suzaki S, Maeda H, Ishii Y, Kyuwa S, Kuroda Y, Yoshikawa Y. Behavioral alterations in response to fear-provoking stimuli and tranylcypromine induced by perinatal human exposure to bisphenol A and nonylphenol in male rats. Environ Health Perspect 2004;112:1159–1164.

Newbold RR, Padilla-Banks E, Snyder RJ, Phillips TM, Jefferson WN. Developmental exposure to endocrine disruptors and the obesity epidemic. Reprod Toxicol 2007a;23:290–296.

Newbold RR, Jefferson WN, Padilla-Banks E.(2007b) Long-term adverse effects of neonatal exposure to bisphenol A on the murine female reproductive tract. Reprod Toxicol 24(2): 253–258. Jul 27.

Noma Y, Takigami H, Kajiwara N. Photolysis studies of technical Decabromodiphenyl ether (DecaBDE) and ethane (DeBDethane) in plastics under natural sunlight. Environ Sci Technol 2008;42 (12):4404–4409.

Ohtani, H., Miura, I., Ichikawa, Y. (2000 Dec) Effects of Dibutyl Phthalate as an environmental endocrine disruptor on gonadal sex differentiation of genetic males of the frog Rana rugosa. Environ Health Perspect 108(12) .

Oldring PKT, Nehring U. Packaging Materials 7. Metal Packaging for Foodstuffs. Brussels: International Life Sciences Institute; 2007.

Otaka H, Yasuhara A, Morita M. Determination of bisphenol A and 4-nonlyphenol in human milk using alkaline digestion and cleanup by solid-phase extraction. Anal Sci 2003;19:1663–1666.

Pickering AD, Sumpter JP. Comprehending endocrine disrupters in aquatic environments. Environ Sci Technol 2003;37:331A–336A.

Pinto B, Reali D. Screening of estrogen-like activity of mineral water stored in PET bottles. Int J Hyg Environ Health 2009;212 (2):228–232.

Plastics Europe. 2010. The plastics portal. Downloaded from http://www.plasticseurope.org/ cited Nov 3, 2010.

Prüss-Üstün A, Corvalán C. Analysis of estimates of the environmentally attributable fraction, by disease. In: Preventing Disease Through Healthy Environments: Towards an Estimate of the Environmental Burden of Disease. Geneva: World Health Organization; 2006.

Rani M, Shim WJ, Han GM, Jang M, Song YK, Hong SH. n-hexabromocyclododecane in polystyrene based consumer products: an evidence of unregulated use. Chemosphere 2014;110:111–119.

Renner R. What fate for brominated fire retardants. Environ Sci Technol 2000;34 (9):222A–226A.

Ropero AB, Alonso-Magdalena P, García-García E, Ripoll C, Fuentes E, Nadal A, (2008). Bisphenol-A disruption of the endocrine pancreas and blood glucose homeostasis. Int J Androl; 31(2):194–200.

Rudel RA, Camann DE, Spengler JD, Korn LR, Brody JG. Phthalates, alkylphenols, pesticides, polybrominated diphenyl ethers, and other endocrine disrupting compounds in indoor air and dust. Environ Sci Technol 2003;37:4543–4553.

vom Saal F, Hughes C. An extensive new literature concerning low-dose effects of bisphenol A shows the need for a new risk assessment. Environ Health Perspect 2005;113:926–933.

vom Saal FS, Cooke PS, Buchanan DL, Palanza P, Thayer KA, Nagel SC, Parmigiani S, Welshons WV. "A physiologically based approach to the study of bisphenol A and other estrogenic chemicals on the size of reproductive organs, daily sperm production, and behavior,". Toxicol Ind Health 1998;14 (1–2):70–76.

Sajiki J, Yonekubo J. Leaching of bisphenol A (BPA) to seawater from polycarbonate plastic and its degradation by reactive oxygen species. Chemosphere 2003;51:55–62.

Sajiki J, Miyamoto F, Fukata H, Mori C, Yonekubo J, Hayakawa H. Bisphenol A (BPA) and its source in foods in Japanese markets. Food Addit Contam Part A Chem Anal Control Expo Risk Assess 2007;24 (1):103–112.

Sánchez-Martínez M, Pérez-Corona T, Cámara C, Madrid Y. Migration of antimony from PET containers into regulated EU food simulants. Food Chem 2013;141 (2):816–822.

Sax L. Polyethylene Terephthalate may yield endocrine disruptors. Environ Health Perspect 2010;118 (4):445–448.

Schecter A, Papke O, Tung KC, Joseph J, Harris TR, Dahlgren J. Polybrominated diphenyl ether flame retardants in the U.S. population: current levels, temporal trends, and comparison with dioxins, dibenzofurans, and polychlorinated biphenyls. J Occup Environ Med 2005;47:199–211.

Schettler T. Human exposure to phthalates via consumer products. Int J Androl 2006;29(1): 134–139; discussion 181–185.

Schmidt CW. Face to face with toy safety. Environ Health Perspect 2008;116:A71–A76.

Schonfelder G, Wittfoht W, Hopp H, Talsness CE, Paul M, Chahoud I. Parent bisphenol A accumulation in the human maternal-fetal-placental unit. Environ Health Perspect 2002;110:A703–A707.

Schönfelder G, Wittfoht W, Hopp H, Talsness E, Paul M, Chahoud I. Parent bisphenol A accumulation in the human maternal-fetal-placental unit. Environmental Health Perspectives 2002;110 (11):703–708.

Scientific Opinion on Bisphenol A (2010): Evaluation of a study investigating its neurodevelopmental toxicity, review of recent scientific literature on its toxicity and advice on the Danish risk assessment of bisphenol A. Via Carlo Magno, Parma, Italy.

Sharpe RM. Is it time to end concerns over the estrogenic effects of Bisphenol A? Toxicol Sci 2009;114 (1):1–4.

Shea KM. Pediatric exposure and potential toxicity of Phthalate plasticizers. Pediatrics 2003;111 (6):1467–1474.

Shotyk S, Krachler M. Contamination of bottled waters with antimony leaching from polyethylene terephthalate (PET) increases upon storage. Environ Sci Technol 2007;41 (5):1560–1563.

Shotyk S, Krachler M, Chen B. Contamination of Canadian and European bottled waters with antimony from PET containers. J Environ Monit 2006;8:288–292.

Shuk-Mei H, Tang W, Frausto J, Prins G. Developmental exposure to Estradiol and Bisphenol A increases susceptibility to prostate carcinogenesis and epigenetically regulates phosphodiesterase type 4 variant 4. Cancer Res 2006;66 (11):5624–5632.

Smith KR, Corvalán CF, Kjellström T. How much global ill health is attributable to environmental factors? Epidemiology 1999;10:573–584.

Stafford CM, Harrison C, Beers KL, Karim A, Amis EJ, VanLandingham MR, Kim H-C, Volksen W, Miller RD, Simonyi EE. A buckling-based metrology for measuring the elastic moduli of polymeric thin films. Nat Mater 2004;3:545–550.

Stahlhut RW, van Wijngaarden E, Dye TD, Cook S, Swan S. Concentrations of urinary phthalate metabolites are associated with increased waist circumference and insulin resistance in adult U.S. males. Environ Health Perspect 2007;115 (6):876–882.

Sun Y, Wada M, Al-Dirbashi O, Kuroda N, Nakazawa H, Nakashima K. High-performance liquid chromatography with peroxyoxalate chemiluminescence detection of bisphenol A migrated from polycarbonate baby bottles using 4-(4,5-diphenyl-1H-imidazol-2-yl)benzoyl chlorine as a label. J Chromatogr B 2000;749:49–56.

Sun Y, Irie M, Kishikawa N, Wada M, Kuroda N, Nakashima K. Determination of bisphenol A in human breast milk by HPLC with column-switching and fluorescence detection. Biomed Chromatogr 2004;18:501–507.

Susiarjo, M., Hassold, T.J., Freeman, E., Hunt, P.A. (2007). Bisphenol A exposure in utero disrupts early oogenesis in the mouse, PLoS Genet, 3(1), 5 10.1371/journal.pgen.0030005.

Swan SH. Prenatal Phthalate Exposure and Anogenital Distance in Male Infants. Environ Health Perspect 2006;114 (2):A88–A89.

Swan SH, Elkin EP, Fenster L. The question of declining sperm density revisited: an analysis of 101 studies published 1934–1996. Environ Health Perspect 2000;108:961–966.

Takai Y, Tsutsumi O, Ikezuki Y, Kamal Y, Osuga Y, Yano T, Taketan Y. Preimplantation exposure to bisphenol A advances postnatal development. Reprod Toxicol 2000; 15:71–77.

Takao Y, Chul Lee H, Ishibashi Y, Kohra S. Tominag screening method for bisphenol A in environmental water microextraction (SPME). J Health Sci 1999;45 (39).

Takigami H, Suzuki G, Hirai Y, Sakai S. Transfer of brominated flame retardants from components into dust inside television cabinets. Chemosphere 2008;73:161–169.

Talsness, C.E., Andrade, A.J., Kuriyama, S.N., Taylor, .JA., vom Saal, F.S. (2009). Components of plastic: experimental studies in animals and relevance for human health. Philos Trans R Soc Lond B Biol Sci, 364(1526), 2079–9.

Teeguarden J. 2012. Estrogen receptor activation potential of internal concentrations of BPA in humans. Paper presented at the AAAS Meeting, February 14, 2012. Boston, MA .

Teeguarden JG, Hanson-Drury S. A systematic review of bisphenol A "low dose" studies in the context of human exposure: a case for establishing standards for reporting "low-dose" effects of chemicals. Food Chem Toxicol 2013;62:935–948.

Teeguarden JG, Calafat AM, Ye X, Doerge DR, Churchwell MI, Gunawan R, Graham MK. Twenty-four hour human urine and serum profiles of bisphenol a during high-dietary exposure. Toxicol Sci 2011;123 (1):48–57.

Terasaki M, Shiraishi F, Fukazawa HI, Makino M. Occurrence and estrogenicity of phenolics in paper-recycling process water: pollutants originating from thermal paper in waste paper. Environ Toxicol Chem 2007;26:2356–2366.

Teshima R, Nakamura R, Nakamura R, Hachisuka A, Sawada J-I, Shibutani M. Effects of exposure to decabromodiphenyl ether on the development of the immune system in rats. J Health Sci 2008;54:382–389.

Tinwell H, Haseman J, Lefevre PA, Wallis N, Ashby J. Normal sexual development of two strains of rat exposed in utero to low doses of bisphenol A. Toxicol Sci 2002;68: 339–348.

Trasande L, Attina TM, Blustein, J. (Sep 19, 2012). Association between urinary bisphenol A concentration and obesity prevalence in children and adolescents. JAMA; 308(11), 1113–1121.

Turyk ME, Persky VW, Imm P, Knobeloch L, Chatterton R, Anderson HA. Hormone disruption by PBDEs in adult male sport fish consumers. Environ Health Perspect 2008;116 (12):1635–1641.

Tyl RW, Myers CB, Marr MC, Thomas BF, Keimowitz AR, Brine DR, Veselica MM, Fail PA, Chang TY, Seely JC, Joiner RL, Butala JH, Dimond SS, Cagen SZ, Shiotsuka RN, Stropp GD, Waechter JM. Three-generation reproductive toxicity study of dietary bisphenol A in CD Sprague-Dawley rats. Toxicol Sci 2002;68:121–146.

Tyl RW, Myers CB, Marr MC, Fail PA, Seely JC, Brine DR, Barter RA, Butala JH. Reproductive toxicity evaluation of dietary butyl benzyl phthalate (BBP) in rats. Reprod Toxicol 2004;18 (2):241–264.

U.S. Agency for Toxic Substances and Disease Registry (ATSDR). 2002. Toxicological profile for Di(2-ethylhexyl)phthalate (DEHP). Downloaded from http://www.atsdr.cdc.gov/toxprofiles/tp9.html. Accessed October 7, 2014.

U.S. Environmental Protection Agency. 2012. Endocrine disruptor screening program (EDSP). Downloaded from http://www.epa.gov/endo/index.htm. Accessed March 11, 2013.

U.S. National Toxicology Program, Center for the Evaluation of Risks to Human Reproduction. 2007. NTP-CERHR reports and monographs. National Institute of Health. Downloaded from http://ntp.niehs.nih.gov/ntp/ohat/bisphenol/bisphenol.pdf. Accessed October 7, 2014.

U.S. Environmental Protection Agency. Ambient water quality criteria—nonylphenol—final (PDF). 290K, (EPA-822-R-05-005). USEPA; 2006. p. 96.

Van den Eede N, Dirtu AC, Covavi A. Analytical developments and preliminary assessment of human exposure to organophosphate flame retardants from indoor dust. Environ Int 2011;37:454–461.

Van der Ven LTM, van de Kuil T, Leonards PEG, Slob W, Cantón RF, Germer S, Visser TJ, Litens S, Håkansson H, Schrenk D, van den Berg M, Piersma AH, Vos JG, Opperhuizen A. A 28-day oral dose toxicity study in Wistar rats enhanced to detect endocrine effects of decabromodiphenyl ether (decaBDE). Toxicol Lett 2008;179:6–14.

Van Vliet ED, Reitano EM, Chhabra JS, Bergen GP, Whyatt RM. A review of alternatives to di (2-ethylhexyl) phthalate-containing medical devices in the neonatal intensive care unit. J Perinatol. (2011) Aug;31(8):551–560.

Vandenberg LN, Hauser R, Marcus M, Olea N, Welshons WV. Human exposure to bisphenol A (BPA). Reprod Toxicol 2007;24:139–177.

Vandenberg LN, Maffini MV, Schaeberle CM, Ucci AA, Sonnenschein C, Rubin BS, Soto AM. Perinatal exposure to the xenoestrogen bisphenol-A induces mammary intraductal hyperplasias in adult CD-1 mice. Reprod Toxicol 2008;26:210–219.

Vandenberg LN, Maffini MV, Sonnenschein C, Rubin BS, Soto AM. Bisphenol-A and the great divide: a review of controversies in the field of endocrine disruption. Endocr Rev 2009;30:75–95.

Vaughn BC, editor. Bisphenol A and Phthalates: Uses, Health Effects and Environmental Risks. In: *Chemical Engineering Methods and Technology*. New York: Nova Science Publishers; 2010.

Versar, Inc. and Syracuse Research Corporation. 2010. Progress Report: Development of Lead and Phthalate Standard Reference Materials for Use in Testing of Children's Products to Ensure CPSIA Compliance. Prepared for U.S. Consumer Products Safety Commission. Available from http://www.cpsc.gov/en/Research–Statistics/-Technical-Reports-/. Accessed October 7, 2014.

Vogel SA. The politics of plastics: the making and unmaking of Bisphenol A "safety". Am J Public Health 2009;99 (S3):S559–S566.

Volkel W, Colnot T, Csanady GA, Filser JG, Dekant W. Metabolism and kinetics of bisphenol A in humans at low doses following oral administration. Chem Res Toxicol 2002;15:1281–1287.

Von Goetz N, Womouth M, Scheringer M, Hungerbuhler K. Bisphenol A: how the most relevant exposure source contributes to total exposure. Risk Anal 2010;30:473–487.

Wagner M, Oehlmann J. Endocrine disruptors in bottled mineral water: total estrogenic burden and migration from plastic bottles. Environ Sci Pollut Res Int 2009;16:3278–3286.

Wagner M, Oehlmann J. Endocrine disruptors in bottled mineral water: estrogenic activity in the E-Screen. J Steroid Biochem Mol Biol 2011;127 (1–2):128–135.

Wargo J. *Plastics that may be Harmful to Children And Reproductive Health*. North Haven: Environment & Human Health, Inc; 2008.

Watanabe W, Tomomi S, Sawamura R, Hino A, Kurokawa M. Effects of decabrominated diphenyl ether (DBDE) on developmental immunotoxicity in offspring mice. Environ Toxicol Pharmacol 2008;26:315–319.

Wegelin M, Canonica S, Alder AC, Marazuela D, Suter MJ-F, Bucheli TD, Haefliger OP, Zenobi R, McGuigan KG, Kelly MT, Ibrahim P, Larroq M. Does sunlight change the material and content of polyethylene terephthalate? J Water Supply Res Technol 2001;50 (3):125–133.

Welle F, Franz R. Migration of antimony from PET bottles into beverages: determination of the activation energy of diffusion and migration modeling compared with literature data. Food Addit Contam 2011;28:115–126.

Welshons WV, Thayer KA, Judy BM, Taylor JA, Curran EM, vom Saal FS. Large effects from small exposures. I. Mechanisms for endocrine-disrupting chemicals with estrogenic activity. Environ Health Perspect 2003;111 (8):994–1006.

Welshons WV, Nagel SC, vom Saal FS. Large effects from small exposures. III. Endocrine mechanisms mediating effects of bisphenol A at levels of human exposure. Endocrinology 2006;147:S56–S69.

Westerhoff P, Prapaipong P, Shock E, Hillaireau A. Antimony leaching from polyethylene terephthalate (PET) plastic used for bottled drinking water. Water Res 2008a;42 (3):551–556.

Westerhoff P, Prapaipong P, Shock E, Hillaireau A. Antimony leaching from polyethylene terephthalate (PET) plastic used for bottled drinking water. Water Res 2008b;42 (3):551–556.

Wetherill YB, Petre CE, Monk KR, Puga A, Knudsen KE. The xenoestrogen Bisphenol A induces inappropriate androgen receptor activation and mitogenesis in prostatic adenocarcinoma cells. Mol Cancer Ther 2002;1:515–524.

WHO/FAO. 2010. Joint FAO/WHO expert meeting to review toxicological and health aspects of bisphenol A and stakeholder meeting, Ottawa, Canada, November 1–5, 2010. Geneva: World Health Organization.

Wong KO, Leo LW, Seah HL. Dietary exposure assessment of infants to bisphenol A from the use of polycarbonate baby milk bottles. Food Addit Contam 2005;22:280–288.

Wormuth M, Scheringer M, Vollenweider M, Hungerbuhler K. What are the sources of exposure to eight frequently used phthalic acid esters in Europeans? Risk Anal 2006;26:803–824.

Wozniak AL, Bulayeva NN, Watson CS. Xenoestrogens at picomolar to nanomolar concentrations trigger membrane estrogen receptor-α mediated Ca++ fluxes and prolactin release in GH3/B6 pituitary tumor cells. Environ Health Perspect 2005;113:431–439.

Zalko D, Jacques C, Duplan H, Bruel S, Perdu E. Viable skin efficiently absorbs and metabolizes bisphenol A. Chemosphere 2011;82 (3):424–430.

Zhou T, Ross DG, DeVito MJ, Crofton KM. Effects of short-term in vivo exposure to polybrominated diphenyl ethers on thyroid hormones and hepatic enzyme activities weanling rats. Toxicol Sci 2001;61:76–82.

Zhou T, Taylor MM, DeVito MJ, Crofton KM. Developmental exposure to brominated diphenyl ethers results in thyroid hormone disruption. Toxicol Sci 2002;66:105–116.

Zhou J, Yu Y, Chen D, Zhong Y. Effects of maternal oral feeding of decabromodiphenyl ether (PBDE 209) on the humoral immunity of offspring rats. J Trop Med (Guangzhou) 2010;10 (3):276–279.

8

PLASTICS AND HEALTH IMPACTS

Packaging is the leading application sector for plastics, accounting for 35–40% of the global consumption of resin. The largest segment of the approximately $500 billion global packaging market is food packaging, with beverage packaging a distant second. Food and beverage packaging comprises 55–65% of the packaging in the United States (Brody, 2008), and in 2008, 38% of all food packaging was made of plastic (with paperboard claiming another ~30% of the market share). In the packaging sector, rigid plastic packaging is the fastest growing segment of the market. Even paperboard packaging and metal cans often incorporate a plastic film on the interior surface in contact with food. Paper-based multilayer packs, such as Tetrapak used to package beverages, also rely on layers of plastic in the construct. With the shift over to recyclable material in packaging plastics benefit as they are easily recyclable and are a cost-effective material with exceptional performance. Practical sustainability guidelines consistent with the elements in the discussion on sustainability (Chapter 2) can be derived for plastic-based food packaging (van Sluisveld, 2013).

Food packaging and food service plastics have the shortest service lifetime of any plastic product manufactured and end up in the waste or litter at a much faster rate compared to other plastic products. Packaging therefore has implications on sustainable development for several reasons. First, being a high-volume-use sector, plastic packaging produces waste that can end up in urban litter and even as ocean debris (see Chapter 10). There is no efficient mechanisms to collect, recycle, or dispose of most plastic litter. Secondly, some of the primary plastic packaging is used in contact with food or beverages, and the possibility of additives (and residual monomer)

Plastics and Environmental Sustainability, First Edition. Anthony L. Andrady.
© 2015 John Wiley & Sons, Inc. Published 2015 by John Wiley & Sons, Inc.

leaching out of the package into the contents via slow diffusion is a concern as some of these compounds, even at low concentrations can be toxic. Also, the packaging sector places a significant demand on fossil fuel resources with only very modest recycling rates achieved in plastic food packaging; the best recycling rates in the US are obtained for PET beverage bottles.

8.1 PACKAGING VERSUS THE CONTENTS

As the primary function of a package is to protect the contents (food or beverage), it is reasonable to expect a "cost" in terms of an environmental footprint to be incurred in its use. Affording mechanical protection, maintaining the shelf-life of food, reducing food waste, and the controlling chemical or biological contamination of the contents ensured by plastic packaging often justify this environmental cost. However, the costs need to be in proportion to the value of the product being protected. In addition to the material and labor expended in fabrication, the "cost" of energy and environmental impacts associated with preparing a product for the market, need to be considered. Generally, the embedded energy and environmental impacts of packaging turn out to be much smaller than those for the contents, making most plastic packaging a reasonable strategy. High-value items such as medication, cosmetics, or electronic goods are packaged at a minute fraction of their costs.

A European study on various food packages (of rye bread, ham, and soy yogurt products) found that only 2–5% of the carbon footprint of the product could be attributed to packaging (Silvenius et al., 2011). The carbon footprint is indirectly related to the embodied energy of the package. There can, however, be exceptions (Williams and Wikström, 2011) as recently reported for several common food items (see Table 8.1).

TABLE 8.1 The Embedded Energy and GWP of Selected Packaged Food Items (1 Kg Portions)

Food item	Ketchup	Bread	Cheese	Beef
Energy (MJ/kg of food) **F**	11	7.8	38	48
Energy (MJ for package) **T**	5.7	0.77	0.65	3.1
Ratio **T/F**	0.53	0.10	0.02	0.07
GWP (kg CO_2 equiv./kg food) **F***	790	610	8500	14000
GWP (kg CO_2 equiv./package) **T***	260	28	44	150
Ratio **T*/F***	0.33	0.05	0.01	0.01
Plastic material used in packaging	PP, LDPE	LDPE, PS[a]	LDPE, PA	PA, LDPE, EVOH[b]

Source: Williams and Wikström (2011).
Data specific to the types of primary and secondary packaging used in retail products at a selected retail location.
[a] Plastic closure tag.
[b] Ethylene vinyl alcohol copolymer film.

The parameters of interest in Table 8.1 are the ratio of the energy demand for package and contents (T/F) and the ratio of GWP for the same (T'/F'). The carbon footprint is indirectly related to the embedded energy of the package. For items such as cheese or beef, the energy cost of the package is only 2 or 7% of that of the food item. The carbon emission attributable to package is also only around 1%, making the package a modest and very reasonable investment to ensure optimal shelf life and consumer safety. By contrast, the ketchup package, the multilayer plastic bottle, takes over 50% of the energy it takes to manufacture the contents (and a third of the carbon footprint as well). LCA studies that compare different beverage packages should be evaluated with caution as transportation energy is often responsible for the significant differences between the choices compared. In general, using larger plastic packages (bulk packaging) for food results in a smaller environmental footprint (Pasqualino et al., 2011). Given the high embedded energy of manufacture, reuse and recycling contribute to economy in energy and material use in packaging.

An extreme example of a disproportionately expensive (in terms of energy) package is bottled water. Global average per capita consumption of bottled water is approximately 25 l, second only to that of carbonated beverages (~31 l) (Rexam Consumer Packaging Report, 2011). It is a popular product because of the convenience it delivers.[1] The energy cost of filtration, disinfection, reverse osmosis, and filling the bottles of locally acquired water (at times transported over short distances) is estimated to be only about 0.02 MJ/l. However, that of manufacturing the plastic bottle is over 4.0 MJ/l, even not counting the energy needed to transport it to the consumer (Gleick and Cooley, 2009)! The package uses at least 200 times the energy needed to produce the water it carries. Depending on transportation costs, this value can be much higher (reaching ~400–500 times that for the water). Environmental impact of treating (non-carbonated) water to be bottled was recently estimated to be 90–1000 times that of making potable tap water (Jungbluth, 2006). There is no significant additional health or safety benefits afforded by the expensive bottled water compared to tap water that uses only 0.005 MJ/l to produce![2] In fact, some studies suggest that endocrine disruptors (ED) (Sb and brominated flame retardants) incorporated into the plastic raw materials during manufacture may leach out into the bottled water (Andra et al., 2012; Fig. 8.1).

Bottled water is generally safe for drinking purposes, but a recent study of over 1900 samples (Marcussen et al., 2013) showed some to contain potentially toxic ion concentrations (e.g., As) that exceed the EU or WHO limit values for drinking water. Growing exports of bottled water by countries such as France, China, and Italy into countries such as the United States, Japan, and Germany add to the ecological footprint of the product because of transportation costs.

Globally, 60% of the carbonated beverages are packed in plastic compared to only 20% in cans (Rexam Consumer Packaging Report, 2011); beer is excluded from this estimate as only approximately 5% of it is packaged in plastic. The embodied energy for manufacture of carbonated drinks is not very different from that for

---

[1] A recent innovation is "boxed water" packaged in paperboard cartons (Boxed Water is Better, 2014).
[2] In fact, 40–60% of bottled water in United States is reprocessed tap water! Unlike the tap water, which is regulated by the USEPA, bottled water is regulated by the FDA, under less stringent guidelines.

FIGURE 8.1   Bottled water sales in the United States is on the increase with a per capita consumption of 29 US gallons in 2011.

bottled water (~0.01 MJ/l) (the carbon footprint for manufacturing the beverage, however, itself will be considerably higher than that for water). Unlike for water, there is no convenient alternative packaging for carbonated beverage except for bottles and metal cans accounting for over a third of the retail value of these products. As with water bottles, recycling the polyester (PET) soft drink bottles can vary significantly but bring down the energy costs and, to a lesser extent, the carbon emission load associated with the product. Plastic (PET) packaging is generally found to be competitive over the aluminum or glass counterparts in comparative cradle-to-grave LCA studies. A significant factor in packaging beverages is the water footprint associated with the process; estimated water footprint in producing 0.5 l PET bottle was estimated to be 150–300 l (Ercin et al., 2011)! A 2009 study comparing PET to glass and aluminum cans (funded by PETRA, the industry trade association) and the results were already discussed in Chapter 5.

8.1.1   Packaging Milk in HDPE

The LCA studies that compare different packaging materials for 1 l milk containers generally show HDPE plastic or PP-coated paperboard cartons to be more energy efficient compared to comparable glass or aluminum packaging. The greenhouse gas (GHG) emissions are also relatively lower for plastics packaging. Generally, both the embodied energy and carbon footprint of the package are dominated by the fabrication and use phases (possibly by transportation as well). A typical such comparison is shown in Figure 8.2 (Ghenai, 2012). As the transportation costs are variable, they are not included in the figure.

This clear LCA picture changes, however, when reuse of containers or material recycling is included in the calculation. Using realistic end-of-life options, including recycling (as well as all the related transport costs), is critical in meaningful LCA studies on packaging as the findings are especially sensitive to these variables. For instance, using recycled aluminum in place of the virgin metal results in a 36% saving in energy (making it competitive with glass) and a 40% saving in GHG

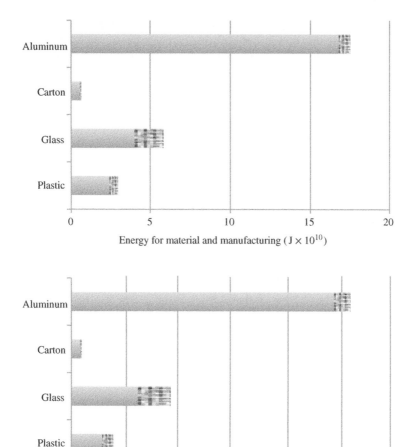

FIGURE 8.2 The energy use and GWG emissions associated with the production of material and fabrication of containers for milk (~1 l). The first segment of bar is for material production, and the second is for manufacturing. Drawn from data in Ghenai (2012).

emissions, in the previous data. Reusing glass bottles recaptures approximately 97% of the original embedded energy.

Another study (Franklin Associates Study, 2008) compared glass bottles, paperboard cartons, HDPE jugs and biodegradable polylactic acid (PLA) containers for packaging of milk. The reusable glass bottles were shown to have the least energy cost and the biodegradable PLA containers, the highest. Despite the high transportation energy (given the relatively higher weight of glass containers), a low energy value is found for the glass bottle because once fabricated, these were assumed to be reused eight times. The energy costs of cleaning to enable reusability, however, were not fully taken into account, in the analysis. A recent packaging study comparing PET vs. glass found that half the environmental impact of glass bottles was in fact associated with the bottling process (Doublet, 2012). Given the approximations in the methodology used, paperboard carton and the HDPE milk jug were estimated to have about the same energy cost. However, the energy for HDPE is almost entirely derived from nonrenewable fossil fuel, while only about 40% of the total energy cost of paperboard is derived from fossil fuel.

The lowest GHG emissions were also associated with the HDPE milk jug, whereas both glass and PLA package showed relatively higher values (both about the same within the margin of error.) With heavier glass bottles, it is the transportation cost that is responsible for most of the emissions. Milk production itself accounts for only approximately 4% of global GHG emissions (FAO, 2010; Gerber et al., 2010), and producing half a gallon of milk results in emission of approximately 4.3 kg CO_2 equivalent. Though the study included estimates of waterborne emissions as well, these were expressed on a simple gravimetric basis; the toxicity or the bioavailability of chemical species in these emissions can vary widely.

The environmental profile of milk packaging systems was also developed in 2008 (Franklin Associates, 2008) where the HDPE milk jug was compared to gable-top paperboard/LDPE carton. Quantifying the highly variable environmental health impacts of each of these is complicated. However, HDPE milk jug (128 oz) was found to have a better environmental profile based on selected impact criteria, compared to paperboard cartons even at a zero rate of recycling. However, for smaller-volume (64 oz) packages, both had about the same embodied energy, but atmospheric emissions, waterborne waste, and solid waste were still significantly lower for the HDPE bottle compared to the paperboard carton. Including recycling in the analysis will further change the profile even more in favor of HDPE. In this study (and also in general), paperboard cartons were assumed non-recyclable. However, in reality product is recyclable[3] into other paperboard packaging.

8.1.2 Overpackaging

Using an amount of packaging materials in excess of what is needed to maintain the full functionality of the package, is overpackaging. With some produce, there is no need for a man-made package; tubers, corn (in husk), apples, or oranges are protected well enough in their natural form. Others such as berries or grapes do need additional packaging in punnets; otherwise, the amount of waste during transportation and retail will be unacceptable. Even in these cases, reducing packaging without compromising the shelf life of contents is often possible (Licciardello et al., 2014). In general, the environmental impact of packaging (including its impact on prevention of food waste) shows a U-shaped curve when plotted against the amount of packaging used in a product. The "optimum packaging" is fully functional but uses the minimum

---

[3] Carton recovery rate is low as it is generally mixed with food waste. In curbside collection, cartons should be placed for collection with plastics/glass and metal, not waste paper.

amount of material for the purpose. Sustainable growth favors economical use of materials, especially non-renewables, with the lowest practical environmental footprint for application including packaging of food, beverage, and medication. Material substitution, packaging portions, reducing voids in packaging and downgauging can make the package more efficient. But, always with the important proviso: "as long as full functionality, especially consumer safety, can be ensured."

8.2 PACKAGE–FOOD INTERACTIONS

Ideally, the package should be an inert container that has no interactions at all with food. But with plastic packaging, a range of such interactions are possible (Piringer and Ruèter, 2000), and the main ones are summarized in Figure 8.3. All these invariably affect the sensory quality and potentially the safety of food:

(a) Permeation: The transfer of gases or vapors, especially oxygen and moisture, across the package (Hotchkiss, 1997).

(b) Migration of additives and printing ink constituents in the plastic packaging into the food or beverage can result in off-flavors and potential toxicity.

(c) Packages can potentially sorb pollutants in the environment via partition. These, once in the plastic matrix, can migrate or interact with the contents.

(d) Scalping of constituents in food by the plastic package can result in altered flavor and aroma of food (Hernandez-Muñoz et al., 2001).

Of these, the permeation of O_2 and humidity affect the shelf life, while migration of additives and scalping of flavors from food affects the sensorial quality and safety of the food.

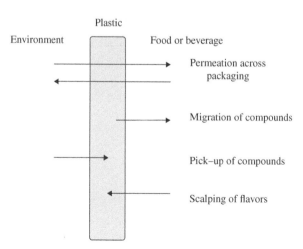

FIGURE 8.3 Summary of interactions between plastic packaging and the food or beverage contents.

8.2.1 Oxygen and Water Permeability

Ingress of oxygen into the package is responsible for slow oxidation of food contents especially at elevated storage temperatures, with the development of off-flavors such as rancidity in fat-containing food items. Moisture in a package can cause slow hydrolytic degradation of food as well as permit microbial growth in non-asceptic packages. Either of these can alter the delicate sensorial qualities of the edible contents. Gas transport properties of packaging therefore often dominate the shelf life of food items.

Gas permeability requirements of the packaging (for O_2, CO_2, and humidity) depend on the contents being preserved. High permeability to humidity is essential in packaging of fresh produce or bread to avoid condensation of water inside the package potentially causing mold growth. Hygroscopic foods such as sugars, milk powder, and dessert mixes on the other hand require low-permeability packaging to avoid moisture ingress that can cake the free-flowing powders. With fresh produce, good gas exchange (for respiration) is critical to maintain their freshness. Specialized products (cooked food and some cheeses, for instance) use modified atmosphere packaging (MAP) where some of the air in the package has been replaced by N_2 or CO_2 and where gas environment must be maintained within the package to ensure maximum shelf life.[4] MAP controls microbial growth within the package (Hotchkiss and Banco, 1992) possibly due to the presence of carbonic acid from reaction of CO_2 with water. Plastic packaging can address these diverse permeability specifications economically with the available wide range of plastic resins and their combinations. Oxygen transmission rate (OTR) and water vapor transmission rate (WVTR) of different plastics in common units are tabulated in Table 8.2.

TABLE 8.2 Selected Barrier Properties of Common Plastic Packaging Films (25 μm Thick) Measured at 38°C

Polymer	Oxygen transmission rate (OTR) at 25°C and 0% RH (cm^3/m^2/day)	Water vapor permeability at 38°C and 90% RH (g/m^2/day)
LDPE	7750	16–31
HDPE	1550–3100	6
PET	50–90	16–23
PP	1550–2480	6
PS	3100–3500	109–155
PC	2480	140
PVDC	2–16	0.8–5
PVC	465–9300	31–155
PA	15–30	155
PET/Al foil/PE	0	0

Source: Adapted from Bhunia et al. (2013).

[4] Some examples are the packaging of red meat packed in 70:30 mixture of O_2: CO_2, ready-to-eat meals in 70:30 mixture of N_2:CO_2, and seafood including white fish in 30:40:30 mix of O_2:CO_2:N_2 (*Air Products (1995): The freshline guide to MAP*).

TABLE 8.3 Some Examples of Multilayer Films Used in Food Packaging

		Layer (%)	Microns
Ground beef	LLDPE/tie/EVOH/nylon	75/5/10/10	150–200
Sausage	ION/tie/EVOH/nylon	75/5/5/15	150–200
	LLDPE/tie/EVOH/o-PET	60/15/10/15	50–100
Bacon	ION/PE/nylon	15/50/35	50–100
Deli meat casing	LLDPE/o-nylon/PVDC	70/20/10	50–80
Fish	LLDPE/tie/EVOH/nylon/tie ION	25/10/10/10/10/25	40–150
Potato chips	o-PP/LDPE/met-o-PP	25/50/2520-60	20–60
Pretzels	o-PP/LDPE/o-PET	25/50/25	30–80
Cereal bag (in box)	HDPE/tie/nylon/tie/EVA	60/10/10/10/10	40–60
Snack nuts	o-PP/LDPE/foil/LDPE	10/20/5/65	40–60
Bakery goods	HDPE/EAA/nylon/EAA	70/10/10/10	30–70
Cheese slices	PP/EVA	20/80	35–40

Data based on Ebnesajjad (2013).
Foil, generally used is aluminum; ION is, ionomer film; Met, refers to metalized film; O, refers to oriented film.

Often, there is no single cost-effective plastic material that meets the combination of barrier properties as well as the mechanical integrity demanded by a specific packaging application. Multilayer packaging addresses this problem; in fact, approximately 17% of all packaging laminates are multilayer constructs. For instance, ethylene vinyl alcohol (EVOH) copolymer resins offer an excellent high-barrier to gases, but only under dry conditions. Therefore, it is generally sandwiched between layers of hydrophobic plastics; multilayer films with PET/EVOH/EVA or LLDPE/EVOH/LLDPE are used in food packaging. Plastic beer bottles, for instance, use PET/EVOH/PET constructs (Bucklow and Butler, 2000). Other useful properties such as heat sealability (at 110–125°C) are often built into multilayer films by adding a copolymer (ethylene-*co*-propylene) or terpolymer (ethylene-*co*-propylene-*co*-butene) layer on either side of polyolefin or multilayer film. Composite polymer/paper/foil packages used for beverages or juices are even more complex in design. The specialized design of these affords excellent protection to the beverage but recycling these can often be a problem. Examples of films used in food packaging (after Ebnesajjad, 2013) are shown in Table 8.3 to illustrate the diversity of the multilayer constructs used in food packaging.

The level of complexity involved in multilayer constructs illustrates the difficulty in post-consumer recycling of these materials. Unlike single-component packaging, the multilayer materials are difficult to separate, identify, and recover as recyclable resin. For instance, a layer of aluminum is often used to improve barrier properties of multilayer packaging films. Aluminum content has to be extracted with dilute alkali or other reagent prior to recycling the plastic. Even then, automated sorting or density-based methods are unable to identify and separate these films at material recycling facilities (MRFs). The only option is to blend the different layer into a mixed plastic recyclate of variable quality. Where a single type of layered film (such as PE/nylon) waste is recovered, the recyclate might be blended with controlled amounts of virgin PE and filler to obtain a plastic blend that can replace the original packaging material (Tartakowski, 2010).

Nanoscale fillers such as clays are used to improve the barrier properties of plastic films. The platelike fillers, in particular, with the relatively higher tortuosity deliver lower gas permeabilities (Alperstein, 2005). At a volume fraction Φ of the nanofiller, the permeability P of a gas through the polymer (Picard et al., 2007), depends on the value of tortuosity τ of the nanoscale filler:

$$\frac{P}{P_o} = \frac{1 - \Phi}{\tau}$$

The magnitude of tortuosity τ depends on the geometric features of the nanofiller material.

Transparent high-barrier packaging consisting of a plastic film with a thin surface layer of vacuum- or plasma-deposited glass has been described (Landau, 2007; Lange and Wyser, 2003). While the silica layer impermeable in theory, micro-cracking of the glass surface can occur during processing, and a measurable permeability is often observed with these. There is a need for ultrahigh barrier plastics that are also transparent in both food and other[5] applications.

8.2.2   Additive Migration and Toxicity

Common polymers including plastics are not toxic in the conventional sense. The long-chain molecules of high polymer are too large in size to be absorbed via the wall of digestive tract, and enzymes that can digest polymers into smaller molecules that can be absorbed via the gut are simply not available in humans. Clean plastics therefore do not pose significant toxicity to humans; however, this applies only to uncontaminated plastics. In practice, plastics contain numerous impurities and additives. Compounded plastics may display toxicity via ingestion because of additives (low-molecular-weight chemical species) or unintentional contaminants in the material. Toxicity related to plastics is attributed to the following classes of contaminants:

(a) Catalyst residues, residual monomer (e.g., BPA, styrene, and vinyl chloride (VC)), and very-low-molecular-weight oligomers present in the plastic resin from polymerization process (Bomfim et al., 2011)

(b) Additives (such as phthalates, antioxidant, slip agents) deliberately added to the plastic during compounding and its processing into useful products (Bonini et al., 2008)

(c) Products from degradation of the polymer or its additives on exposure to heat or solar ultraviolet radiation yielding toxic products

(d) Organic chemicals picked up from the environment (air, water, or contents in a package) and concentrated in the polymer through partition

---

[5]Photovoltaic modules incorporate materials that need to be maintained under very low humidity over extended durations. Transparent plastic films with WVTR approximately 5×10^5 (g/m^2/day) are available for the application.

TABLE 8.4 Some Common Additives in Plastics Used for Packaging Food and Beverages

Additive	Example	Function
Antiblock agent	Talc, mica, ceramic spheres	Prevents sticking of films
Antifog additive	Sorbitan and glycerol esters	Prevents condensation inside package
Antistatic agents	Carbon and metallized fillers	Controls static buildup
Antioxidants	Hindered amines, light absorbers	Controls light-induced oxidation
Biocides	Ethoxylated amines	Controls microbial contamination of package surface
Lubricants and slip agents	Waxes and amides	Ease of processing
Heat stabilizers		Prevents degradation and discoloration during processing
Plasticizers	Epoxidized oils, phthalates	Makes the plastic softer and pliable

Source: Adapted from Sing et al. (2012).

The items (c) and (d) are also referred to as "unintentionally added substances" in plastics. All these chemicals remain dissolved in the plastic matrix. Being not covalently bonded to the polymer molecules, these can leach out of the plastic and in the case of food-contact plastics present a source of contamination of the food. It is important to realize, however, that such migration in most instances results in very low levels of the chemicals in the food, often too low to have any observable adverse effect. But with some classes of compounds, such as the EDs discussed in Chapter 7, this can still be a concern.

The functional additives that are commonly employed in plastics food packaging are listed in Table 8.4 (Singh et al., 2012).

The kinetics of migration of additives from the package into the (food) contents depends on the characteristics of the plastics such as density or free volume, crystallinity, glass transition temperature (Tg), as well as the polarity, molecular mass, and boiling point of the migrant species. Migration rates also depend on the temperature, relative humidity, pH value, and the composition of the food contents (Sajilata et al., 2007). Given the number of variables that can affect the results, reported data must be compared cautiously.

It is the differential solubility of these compounds that results in their being partitioned between the food and plastic, achieving an equilibrium concentration.

Diffusion of chemicals in and out of the plastic film conforms to Fick's second law:

$$\frac{dc}{dt} = -D_p \frac{d^2c}{dx^2}$$

where c is the concentration of the migrating chemical in the packaged (food) content, t is the time and x is the distance across the package wall, and D_p is the diffusion coefficient. Since the packages remain on the shelf or in storage for long durations, it is the equilibrium concentration of the migrant chemical in the contents, rather than its rate of diffusion that is of interest. The partition coefficient K_{sp} quantifies this concentration and is defined as the ratio of the equilibrium concentration of migrant in the food or simulant, C_s, to that in the plastic packaging material, C_p (the subscripts s and p refer to simulant and plastic, respectively):

$$K_{s/p} = \frac{C_s}{C_p}$$

The value of K_{sp} varies with the temperature, the chemical species, as well as the type of plastic used in the package. The package also has a second key interface: that between air/plastic with a value $K_{a/p}$.

In lab assessments of migration of additives, food simulants such as dilute (3% v/v) acetic acid, 10% and 20% (v/v) ethanol, olive oil and distilled water are commonly used in place of actual food items. Foods that are high in lipids may partition hydrophobic additive migrants at concentrations higher than indicated in standard tests (Sanches Silva et al., 2007). The levels of such migrating compounds will be low, especially in aqueous extracts, and analytical protocols need to be followed closely to quantify them (Nerin et al., 2013). In typical laboratory testing, the food simulant is allowed to be in contact with the packaging for fixed durations at a constant temperature, and the migrant concentration in food and/or package is analyzed. Bhunia et al. (2013) has compiled a detailed list of migration studies for common plastic packaging systems.

Typical migration data for eight additives, mostly plasticizers in several different commercial packages, illustrates typical data (Fasano et al., 2012) and are summarized in Table 8.5. As phthalates are EDCs, their migration into packaged food is a particular concern. The following were studied: di-(2-ethylhexyl)phthalate (DEHP), diethylphthalate (DEP), di-n-butylphthalate (DBP), butylbenzylphthalate (BBzP), di-iso-nonylphthalate (DiNP), di-n-octylphthalate (DnOP), diisododecylphthalate (DiDP), nonylphenol (NP), and bisphenol A (BPA).

Rates of diffusion of molecules in a polymer will generally be faster at higher temperatures. The Arrhenius equation gives the dependence of the change in magnitude of the diffusion coefficient D_p with temperature:

$$D_p = A \exp^{\left(-\frac{\Delta E}{RT}\right)}$$

where ΔE is the activation energy and A is the preexponential factor. Log (D_p) is inversely proportional to the temperature T (K). Diffusion coefficients at two temperatures, T_1 and T_2, can be used to calculate the value of ΔE:

$$\Delta E = -2.3R \left\{ \log D_1 - \log D_2 \right\} \Big/ \left\{ \frac{1}{T_1} - \frac{1}{T_2} \right\}$$

TABLE 8.5 Levels of Phthalates, OP, NP, BPA and DEHA (mean ± sd) in ng/L in the Different Food Packaging Items Considered in this Study[a]

Packaging	DMP mean ± sd	DBP mean ± sd	BBP mean ± sd	DEHP mean ± sd	OP mean ± sd	NP mean ± sd	BPA mean ± sd	DEHA mean ± sd
Oil tuna can	66 ± 20	<LOD	<LOD	2668 ± 591	NR	NR	NR	<LOD
Natural tuna can	106 ± 12	<LOD	39 ± 4	<LOD	<LOD	3912 ± 459	824 ± 148	116 ± 58
Marmalade cup	338 ± 76	<LOD	275 ± 67	7921 ± 738	<LOD	<LOD	<LOD	340 ± 71
Film	<LOD	30 ± 8	<LOD	<LOD	17 ± 5	57 ± 28	<LOD	<LOD
Bread bag	19 ± 3	65 ± 15	<LOD	436 ± 35	338 ± 163	238 ± 128	189 ± 46	20 ± 10
Tetrapak	12 ± 5	40 ± 15	<LOD	<LOD	12 ± 6	24 ± 6	<LOD	73 ± 59
Yogurt packaging	58 ± 37	<LOD	<LOD	<LOD	NR	NR	NR	1989 ± 539
Polystyrene dish	19 ± 6	<LOD	<LOD	701 ± 479	39 ± 4	184 ± 51	<LOD	40 ± 10
Teat	<LOD	<LOD	18 ± 4	<LOD	<LOD	378 ± 130	<LOD	<LOD
Baby's bottle	<LOD	121 ± 33	<LOD	<LOD	33 ± 10	154 ± 86	404 ± 69	<LOD
Plastic wine top	273 ± 111	1684 ± 281	274 ± 108	14176 ± 3360	26629 ± 1185	958 ± 490	NR	1099 ± 348
All packaging	85 ± 116	363 ± 640	116 ± 135	5086 ± 5348	6619 ± 20910	720 ± 1251	315 ± 300	651 ± 875

[a] Data based on Fasano et al. (2012).
LOD, limit of detection; NR, not recovered (OP, NP, and BPA not recovered in acetic acid 3%).

This effect of temperature is also the basis for the precaution against heating food contained or wrapped in plastic materials that contain additives or residual monomer. Several studies (Galotto and Guarda, 2004; Rijk and Kruijf, 1993) suggest that lower migration rates are obtained with microwave heating compared to oven (thermal) heating of fatty foods in contact with plastics, possibly because of the shorter durations of contact.

8.2.3 Residual Monomer in Packaging Resin

PE and PP being based on gaseous monomers do not result in any trapped residual monomers. Generally, only traces of residual Ziegler–Natta catalysts used in low-pressure polymerization of polyolefins are usually present in the resin and that too at low (<100 ppm) concentrations. Other plastics such as polystyrene (PS), acrylics (PMMA), poly(ethylene terephthalate) (PET), poly(vinyl chloride) (PVC), and polycarbonate (PC) as well as thermoset polymers such as epoxy and polyurethane generally carry traces of residual monomer as well as by-products of the polymerization reaction (Araújo et al., 2002). Efforts are continually being made to reduce residual monomer levels, and therefore, only those reported in the recent literature are reliable. Also, the reported residual monomer levels depend primarily on the specific type of polymerization used (e.g., bulk, solution, emulsion) in manufacture as well as on the sensitivity of analytical methods used to estimate them. The values reported varied by as much as 100% depending on the analytical method used to estimate styrene in PS resin (Garrigós et al., 2004). Recent studies (Borrelli et al., 2005) on PVC, for instance, the residual VC values of 2–5 ppb for food packaging, well below the acceptable levels (by the FDA). Residual monomer is more common in thermoset plastics compared to common thermoplastics. The most studied system for residual monomer is dental acrylic resin that is UV-cured *in situ* on the tooth with inevitable monomer residue left over in the buccal cavity (Kawahara et al., 2004; Lung and Darvell, 2005).

The significance of residual monomers depends not only on their levels in the plastic resin but also on how readily they leach out and their toxicity. A hazard rating for monomers based on their modes of toxicity has been developed as shown in Table 8.6 (Lithner et al., 2011). However, the risk from residual monomer toxicity is ultimately the product of its hazardousness and the likelihood of human exposure to the leached monomer via ingestion or transdermal delivery.

8.2.4 Scalping of Flavor Components

Flavor scalping is the loss of sensorial quality of a packaged food due to either its aroma or flavors being absorbed by the packaging material. Perhaps the most studied case is the scalping of a citrus flavor component, limonene, from orange juice packages (Moshanos and Shaw, 1989). In laboratory studies with model solutions of the terpene, the extent of scalping was reported to vary with the plastic as follows: LDPE > polyamide > PS. The solubility parameter of the specific flavor or aroma component and that of the plastic in contact with food determines the extent of scalping of compounds from the food content by the plastic package. Where Hildebrand solubility parameters are

TABLE 8.6 Toxicity Levels of Monomers in Common Plastics

Monomer	Hazard rating	Plastic	Mode of toxicity	Residual monomer (ppm)	
Acrylonitrile	11,521	PAN	C, skin, eye	20–1,000	Araújo et al. (2002)
Vinyl chloride	10,001	PVC	C	<50–30,000	Araújo et al. (2002)
Vinylidene chloride	111	PVDC	C	0.001–0.020	Ohno et al. (2006)
Styrene	30	PS		2,300–6,304	Garrigós et al. (2004)
BPA	1210	PC, epoxy	Skin	153–458	American Chemical Society (2013)
Epichlorohydrin		epoxy	C, skin	200–25,000	Araújo et al. (2002)
Methyl methacrylate	1021	PMMA	Skin	20,000–25,000	Vallo et al. (1998)

Endocrine disruptor effects were not taken into account in estimating the toxicity of monomers. Modified from Lithner et al. (2011).
E, Eemulsion polymerization; S. suspension polymerization.
Residual monomer values selected from Araújo et al. (2002).

TABLE 8.7   Percentage of Component in Cold-pressed Orange Oil
Sorbed by Different Plastics it was in Contact with for a 4-day Period

	LDPE	HDPE	PP	Surlyn
Limonene	68	30	28	53
Pyrene	58	21	20	49
Myrcene	66	28	22	78
Decanal	18	6	6	11
Dodecanal	22	7	9	16

about the same for the plastic material and the contents, the potential for scalping is relatively high (Sajilata et al., 2007). Also, flavor compounds generally tend to diffuse into
and dissolve at a faster rate in plastics that are above their glass transition temperature,
Tg, at the use temperature, such as with PE and PP. The equilibrium concentration of the
migrant in the plastic will also be influenced by the percent crystallinity of the matrix.
Lower percent crystallinity favors higher solubility of migrants as only the amorphous
regions in a semicrystalline polymer (Nielsen and Jägerstad, 1994) can dissolve the
migrant species. In some instances, the sorption of compounds that plasticize the polymer can increase the gas permeability of the packaging (Van Willige et al., 2000).

Examples of the dependence of aroma scalping on plastic/migrant combination
are illustrated by the data on partition of key orange oil aroma constituents into four
types of common plastics (see Table 8.7) (Charara et al., 1992). The nonpolar
limonene is readily absorbed by LDPE; a polar polymer such as PET would absorb
much less of the compound.

8.3   STYRENE AND EXPANDED POLYSTYRENE FOOD SERVICE MATERIALS

In 2012, the world production of styrene was 26.4 MMT with China, the United
States, and Japan accounting for approximately 44% of the production. Styrene ranks
16th in air emissions of volatile chemicals in the United States. Not surprisingly, the
USEPA found 100% of the human fat tissue samples collected in a 1982 study, to
contain styrene.[6] Most of the styrene emissions are not from food packaging but from
automobile exhaust and industrial uses (e.g., reinforced composites) of styrene. Not
generally taken into account is the exposure inside automobiles (Ayoko et al., 2008).
Styrene ranks 4th in mean concentration of volatile constituents inside the cab and is
one of the five compounds in the mix at a concentration greater than 40 μg/m³. Also,
cigarette smoke (~147 μg/cigarette; Cohen et al., 2002) contributes significantly to
the intake. It is readily oxidized in air and outdoor measurements usually find concentrations of about 1 ppb (National Toxicology Program, 2006).

Expanded polystyrene (EPS) foam widely used in the food service industry is made
by steam expansion of PS beads containing pentane or other blowing agents. The foam is

---

[6] The National Human Adipose Tissue Survey for 1986 identified styrene residues in 100% of all samples
of human fat tissue taken in 1982 in the United States.

molded during the expansion (or subsequently extruded through a die to obtain a sheet that can be thermoformed). Polystyrene products may contain up to 153–458 ppm of residual styrene monomers (American Chemical Council, 2013), dimers, and trimers (Choi et al., 2005) as well as traces of the 7,8-oxide of styrene. With food service products, EPS comes into direct contact with hot/cold food or beverages that can extract residual styrene from packaging particularly efficiently. This is an obvious route of ingestion and the amounts migrating into food can be estimated using the migration kinetics of styrene reported by Lickly in 1995 (Lickly et al., 1995; Tang et al., 2000). For instance, eggs packaged in PS cartons for 2 weeks had seven times higher levels of styrene and ethyl benzene compared to fresh farm eggs (Matiella and Hsieh, 2006). Migration of the monomer from EPS foam cups into water (pH ~ 5) at 80°C over a 35–120 min of exposure was as high as 0.3–4.2 µg/l (Tawfik and Huyghebaert, 1998). Though reliable quantitative data is not available, there is little doubt that fatty foods such as meat or dairy products in contact with PS containers also pick up residual monomer.

However, the levels of migrated styrene and other compounds in packaged food are low compared to maximum permitted levels. Exposure to styrene at non-occupational levels causes primarily mild irritation of the eyes and the respiratory tract. Ingestion or inhalation of styrene results in low acute toxicity and neither styrene nor its dimer or trimer shows any ED effects (Date et al., 2002; Hirano et al., 1999). The monomer released into air is photooxidized readily and has a short half-life of only 3.5–9.0 h (Alexander, 1990). In any event, some common foods are naturally rich in styrene with levels that far exceed those resulting from migration via packaging! Examples include cinnamon (524 ppm) (Lafeuille et al., 2009), some cheeses (up to 5000 ppb) (Tang et al., 2000), or raw avocado. Also, cigarettes contain up to 147 µg styrene/cigarette (Cohen et al., 2002), and smoking is a major route of personal exposure to styrene.

However, an expert panel of scientists convened by the National Toxicology Program (NTP) recently recommended that styrene be listed in the 12th Report as being "reasonably anticipated to be a human carcinogen" (National Toxicology Program, 2008). Their process relied on a review of published studies in arriving at this classification.[7] There is evidence from occupational exposures (Budroni et al., 2010; Cocco et al., 2010; Huff and Infante, 2011; Infante and Huff, 2011) and animal studies (Huff, 1984) to support the potential carcinogenicity of the monomer. Also, styrene 7,8-oxide, the primary oxidation product of the monomer, is mutagenic (Filser et al., 1993). Occupational exposure to styrene increases the risk of lymphatic and hematopoietic carcinoma. Studies on occupational exposure did show increased cancer risk due to exposure (Delzell et al., 2006; Kolstad et al., 2012). In contrast, a 2013 study (Collins et al., 2013) on 16,000 cases of occupational exposure to styrene found no credible correlation between exposure and cancer. While the evidence is not always consistent, there appears to be enough credible data to seriously pose the question of carcinogenicity of styrene and related compounds. Despite the level of uncertainty in the reported data, the precautionary stance adopted by the NTP is warranted and styrene does deserve closer attention at least until the issue of carcinogenicity is resolved.

[7] In 2011, the US Department of Health and Human Services categorized styrene as a "reasonably anticipated to be a human carcinogen." Chemical industry legally challenged the classification without success.

8.3.1   Exposure to Styrene from Packaging

The adverse impacts of human exposure depend on the total daily intake (TDI). Styrene exposure in US populations was estimated to be less than 0.3 (mcg/kg body weight/day) in nonsmokers and <3.51 and <2.86 (mcg/kg body weight/day) in smokers ages 12–19 years and 20–70 years, respectively (NTP, 2006). Tang et al. (2000) arrived at TDI values of 18.2–55.2 μg/person/day (6.7–20.2 mg/person/year) of styrene. However, most (~98%) of this intake was attributed to inhalation rather than to ingestion with food (amounting to only 0.2–1.2 μg/kg bw/day). But a Canadian study (Newhook and Caldwell, 1993), by contrast, found food to be the leading source of styrene (0.11 μg/kg bw/day) making up the TDI. An industry-sponsored study estimated the food-related exposure to be 6.6 μg/person/day. All these values, however, are well below the generally accepted tolerable or "safe" levels proposed for styrene. The WHO reported (WHO, 2003) a tolerable daily intake (TDI) of 7.7 (μg/kg bw/ day) or 0.46 mg/person/day for a 60 kg individual. The REACH regulations in EU proposed (IUCLID, 2013) a derived no effect levels (DNELs) for long-term exposure for styrene of 2.1 (mg/kg bw/day). Safe levels established for chemicals are often based on limited experimentation (as with ED chemicals discussed in Chapter 7), and these may be revised as more information emerges as to different modes of toxicity. Given the possible carcinogenicity associated with styrene, a precautionary stance is reasonable.

8.3.2   Leachate from PET Bottles

The polyester PET was reported to have 4,024–11,576 mg/kg of extractible material (Kim and Lee, 2012b) consisting mainly of cyclic oligomers along with 1.3–2.8% of linear oligomer. Terephthalic acid (TPA), hydroxyethyl terephthalates, and *bis*-hydroxyethyl terephthalate were also found at levels of 3.0–28.2 mg/kg, 16.8–118.2 mg/ kg, and 3.9–26.7 mg/kg, respectively. However, these were extracted using organic solvent mixtures and are generally not extractible by aqueous food or beverage.

In studies with nonfatty food simulants in contact with PET, the leachate at 40°C over 10 days was only 0–0.94 μg/dm^2, and with fatty food simulant,[8] 0.36–1.05 μg/dm^2 of cyclic oligomer leached at 20°C over 48 h of contact (Kim and Lee, 2012a). Migration of the three monomers (TPA, IPA, and DMT)[9] into food simulants (water, 4% acetic acid, 20% alcohol, and *n*-heptane) from 56 PET containers collected from open markets showed no extracted monomer at a detection level of 0.1 ppm (Park et al., 2008). The TDI associated with this level of migrants is well below the accepted safety levels. Values for water extracts of migrants reported in different studies vary widely because of differences in experimental conditions, specifically the duration of contact or storage, temperature, and pH of contents as well as the standard test protocol used (Kim and Lee, 2012a). The published data does not support a precautionary stance for the use of PET in beverage bottle or in other food-contact applications.

---

[8] A standard set of simulants are commonly used to assess migration of compounds from packaging. These are 3% acetic acid, 10% alcohol/water, 50% ethanol/water and olive or sunflower oil.
[9] Terephthalic acid (TPA), isophthalic acid (IPA), and dimethyl terephthalate (DMT).

8.4   RANKING COMMON PLASTICS

A reasonable approach to facilitating environmental sustainability is to recognize particularly unsustainable plastic material in specific applications and substitute these with more sustainable ones. Ranking the main classes of plastics on the basis of their energy costs and fossil fuel use as well as externalities generated during their manufacture (with special attention paid to toxicity) is a useful starting point in assessing their environmental sustainability. While the rankings are based on LCAs, their limited scope and emphasis of selected environmental impacts such as recyclability or impacts on human health must be appreciated. The available rankings are material-specific and are not directly linked to specific application areas and therefore cannot be easily used in making informed choices in material selection. Attempts at ranking the common classes of thermoplastics (Greenpeace, 1998; Institute for Agriculture and Trade Policy, 2008; Miller et al., 2006; Van der Naald and Thorpe, 1998; Yarwood and Eagan, 1998) including some by manufacturers of consumer goods (Henderson et al., 2009; Opel Automaker, 2002) have been reported. Despite the differences in the scope of LCAs these were based on and that they did not take specific applications into account, the rankings obtained are very similar.

Early attempts yielded the highly publicized plastic pyramid (Greenpeace International) shown in Figure 8.4. In 2005, a similar ranking was developed and presented as a plastic spectrum (Rossi et al., 2005), where the ranking based on environmental hazard was supplemented also by the recyclability of the plastic. Along the same lines is the development of a "plastic score card" where the resins are given a score ranging from A+ to F based on three core considerations of sustainable feedstocks, green chemistry, and closed-loop system design. The general rankings of resins in all studies were not that different from that in the original plastic pyramid in Figure 8.4.

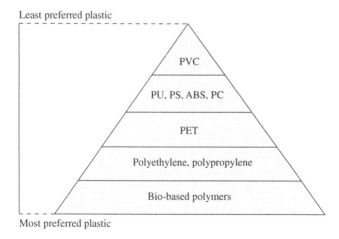

FIGURE 8.4   Plastic pyramid originally proposed in 1998 by Van der Naald and Thorpe.

A recent report (Tabone et al., 2010) presents rankings based on two sets of criteria, the principles of green chemistry or engineering and the standard LCA methodology for assessing environmental impacts, for common plastics. The two rankings derived were different, and those from green design principles are given in the following:

$$\text{PLA/PHA} < \text{HDPE} < \text{PET} < \text{LDPE} < \text{Bio-PET} < \text{PP} < \text{PS} < \text{PVC} < \text{PC}$$

However, the LCA rankings of these were based on had a limited scope, and were assessed on a volume basis (as opposed to a functional unit such as a number of containers of specified capacity). The footprint of a product must be taken together with its functionality or the contribution the product makes to human lifestyle. This requires the footprint be assessed on a functional unit basis. The Institute for Agriculture and Trade Policy (IATP) (2005) study on food-use plastics also ranked PVC as a hazardous plastic and recommends avoiding its use (plasticized with DEHA or other plasticizer). It advocates avoiding PS (which can leach residual styrene) and PC (which can leach BPA) plastics as well, with any food-contact products.

A 2011 study by Lithner et al. (2011) assessed the environmental and health hazards posed by plastics, based on the toxicity of their monomers. The classification is not inherent to the polymer as it is based primarily on residual monomer (with selected additives, plasticizer, and flame retardants). The ranking (see Table 8.8) can have relevance only for occupational exposures and in some food-contact uses of plastics. Also, future advances in residual monomer reduction technology and green substitution of additives can change the status of a polymer in this assessment. Where "recyclability" is used as a ranking criterion, it generally refers to technical recyclability that has little to do with if the resin will in fact be recycled in practice.

While there is some justification for using such rankings as a general guide to material selection, there are shortcomings in the approach. The most obvious is that

TABLE 8.8   Hazard Levels of Common Plastics Estimated from Monomer Characteristics

Rank	Plastic material	Rank	Plastic material
V	Flexible PU foam	IV	Phenol-formaldehyde resin
	Plasticized PVC (worst)		Unsaturated polyester (with MMA)
	Rigid PVC		Polycarbonate (PC)
	Rigid PU foam		Unsaturated polyester (with styrene)
	Plasticized PVC		Poly(methyl methacrylate)
	High-impact PS (HIPS)		Thermoplastic polyurethane
III	Nylons	II	LDPE, LLDPE, HDPE
	Expanded PS foam		PET
I	PP		

Source: Data based on Lithner et al. (2011).
Endocrine disruptor effect was not taken into account here; otherwise, PC would possibly be considered more hazardous (Rank V).

comparisons are made on an equal weight basis. For example, material substitution from rigid PVC to a polyolefin, for pressure pipe application, might be considered desirable, as a Rank V plastic is being replaced by a Rank I plastic. But, the wall thickness and therefore the weight of resin used in the two pipes to achieve the same performance (pressure rating) will be very different. Also, the health impacts (especially those from migrants into food in contact with plastic) cannot be assessed on the basis of broad classes of polymers. Substituting a plastic lower in rank with one that is higher does not always result in a choice that is better for the environment. When all factors are taken into account PVC, the worst rated resin, may in fact be desirable in specific medical applications.

A second limitation is that additives used in its formulation often determine the environmental desirability of a plastic material used in a given application. While the pure polymers are nontoxic, the toxicity of leachates from compounded plastics must also be considered in developing rankings. Additives and residual monomers are responsible for toxicity outcomes that vary widely within a single class of plastic. Their health impacts often overshadow those due to residual monomers. Phthalates are used in some PVC products such as medical tubing but not in others (such as potable water pipes), just as brominated flame retardants are used only in some formulations of PS foam (EPS) but not in injection-molded PS cutlery. A practical assessment of the acute toxicity of water extracts of various plastic products (1–3 days at 50°C) to *Daphnia magna* sp. shows that for 42% of the 48-hour extract tested, the EC50 values (the concentrations causing toxicity effects in 50% of the test organisms) were below 250 g plastic/l (Lithner et al., 2012). All extracts from plasticized PVC (but not rigid PVC) were toxic at concentration of only 17–24 g plastic/l illustrating the toxicity of plasticizers. The exposure of plastics at high temperatures for extended periods of time is not common in everyday use, and the test is an accelerated leaching test. However, the mortality of *Daphnia* stored at ambient temperatures in water contained in food-grade LDPE and HDPE was found to be significantly higher than in glass container controls (Rkman-Filipovic et al., 2012).

Overall, the use of ranking systems regardless of the life cycle analysis it is based on, can be misleading. Sustainability assessments cannot be made on broad classes of plastics or even categories of their compounds. In each case, the environmental cost of using plastics must be balanced with the societal benefits afforded by the products. What is particularly useful is a ranking of different plastics or materials of construction for a specific application area such as potable water pipes, greenhouse glazing, insulation of homes, or packaging of classes of food. Ultimately, the sustainable choice of a plastic for a given application is one that delivers the required functional attributes at the minimum ecological footprint. At times, this may unavoidably indicate a generic plastic with a poor environmental rating. High environmental costs are often accommodated for indispensable societal services. If the societal benefit of the product is high enough, it will still be manufactured and used. Sustainable growth in these instances still demands research to find an adequate replacement and strategies to mitigate the hazard associated with the product, not its substitution with a product of reduced functionality.

8.4.1   PVC

Most comparative assessments of plastics find PVC to be a particularly low-ranking, therefore an undesirable, plastic material (Lindahl et al., 2014) that needs to be phased out, especially in packaging and building uses. Its production, use, and disposal are in fact beset with a range of adverse environmental impacts (Thornton, 2002). There is a large enough body of literature that suggests the use of PVC is associated with a particularly high environmental cost. In some applications, PVC does not offer unique functionality that would justify that high an environmental cost.

The major uses of PVC in buildings include pipes, window frames, siding (in the United States), roofing, and flooring. The US Green Building Council, however, recommends against the use of PVC in sustainable building design. The key environmental issues associated with PVC use are as follows.

(a) Over half the weight of PVC is made up of chlorine! Chlorine production is energy intensive and consumes approximately 1% of the world's electrical energy. Large-scale synthesis of the vinyl chloride monomer (VCM) results in inevitable emission of traces of chlorine, the carcinogenic monomer, and possibly dioxins, into air. The chlorine-rich waste (~1 million tons annually) from the process is incinerated, potentially yielding hazardous (e.g., dioxin) emissions.[10] (Dioxins are highly toxic chemicals and the acceptable daily exposure set by EPA is only 0.7 pg/kg of body weight.[11]) This creates an occupational safety concern, potential for accidental spills where transport of VCM and Cl_2 is involved, as well as an air pollution issue. Mercury chloride was used as catalysts in vinyl chloride manufactured by hydrochlorination of acetylene, an obsolete process used rarely in modern times. Mercury can be accidentally or otherwise released into the environment (as in the case of pollution in Minamata Bay, Japan, leading to widespread regional neurotoxic effects).

(b) Soft PVC products use phthalate plasticizers at high weight fractions (sometimes as high as 50–60% by weight). Over 6 MMT of phthalates (in 2011) is used annually in PVC products. Phthalates are released into the environment during production use and disposal of plasticized PVC products. Phthalates are endocrine disrupters and a health hazard as extensively discussed in Chapter 7.

(c) Lead and cadmium compounds are often used as thermal stabilizers in PVC. The presence of these metal residues complicates disposal of PVC waste. The industry is moving to phase out the use of lead. About a third of the production uses a process that uses and releases mercury into the environment.

---

[10] In 2012, the USEPA issued new restrictions on highly toxic emissions from PVC manufacturing facilities. The annual emission reductions as a result of these from major sources are estimated to be 238 tons of total air toxics including 21 tons of hydrogen chloride.

[11] Dioxins are persistent and bioaccumulative. Once in the food web it can partition and bioconcentrate, resulting in exposure (especially of children) to even higher levels than the 0.7 pg/kg level.

(d) There is no infrastructure to effectively recycle PVC waste. Even by 2020, its mechanical recycling will only reach a maximum of 18% (Leadbitter, 2002) under the best conditions. It cannot be easily incinerated as HCl is formed during the combustion and has to be sequestered in lime. Also, there is the possibility of dioxin formation during incineration.

In a recent study (Technical Science Advisory Committee of the U.S. Green Building Council, Altshuler et al., 2007), PVC was compared to competing materials in windows (aluminum and wood), pipe (cast iron and ABS), siding (aluminum and fiber cement), and resilient floor (sheet vinyl, linoleum, and cork). Impacts assessed for each product category were as follows: carcinogenicity, total health hazard, and adverse environmental impacts (acidification, eutrophication, smog, ozone depletion, global climate change, fossil fuel depletion, and ecotoxicity). The PVC products either ranked the worst or tied with the worst choice, in human health assessments even when the comparison was based on low-end estimates. It ranked higher than cast iron pipes, aluminum siding, and aluminum windows on overall environmental impact but not significantly different from ABS pipes, wooden windows, wood, or fiber cement siding.

However, there might be low-volume niche applications where PVC does provide unique value. A good example is the use of plasticized PVC in blood bags used in the United States. While alternative plastics might be used (EVA, PU, PE), PVC remains the ideal material for this application and has been used for over 50 years (though the plasticizer has changed from the original tricresyl phosphate to a phthalate, DEHP). Though widely considered a "bad" plastic with trace DEHP plasticizer leaching into the blood (Lozano and Cid),[12] it continues to serve in this niche use where it provides exceptional functionality.

REFERENCES

Alexander M. The environmental fate of styrene. SIRC Rev 1990;1:33–42.

Alperstein D. Experimental and computational investigation of EVOH/clay nanocomposites. J Appl Polym Sci 2005;97:2060–2066.

Altshuler K, Horst S, Malin N, Norris G, Nishioka Y. (2007) Assessment of the technical basis for a PVC-related materials credit for LEED The US Green Building Council. Washington, DC: Advisory Committee on PVC, US Green building Council.

American Chemical Council. PFPG FDA task force report. The Safety of Styrene-Based Polymers for Food-Contact Use. Update 2013. August 29, 2013 Washington, DC: Pira (2013) Ref. 00A11J0015.

Andra SS, Makris KC, Shine JP, Lu C. Co-leaching of brominated compounds and antimony from bottled water. Environ Int 2012;38 (1):45–53.

Araújo PHH, Sayer C, Poco JGR, Giudici R. Techniques for reducing residual monomer content in polymers: a review. Polym Eng Sci 2002;42 (7):1442–1468.

---

[12] In the case of the red blood cell storage, the leached DEHP is believed to stabilized the cell membrane, allowing the extension of storage time up to 49 days (Lozano and Cid, 2013).

Ayoko GA, Blinco JP, Kirk KM, Uhde E. Exposure to airborne organic compounds inside passenger vehicles. In: Proceedings of the 11th International Conference on Indoor Air Quality and Climate, August 22–27. Paper ID–340, Copenhagen; 2008.

Bhunia K, Sablani SS, Tang J, Rasco B. Migration of chemical compounds from packaging polymers during microwave, conventional heat treatment, and storage. Compr Rev Food Sci Food Saf 2013;12:523–545.

Bomfim MVJ, Zamith HPS, Abrantes SMP. Migration of ε-caprolactam residues in packaging intended for contact with fatty foods. Food Contr 2011;22 (5):681–684.

Bonini M, Errani E, Zerbinati G, Ferri E, Girotti S. Extraction and gas chromatographic evaluation of plasticizers content in food packaging films. Microchem J 2008;90 (1):31–36.

Borrelli FE, de la Cruz PL, Paradis RA. Residual vinyl chloride levels in U.S. PVC resins and products: historical perspective and update. J Vinyl Addit Technol 2005;11:65–69.

Boxed Water (2014) Boxed water is better. Grand Rapids (MI) 49503. Available at http://www.boxedwaterisbetter.com. Accessed October 4, 2014.

Brody A. Packaging by the numbers. Food Technol 2008;62 (2):89–91.

Bucklow I, Butler P. Plastic beer bottles. Mater World 2000;8 (8):14–17.

Budroni M, Sechi O, Cesaraccio R, Pirino D, Fadda A, Grottin S, Flore MV, Sale P, Satta G, Cossu A, Tanda F, Cocco PL. Cancer incidence among petrochemical workers in the Porto Torres industrial area, 1990–2006. Med Lav 2010;101:189–198.

Charara ZN, Williams JW, Scmidt RH, Marshall MR. Orange flavor absorption into various polymeric packaging materials. J Food Sci 1992;57:963–968.

Choi JO, Jitsunari F, Asakawa F, Lee DS. Migration of styrene monomer, dimers, and trimers from polystyrene to food simulants. Food Addit Contam 2005;22 (7):693–699.

Cocco P, t'Mannetje A, Fadda D, Melis M, Becker N, de Sanjosé S, Foretova L, Mareckova J, Staines A, Kleefeld S, Maynadié M, Nieters A, Brennan P, Boffetta P. Occupational exposure to solvents and risk of lymphoma subtypes: results from the Epilymph case-control study. Occup Environ Med 2010;67:341–347.

Cohen JT, Carlson G, Charnley G, Coggon D, Delzell E, Graham JD, Greim H, Krewski D, Medinsky M, Monson R, Paustenbach D, Petersen B, Rappaport S, Rhomberg L, Ryan PB, Thompson K. A comprehensive evaluation of the potential health risks associated with occupational and environmental exposure to styrene. J Toxicol Environ Health B Crit Rev 2002;5 (1–2):1–265.

Collins JJ, Bodner KM, Bus JS. Cancer mortality of workers exposed to styrene in the U.S. reinforced plastics and composite industry. Epidemiology 2013;24 (2):195–203.

Date K, Ohno K, Azuma Y, Hirano S, Kobayashi K, Sakurai T, Nobuhara Y, Yamada T. Endocrine-disrupting effects of styrene oligomers that migrated from polystyrene containers into food. Food Chem Toxicol 2002;40:65–75.

Delzell E, Sathiakumar N, Graff J, Macaluso M, Maldonado G, Matthews R. (2006). Health effects institute: an updated study of mortality among North American synthetic rubber industry workers. Res Rep Health Eff Inst, 132, 1–63. discussion 65–74.

Doublet G. LCA of Rivella and Michel soft drinks packaging [Master Thesis]. Zürich: Institute of Environmental Engineering (IFU) at Swiss federal institute of technology; (2012). Available at http://www.esu-services.ch/fileadmin/download/doublet-2012-masterarbeit.pdf. Accessed October 4, 2014.

Ebnesajjad S. Plastic Films in Food Packaging: Materials Technology and Applications. Waltham: Elsevier; 2013.

Ercin AE, Aldaya MM, Hoekstra AY. Corporate water footprint accounting and impact assessment: the case of the water footprint of a sugar-containing carbonated beverage. Water Resour Manage 2011;25:721–741.

Fasano E, Bono-Blay F, Cirillo T, Montuori P, Lacorte S. Migration of phthalates, alkylphenols, bisphenol A and di(2-ethylhexyl)adipate from food packaging. Food Contr 2012;27 (1):132–138.

Filser JG, Schwegler U, Csanády GA, Greim H, Kreuzer P, Kessler W. Species-specific pharmacokinetics of styrene in rat and mouse. Arch Toxicol 1993;67:517–530.

FAO. (2010). Greenhouse gas emissions from the dairy sector—a life cycle assessment. Rome: Food and Agriculture Organization. Available at www.fao.org/agriculture/lead/themes0/climate/emissions/en/. Visited in March 2011.

Franklin Associates. *Resource and Environmental Profile Analysis of Polyethylene Milk Bottles and Polyethylene-coated Paperboard Milk Cartons.* Prairie Village: Franklin Associates; 2008.

Franklin Associates Study. *LCI Summary for Four Half-Gallon Milk Containers.* Prairie Village: Franklin Associates; 2008.

Galotto MJ, Guarda A. Suitability of alternative fatty food simulants to study the effect of thermal and microwave heating on overall migration of plastic packaging. Packag Technol Sci 2004;17:219–223.

Garrigós MC, Marín ML, Cantó A, Sánchez A. Determination of residual styrene monomer in polystyrene granules by gas chromatography–mass spectrometry. J Chrom A 2004;1061 (2):211–216.

Gerber P, Vellinga T, Steinfeld H. Issues and options in addressing the environmental consequences of livestock sector's growth. Meat Sci 2010;84:244–247.

Ghenai C. Life cycle assessment of packaging materials for milk and dairy products. Int J Therm Environ Eng 2012;4 (2):117–128.

Gleick PH, Cooley HS. Energy implications of bottled water. Environ Res Lett 2009;4:1–6.

Greenpeace. *PVC Plastic: A Looming Waste Crisis.* Amsterdam: Greenpeace International; 1998.

Henderson R, Locke RM, Lyddy C, Reavis C. *Nike Considered: Getting Traction on Sustainability: MIT Sloan Management.* Cambridge: MIT Sloan School of Management; 2009.

Hernandez-Muñoz P, Catala R, Gavara R. Food aroma partition between packaging materials and fatty food stimulants. Food Addit Contam 2001;18:673–682.

Hirano S, Azuma Y, Date K, Ohno K, Tanaka K, Matsushiro S, Sakurai T, Shiozawa S, Chiba M, Yamada T. Biological evaluation of styrene oligomers for endocrine-disrupting effects. J Food Hygienic Soc Jpn 1999;40:36–45.

Hotchkiss JH. Food-packaging interactions influencing quality and safety. Food Addit Contam 1997;14:601–607.

Hotchkiss JH, Banco MJ. Influence of new packaging technologies on the growth of microorganisms in produce. J Food Protect 1992;55 (10):815–820.

Huff J. Styrene, styrene oxide, polystyrene, and beta-nitrostyrene/styrene carcinogenicity in rodents. Prog Clin Biol Res 1984;141:227–238.

Huff J, Infante PF. Styrene exposure and risk of cancer. Mutagenesis 2011;26 (5):583–584.

Infante PF, Huff J. Cancer incidence among petrochemical workers in the Porto Torres industrial area. La Medicina del lavoro 2011;102 (4):382–383.

Institute for Agriculture and Trade Policy (IATP). *Smart Plastics Guide: Healthier Food Uses of Plastics*. Minneapolis: IATP; 2005.

Institute for Agriculture and Trade Policy (IATP). *Smart Plastics Guide: Healthier Food Uses of Plastics*. Minneapolis: IATP; 2008.

IUCLID. 2013. Chemical safety report, styrene. Available at http://echa.europa.eu/. Accessed September 23, 2013.

Jungbluth N. 2006. Comparison of the environmental impact of tap water vs. bottled mineral water, Swiss Gas and Water Association (SVGW). Available at http://www.esu-services.ch/download/jungbluth-2006-LCA-water.pdf. Accessed October 4, 2014.

Kawahara T, Nomura Y, Tanaka N, Teshima W, Okazaki M, Shintani H. Leachability of plasticizer and residual monomer from commercial temporary restorative resins. J Dent 2004;32:277–283.

Kim DJ, Lee KT. Analysis of specific migration of monomers and oligomers from polyethylene terephthalate bottles and trays according to the testing methods as prescribed in the legislation of the EU and Asian countries. Polymer Test 2012a;31 (8):1001–1007.

Kim DJ, Lee KT. Determination of monomers and oligomers in polyethylene terephthalate trays and bottles for food use by using high performance liquid chromatography-electrospray ionization-mass spectrometry. Polymer Test 2012b;31 (3):490–499.

Kolstad HA, Ebbehøj N, Bonde JP, Lynge E, Albin M. Health effects following occupational styrene exposure in the reinforced plastics industry. Ugeskr Laeger 2012;174 (5): 267–270.

Lafeuille JL, Buniak ML, Vioujas MC, Lefevre S. Natural formation of styrene by cinnamon mold flora. J Food Sci 2009;74 (6):M276–M283.

Landau S. The future of flavor and odor release. In the future of caps and closures—latest innovations and new applications for caps and closures. Intertech-Pira Conference; June 20–21, 2007; Atlanta: Pira; 2007.

Lange J, Wyser Y. Recent innovations in barrier technologies for plastic packaging—a review. Packag Technol Sci 2003;16:149–158.

Leadbitter J. PVC and sustainability. Prog Polym Sci 2002;27:2197–2226.

Licciardello F, Cipri L, Muratore G. Influence of packaging on the quality maintenance of industrial bread by comparative shelf life testing. Food Packag Shelf Life 2014; 1 (1):19–24.

Lickly TD, Breder CV, Rainey ML. A model for estimating the daily dietary intake of a Substance from food-contact articles: styrene from polystyrene food-contact polymers. Regul Toxicol Pharm 1995;21:406–417.

Lindahl P, Robèrt KH, Ny H, Broman G. Strategic sustainability considerations in materials management. J Clean Prod 2014;64:98–103.

Lithner D, Larsson Å, Dave G. Environmental and health hazard ranking and assessment of plastic polymers based on chemical composition. Sci Total Environ 2011;409 (18): 3309–3324.

Lithner D, Nordensvan I, Dave G. Comparative acute toxicity of leachates from plastic products made of polypropylene, polyethylene, PVC, acrylonitrile–butadiene–styrene, and epoxy to Daphnia magna. Environ Sci Pollut Res 2012;19 (5):1763–1772.

Lozano M, Cid J. DEHP plasticizer and blood bags: challenges ahead. ISBT Sci Ser 2013;8:127–130.

Lung CYK, Darvell BW. Minimization of the inevitable residual monomer in denture base acrylic. Dent Mater 2005;21 (12):1119–1128.

Marcussen H, Holm PE, Hansen HCB. Composition, flavor, chemical food safety, and consumer preferences of bottled water. Compr Rev Food Sci Food Saf 2013;12:333–352.

Matiella JE, Hsieh TC. Volatile compounds in scrambled eggs. J Food Sci 2006;56 (2): 387–390.

Miller H, Rossi M, Charon S, Wing G, Ewell J. Design for the next generation: incorporating cradle-to-cradle design into Herman Miller products. J Ind Ecol 2006;10:193–210.

Moshanos MG, Shaw PE. Changes in composition of volatile constituents in aseptically packaged orange juice. J Agr Food Chem 1989;37:157–161.

National Toxicology Program (NTP). NTP-CERHR Monograph on the Potential Human Reproductive and Developmental Effects of Styrene. NIH Publication 06-4475. Feb. 2006; 2006.

National Toxicology Program (NTP). Report on Carcinogens (RoC). Federal Register, 73, 174; 2008.

Nerin C, Alfaro P, Aznar M, Domeño C. The challenge of identifying non-intentionally added substances from food packaging materials: a review. Anal Chim Acta 2013;775:14–24.

Newhook R, Caldwell I. Exposure to styrene in the general Canadian population. In: Sorsa M, Peltonen K, Vainio H, Hemminki K, editors. *Butadiene and Styrene: Assessment of Health Hazards*. Lyon: IARC; 1993. *IARC Scientific Publications No 127*; p 27–33.

Nielsen TJ, Jägerstad IM. Flavor scalping by food packaging. Trend Food Sci Technol 1994;5:353–356.

Ohno H, Kawamura Y. Analysis of vinylidene chloride and 1-chlorobutane in foods packaged with polyvinylidene chloride casing films by headspace gas chromatography/mass spectrometry (GC/MS). Food Addit Contam 2006;23 (8):839–844.

Opel Automaker. *Study on Recyclability. Environmental Report 2000/2001*. Rüsselsheim: Adam Opel AG; 2002.

Park H-J, Lee YJ, Kim M-R, Kim KM. Safety of polyethylene terephthalate food containers evaluated by HPLC, migration test, and estimated daily intake. J Food Sci 2008;73:T83–T89.

Pasqualino J, Meneses M, Castells F. The carbon footprint and energy consumption of beverage packaging selection and disposal. J Food Eng 2011;103 (4):357–365.

Picard E, Vermogen A, Gérard J-F, Espuche E. Barrier properties of nylon 6-montmorillonite nanocomposite membranes prepared by melt blending: Influence of the clay content and dispersion state: consequences on modelling. J Membr Sci 2007;292 (1–2):133–144.

Piringer OG, Ruèter M. Sensory problems caused by food and packaging interactions. In: Piringer OG, Baner AL, editors. *Plastic Packaging Materials for Food: Barrier Function, Mass Transport, Quality Assurance, and Legislation*. Sussex: John Wiley & Sons; 2000.

Rexam Consumer Packaging Report: Packaging Unwrapped. 2011. Rexam PLC (London SW1P 3XR). www.rexam.com. Downloaded on October 22, 2014 from http://www.rexam.com/files/pdf/packaging_unwrapped_2011.pdf. Accessed October 28, 2014.

Rijk R, Kruijf ND. Migration testing with olive oil in a microwave oven. Food Addit Contam 1993;10 (6):631–645.

Rkman-Filipovic S, Mourzaeva-Solomonov E, Herrmann O, Rodgers D. It's in the bag! Toxicity identification and evaluation of plastic bag toxicity: implications for aquatic toxicity testing and changes to testing procedures. Integr Environ Assess Manage 2012;8:197–199.

Rossi, M., Griffith, Ch., Gearhart, J., and Juska, C. (2005). Moving towards sustainable plastics. A report card on the six leading automakers. A Report by the Ecology Center, U.S.A. Available at http://www.ecocenter.org/publications/downloads/auto_plastics_report.pdf. Accessed October 4, 2014.

Sajilata MG, Savitha K, Singhal RS, Kanetkar VR. Scalping of flavors in packaged foods. Compr Rev Food Sci Food Saf 2007;6:17–35.

Sanches Silva A, Cruz JM, Sendón García R, Franz R, Paseiro Losada P. Kinetic migration studies from packaging films into meat products. Meat Sci 2007;77 (2):238–245.

Silvenius F, Katajajuuri J, Grönman K, Koivupuro RHS, Virtanen Y. Role of packaging in LCA of food products. In: Finkbeiner M, editor. *Towards Life Cycle Sustainability Management*. Dordrecht: Springer Science+Business Media B.V.; 2011.

Singh P, Saengerlaub S, Wani AB, Langowski H. Role of plastics additives for food packaging. Pigm Resin Technol 2012;41 (6):368–379.

Tabone MD, Cregg JJ, Beckman EJ, Landis AE. Sustainability metrics: life cycle assessment and green design in polymers. Environ Sci Technol 2010;44 (21):8264–8269.

Tang W, Hemm I, Eisenbrand G. Estimation of human exposure to styrene and ethylbenzene. Toxicology 2000;144 (1–3):39–50.

Tartakowski Z. Recycling of packaging multilayer films: new materials for technical products. Resour Conserv Recycl 2010;55 (2):167–170.

Tawfik MS, Huyghebaert A. Polystyrene cups and containers: styrene migration. Food Addit Contam. 1998;15 (5):592–599.

Thornton J. *Environmental Impacts of PVC Building Materials*. Washington, DC: Healthy Building Network; 2002.

Vallo CI, Montemartini PE, Cuadrado TR. Effect of residual monomer content on some properties of a poly(methyl methacrylate)-based bone cement. J Appl Polym Sci 1998;69: 1367–1383.

Van der Naald W. and Thorpe B.G. (1998). PVC Plastic: A looming waste crisis. Greenpeace International. ISBN 90-73361-44-3.

van Sluisveld MAE, Worrell E. The paradox of packaging optimization—a characterization of packaging source reduction in the Netherlands. Resour Conserv Recycl 2013;73:133–142.

Van Willige RWG, Linssen JPH, Voragen AGJ. Influence of food matrix on absorption of flavour compounds by linear low-density polyethylene: proteins and carbohydrates. J Appl Polym Sci 2000;80:1779–1789.

Williams H, Wikström F. Environmental impact of packaging and food losses in a life cycle perspective: a comparative analysis of five food items. J Clean Prod 2011;19:43–48.

WHO. (2003). Styrene in Drinking Water; Background Document for Development of WHO Guidelines for Drinking-water Quality. World Health Organization. Report WHO/SDE/ WSH/03.04/27. Geneva: World Health Organization.

Yarwood J, Eagan P. *Design for Environment Toolkit: A Competitive Edge for the Future*. St. Paul, MN: Minnesota Office of Environmental Assistance and Minnesota Technical Assistance Program 1998.

9

MANAGING PLASTIC WASTE

Postconsumer plastic waste was an inevitable result of high volume use of plastics in consumer applications such as packaging. Over the last two decades, the plastics waste streams have grown in volume, reflecting the volume growth in use of plastics as a material. In fact the fraction of plastics in the waste stream has increased at a rate faster than that of the municipal solid waste (MSW) stream itself. The popularity of disposable plastic goods such as packaging and foodservice items contributes significantly to plastic in MSW. The "open loop" linear consumerism results in the net conversion of fossil fuel and other resources into products and then quickly into waste on a continuing basis. Residential waste collected at curbside and commercial waste collected and removed for processing, referred to as MSW, constitute managed streams of waste. Some of the waste is recovered and recycled. Over half of the US communities have access to recycling via curbside collection and drop off facilities. The waste is allocated to materials recovery facilities (MRFs), landfills, incineration plants, or cogeneration facilities. In 2008 MSW accounted for only 40–50% of the plastic waste in EU countries (Plastics Europe et al., 2009; Villanueva et al., 2010). Agricultural waste, auto reclamation, electronics waste, and ship breaking waste are also managed waste streams. In contrast to MSW, the unmanaged waste stream, specifically urban plastic litter with its negative aesthetic impact, health hazards, and ecological impacts, remains the most contentious aspect of plastic waste management.

This chapter primarily deals with plastics in the MSW stream and the choices available for their recovery and treatment. The discussion here focuses on the

Plastics and Environmental Sustainability, First Edition. Anthony L. Andrady.
© 2015 John Wiley & Sons, Inc. Published 2015 by John Wiley & Sons, Inc.

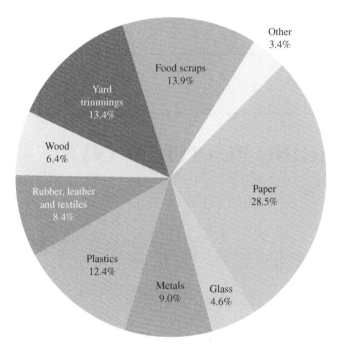

FIGURE 9.1   The composition of the USMSW stream of 250 million tons generated in the year 2010.

sustainability of available waste management choices as they pertain to plastic waste. Environmental sustainability and the principles of circular economy demand that post-use plastics be regarded a potential resource or a raw-material rather than a waste disposal problem inherent to consumer products at the end of their lifecycle. Preferred choices of recovery or reuse must have the optimal triple footprint; energy efficiency, resource conservation, and minimal environmental pollution.

The composition of the MSW stream in the United States is assessed each year by the USEPA (compiled by Franklin Associates) based on a materials flow model. The *BioCycle Journal* (compiled by Columbia University, SC) also reports biennial assessments of the same but is based on data from State reports. Because of differences in methodology and the disparity in items captured in each, the two estimates do not agree and it is difficult to select one as being the better (Tonjes and Greene, 2012). The more popularly quoted data in the literature is by the USEPA and the data for the composition of MSW (USEPA, 2010) are shown in Figure 9.1.

Plastics account for only about 12% by weight of the MSW in the United States (globally it is ~10%), amounting to about 30 MMT annually. This is an average value for the US; data at the state level show values as high as 17% in locations such as New York. With both the population and per capita consumption on the increase,

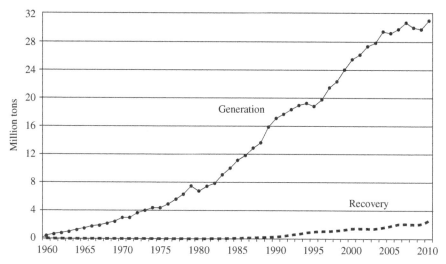

FIGURE 9.2   Generation and recovery of the plastics in municipal solid waste stream in the United States. Source: USEPA.

the amount of plastics waste ending up in the MSW stream will continue to increase in the future.

In the United States, only 8.2% of the plastic in MSW was recovered for reuse in some form in 2010 (Fig. 9.2). On recovery, the waste plastics in the MSW is sorted, cleaned, and reprocessed into useful products. Separating plastics from other waste is carried out in MRFs to obtain a single stream of mixed plastic products that often contains varieties of plastics unsuitable for recycling. It has to be further sorted to separate out popularly recycled varieties such as poly(ethylene terephthalate) (PET) and high density polyethylene (HDPE) from the mix. While hand sorting is possible, most modern MRFs use faster mechanical sorting. Plastic-only streams presorted at source by consumer is also processed by MRFs where machine sorting of plastics into different classes (based on near infra-red spectral signature) is particularly convenient and efficient. With source-separated all-plastics streams, the residuals (unrecyclables) are relatively low and cleaning of the material is easier. However, in the US only resin types such as PET or HDPE where there is a ready market for the recovered resin is recycled; the rest of the plastic waste is landfilled.

Plastic waste is classified into several subcategories in the USEPA compilation and the fraction of specific resins in each, identified. It is instructive to study the data (despite their limited reliability) to identify which types of plastics are prevalent in the MSW stream. Table 9.1 shows that four resins (low density polyethylene (LDPE), polypropylene (PP), HDPE, and PET) dominate the plastics fraction in the MSW. Of these, the PET and HDPE are recycled to a significant extent (at 28% and 6% respectively). Majority of the plastic wastes however is landfilled. The United States lags behind European countries such as Sweden, Austria, Belgium, and Germany as well as Japan in plastics waste recovery.

TABLE 9.1   The Plastic Types Mostly Encountered in the MSW Stream

Categories	Amount × 10⁶ tons	Leading polymer[a]	Recovery (%)
Durable goods	10.9	PP, LDPE, HDPE	6.4
Nondurables	6.4	LDPE, PP, PS	Negligible
Bottles containers	5.3	PET, HDPE	PET ~ 29
			HDPE ~ 27
Bags, sacks, wraps	3.9	LDPE, PP, HDPE	11.5
Other packaging	4.45	LDPE, PP, PET	
Total	31.0	LDPE, PP, HDPE, PET	8.2

Source: From USEPA (2010).
LDPE estimate includes LLDPE.
[a] Indicated in the order of decreasing fraction in the category.

9.1   RECOVERY OF WASTE

Post-consumer plastics are classified as waste not because they have no further intrinsic value, but often because these have lost a single critical aspect of their functionality during use. Food service plastic items illustrate this particularly well. A used cup or bottle is no longer clean or sterile but the physical and mechanical properties, the thermoplastic nature, the bio-inertness and the barrier effectiveness of the "waste" is no different from those of unused item. Clearly, it is still a valuable resource synthesized at a high cost in terms of depleted resources, fossil-fuel energy, and the ecological footprint associated with its externalities. The unrealistic market cost of the item often does not include its environmental cost discouraging the user from re-using an item a number of times. The true cost of the product taking into account all the costs is high enough to easily justify a vigorous effort at reuse and recovery. There is no justification for any plastic product made at a high environmental cost to be deemed a service life of only a few minutes. The primary argument for considering plastic waste as a resource is its high embodied energy and the thermal energy that can be derived from it as a solid fuel. Heating values of plastics in comparison to conventional fuels are shown in the bar diagram (Fig. 9.3). In fact one approach to recovering the energy is to use source-separated plastic waste as fuel in power plants or in cement manufacture.

In essence, sustainable growth demands that the linear unidirectional flow of scarce nonrenewable materials from sources to waste be converted into a cyclic flow to the extent practical. Recovery of waste has to be promoted and implemented as an integral part of this change. The sustainability goal with respect to plastic waste is twofold:

(a) Material economy: extending the useful service life of the material by reuse and recycling
(b) Full utilization of the non-renewable resource: recovery of either the energy and/or feedstock resources from waste destined for storage (in landfills) or destruction (burning).

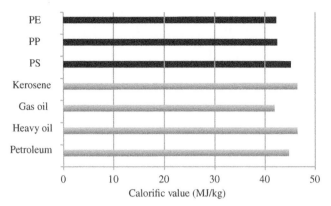

FIGURE 9.3 A comparison of the heating value of plastics and conventional fuels.

Reducing the use of plastics where it makes economic, functional, and environmental sense to do so also conserves the resource and is captured within the first goal. Where reduction is by using alternative materials, it is important to base the decision on an appropriate life cycle analysis (LCA) to make sure that the alternative (usually paper, wood, metal, or glass) in fact has a lower environmental footprint compared to plastics. There is no justification for adopting the extreme position that plastics as a class are undesirable and should be therefore avoided wherever possible in all applications. Multiple use (reuse) of a plastic product prior to its disposal is clearly a superior means of extending its useful life compared to recycling it after a single use. This helps conservation as recycling invariably has some energy and externalities "costs" associated with it.

Recovery of plastics from MSW can be as a material resource (by material recycling), a feedstock (by thermal or chemical degradation), or as energy (by incineration or biological treatment). Different technical methods are available for each of these recovery options.

Estimates for recovery given in Figure 9.4 are average values for the United States; the numbers can be very different at the state level where the choice of disposal is dictated by economic considerations. While states such as MS, OK, or MT with relatively undeveloped recycling programs for materials landfill almost all wastes, others such as CA and OR recycle greater than 10% of the plastics. CT, MA, ME, NJ, and HI rely more on incineration as a means of waste disposal and therefore recover significantly more energy from plastic wastes. Table 9.2 shows the recovery effort for different classes of plastics, based on USEPA data for 2011. The recovery rates for the United States summarized in the table are quite unimpressive compared to those for Western Europe or Japan. Consumer education, incentives, and improved recycling infrastructure can increase these rates significantly.

Another recent waste management trend is to export waste, often sorted plastic waste, for reprocessing. The EU countries for instance export waste plastics (mainly polyethylene (PE) and PP) to member countries (mostly Germany and France) as well as to Asia (mostly China) for reprocessing.

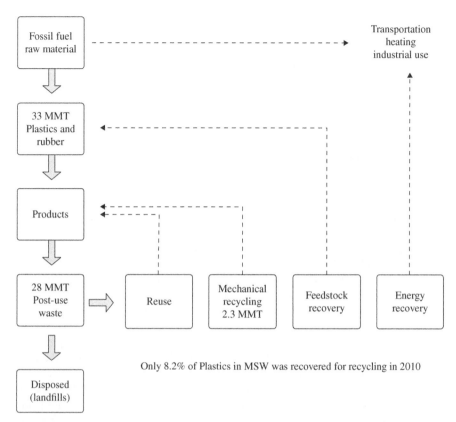

FIGURE 9.4 Waste management options in United States (2010). The numbers in select boxes are for percentage of plastic waste in the MSW.

TABLE 9.2   Breakdown of Different Classes of Plastics in MSW and their Recovery

Plastics	Waste generation short tons	Percent waste generation (%)	Percent recovery of waste (%)	Recovery rate (%)
HDPE	5,450,000	17.6	22.4	10.5
LDPE/LLDPE	7,430,000	23.9	16.5	5.7
PET	3,980,000	12.8	30.6	19.6
PP	7,530,000	24.3	1.6	0.5
PS	2,060,000	6.6	0.8	1.0
PVC	910,000	2.9	0	0

Source: From EPA (2011).

The different recovery options available for plastics waste might be summarized as follows:

9.1.1 Material Recycling

The products are recovered from the MSW stream via curbside collection, separated/sorted at MRFs, and then cleaned and ground into chips at plastic reclaiming facilities to be remelted into recycled resin pellets. The recycled resin is used, mixed with virgin plastics, in the fabrication of plastic products. Mechanical recycling works best when applied to source-separated streams of waste plastics products.

9.1.2 Feedstock Recovery

Waste plastic is changed by heat or chemical agents into chemicals that might be used in the production of new polymers or as general chemical feedstocks (Al-Salem et al., 2010). Monomer recovery works well with source-separated plastics whereas general feedstock recovery is better suited for mixed plastic waste.

9.1.3 Energy Recovery

Energy recovery can be either via incineration or via biological treatment:

(a) Burning of waste material and recovery of a part of the heat energy: applicable to mixed plastic waste
(b) Biological recovery technologies with energy recovery: includes engineered landfills with leachate and/or gas extraction and anaerobic digestion

In the United States, landfill disposal still accounts for the majority of MSW and therefore for plastic waste as well. Mechanical recycling and incineration (waste to energy processes or WTE) are also in use, but to a much lesser extent. In Europe, whatever is not recycled is mostly incinerated for energy recovery. Feedstock recovery, though an attractive option, is not practiced as a large-scale MSW treatment option. Figure 9.5 summarizes the available technical options for recovery of plastics waste.

9.2 PYROLYSIS OF PLASTIC WASTE FOR FEEDSTOCK RECOVERY

9.2.1 Direct Thermolysis

Pyrolysis, also referred to as "cracking," is a relatively low-cost treatment option where waste is converted into gas and liquid fuel. Presorting of MSW to select the plastic fraction and shredding it down to a suitable size (both steps requiring energy) is necessary to ensure good yield of fuel. Plastic waste is heated with or without a catalyst (Al-Salem et al., 2010; de Marco et al., 2009; Kaminsky et al., 1995) in the absence of air and is thermally degraded into small molecules. Using a catalyst (such as silica-alumina, zeolite, zirconia, or clay) allows for pyrolysis at lower

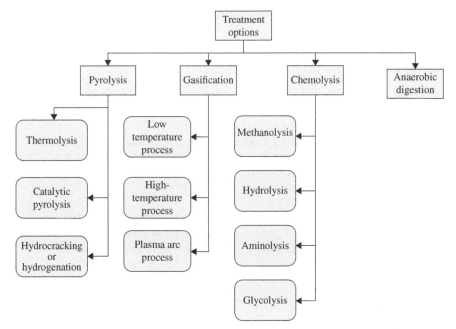

FIGURE 9.5   Different available recovery options for plastics waste.

temperatures and increases the conversion rates into useful products (Ding et al., 1997; Panda et al., 2010). The composition of fuel products as well as the ratio of gas to liquid yield in pyrolysis is determined by the type of catalyst used.

Non-catalytic pyrolysis is carried out at 650–900°C and is designed to maximize the yield of fuel oils (Tukker et al., 1999). The process yields low-grade gasoline or oil and is relatively inefficient. One of the most important of these, the BP process for instance, based on a fluidized bed reactor, is operated at 500°C in the absence of air. About 80% of the plastic is converted into a liquid mix of hydrocarbons (oil) and the remaining (8–10%) into a gas rich in monomers, under these pyrolysis conditions. A broad range of hydrocarbons ranging from C_5 to C_{28} results from high-temperature thermolysis.

Pyrolysis at moderate temperatures (300–500°C) is achieved in catalyzed reactions (Ding et al., 1997; Park et al., 1999) and yields a wax suitable for further catalytic cracking (Lee, 2012) in a second step, to produce useful petrochemicals. These can be converted into a high-grade gasoline substitute with a high content of isoparaffins (Songip et al., 1993). Pyrolysis-based recovery can yield fuels such as the naphtha, furnace oil, and heavy fuel that might also be used without further refining. Oils from low-temperature pyrolysis generally contain less of olefins and more branched hydrocarbon and aromatics.

The most studied and already demonstrated pyrolytic technique is fluidized-bed pyrolysis of waste (Conesa et al., 1997; Williams and Williams, 1999). The process is also carried out in two-steps where the plastics are initially thermally processed

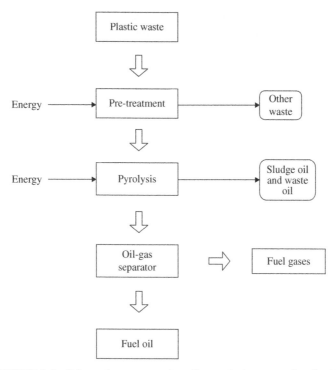

FIGURE 9.6 Schematic representation of a pyrolysis process for plastics.

followed by catalyzed thermal degradation (Horvat and Ng, 2005; Yang et al., 2013). Application of pyrolysis to plastics waste and rubber waste (Al-Salem et al., 2009) has been reported in the literature. A state-of-the-art pyrolysis unit (close to commercialization)[1] is conservatively estimated to convert a ton of waste plastics into three to six barrels (31.5 US gallons per barrel) of fuel oil; the 29 million tons of presently landfilled plastic waste can potentially be converted into 3.6 billion gallons of oil with minimal emissions associated with the process (Themelis et al., 2011). Figure 9.6 illustrates the main features of a pyrolysis process.

In practice, a mixed plastic stream separated out from MSW is pyrolyzed where synergistic effects can sometimes be obtained. The yield and composition of fuel depend on the feed stream (Wong and Broadbelt, 2001) as the pyrolysis reactions are feed dependent.[2] While pyrolysis can in theory handle a range of input streams of mixed plastics, most processes can only tolerate limited amounts of PVC and PET scrap. Both these yield relatively less oil per unit mass, and PVC degradation produces corrosive

---

[1] A state-of-the-art single stream pyrolysis pilot facility is in operation in Akron, Ohio, with a second one planned in Cleveland, OH. Commercially available Agilyx (Beaverton, OR) units, for instance, claim to yield 60 barrels of oil per 10 tons of plastic waste.

[2] Approximate fuel yield in pyrolysis of common plastics are as follows: PS ~ 90%, PP and PE ~ 70%, and PET ~ 30%.

TABLE 9.3   Yield of Products from Pyrolysis of Mixed Plastic Waste[a] at 440°C (dehydrochlorination Step was at 300°C for 30 Min)

Pyrolysis method	Liquid	Gas	Residue
Uncatalyzed pyrolysis	79.3 ± 1.9	17.7 ± 1.9	3.0 ± 0.3
Catalyzed pyrolysis (zeolite ZSM-5)	56.9 ± 3.0	40.4 ± 3.0	3.2 ± 0.2
Uncatalyzed followed by catalyzed	69.0 ± 0.1	29.0 ± 0.1	2.00 ± 0.0
Dehydrochlorination + catalytic	56.8 ± 2.0	41.2 ± 2.0	2.00 ± 0.2

Source: Based on data from Lopez-Urionabarrenechea et al. (2012).
[a]PE = 40%, PP = 35%, PS = 18%, PET = 4, and PVC = 3.

HCl that has to be removed during the process. Though it has never been demonstrated, bio-based plastics such as poly(lactic acid) (PLA) should also be amenable to conventional pyrolysis.

A carbonaceous char residue waste (2–13%), often contaminated with catalyst residue and inorganic fillers, is formed in pyrolysis. This may be either used as solid fuel, activated C, or disposed of in a landfill.

Mechanism of pyrolytic degradation is complex involving a large set of reactions, even for a single class of plastic. Generic degradation reactions such as chain scission, H-transfer, unzipping, disproportionation, and combination occur in the process. Typical mix of products from mixed plastic waste streams with different pyrolysis conditions are shown in Table 9.3.

The rate of loss in weight of the plastic can be described by first-order kinetics (Marcilla et al., 2003), allowing an activation energy for the process to be estimated and is as follows:

$$-\frac{dm}{dt} = km^n$$

$$\text{Ln } k = \text{Ln } A - \left(\frac{E}{RT}\right)$$

where, m is the mass of solid plastic left at time t, n is the order of reaction, A is the preexponential factor, and E is the activation energy. For PE, for instance, $E \sim 250$ kJ/mol and the reaction rate is very significantly affected by the pyrolysis temperature. Industrial-scale pyrolysis is carried out in rotary drum, melt furnace, or fluidized bed reactors.

No commercial scale facilities are operational in North America but numerous companies have demonstrated technical feasibility of pyrolysis.

9.2.2   Hydrogenation or Hydrocracking

Hydrogenation technologies are not that widely used in waste management. It is usually a two-stage process where plastic waste is first subjected to mild cracking by low-temperature pyrolysis and the condensable product is reacted with hydrogen to obtain petrochemicals or diesel fuel (Joo and Guin, 1997). Typically, the process involves reaction of the plastic degradation products with hydrogen over a catalyst in an autoclave at moderate temperatures and pressures (typically 423–673°K and 3–10 MPa hydrogen) (Panda et al., 2010). Catalysts used include oxide- or zeolite-supported transition metals that catalyze both reactions. The cracking reaction is endothermic while the hydrogenation itself is exothermic. Typical feeds include PE, PET, PS, PVC, and mixed plastics or plastic/tire rubber mixes.

Both pyrolysis and hydrocracking are particularly desirable waste management approaches from a sustainability standpoint, for several reasons.

(a) They produce only small amounts of waste to be landfilled and result in minimal air emissions (flue gases and CO_2) or water pollution.

(b) These yield an energy-rich fuel that is ready for use with little or no further processing. Pyrolysis yields about a gallon of ready-to-use fuel per about 7.6 lbs of scrap plastic and the BTU value of the fuel is higher (by about 15%) than that from incinerating the plastic itself as a solid fuel (4R Sustainability, Inc., 2011). However, process energy costs (generally amounting to a few percent of the energy content of waste (Gonçalves et al., 2008) in producing the fuel is significant.

(c) Where monomer recovery in high yield is possible (as with polystyrene waste), it can, at least in theory, provide a route to converting waste plastics into virgin resin (Smolders and Baeyens, 2004).

9.2.3 Gasification

9.2.3.1 Thermal Gasification Pyrolysis of mixed waste at a temperature higher than about 600°C in a controlled atmosphere of air, O_2, or steam yields a mix of gaseous oxygenated products (CO, H_2, and CH_4) (Aznar et al., 2006; Zia et al., 2007). As with pyrolysis proper, this process leaves behind a char residue and yields fuel gas. Plastics are often co-gasified with biomass (Pinto et al., 2002), coal (Aznar et al., 2006), or wood (van Kasteren, 2006) in mixed waste streams. The composition of products obtained depends on the mix of waste used, temperature, and atmosphere as well as the residence time in the reactor (He et al., 2009). The variability in composition of input waste streams does not allow gasification or other advanced thermolytic techniques to deliver a product stream of consistent composition or quality. Gasification of PE (He et al., 2009), PP (Xiao et al., 2007), and PET-containing waste streams (Pohorely et al., 2006) has been studied in some detail. The products include some char, liquid fuel, and fuel gases; at higher temperatures and longer residence times more of the gaseous products are obtained.

The main gasification reactions are shown in Table 9.4. Many other reactions are also involved in gasification (see for instance de Souza-Santos, 2010), but a detailed discussion of their chemistry is beyond the scope of this chapter.

The calorific value of the syngas produced is highest (10–15 MJ/m³) in O_2 atmospheres and lowest (4–7 MJ/m³) in air (Arena, 2012). Unlike with incineration, the

TABLE 9.4   The Main Reactions Involved in Gasification

Solid–gas reactions	Gas–gas reactions
$C + O_2 \rightarrow CO$ (exothermic)	$CO + H_2O \rightarrow CO_2 + H_2$ (exothermic)
$C + O_2 \rightarrow CO_2$ (exothermic)	$CO + 3H_2 \rightarrow CH_4 + H_2O$ (exothermic)
$C + 2H_2 \rightarrow CH_4$ (exothermic)	$C + H_2O \leftrightarrow CO + H_2 + 131\,MJ/kmol$: *Water gas reaction*
$C + H_2O \rightarrow CO + H_2$ (endothermic)	$CO + H_2O \leftrightarrow CO_2 + H_2 - 41\,MJ/kmol$: *Water gas shift reaction*
$C + CO_2 \rightarrow 2CO$ (endothermic)	$C + CO_2 \leftrightarrow 2CO + 172\,MJ/kmol$: *Boudouard reaction*

oxygen-deficient atmosphere used in gasification discourages the formation of any dioxins (Arena et al., 2010). At higher temperatures (>1200°C), any toxic polyaromatic hydrocarbons (PAHs) formed are cracked into safe gasses. The Texaco gasification process, for instance, involves two steps. First, the plastic is thermally cracked into oil and condensable gases at around 700°C. This mix is then passed through a gasifier with air and steam at 1200–1500°C to obtain a mix of CO and H_2 (Syngas) which is then purified.

Some tar or char is invariably formed in the process, reducing the yield of syngas. Using activated charcoal (Kim et al., 2011) or olivine (Mastellone and Arena, 2008) in the feed can reduce the tar fraction. With low-temperature (600–900°C) gasification (Namioka et al., 2011), some liquid oils might also be produced and the residual ash/char may require vitrification or other treatment before disposal. The residue in high-temperature gasification is generally an inert slag that can be landfilled without further processing. Any PVC in the feed stream can present a problem because of the HCl produced on thermal degradation. This dehydrochlorination product is generally absorbed on lime, and the inorganic chloride formed is safely disposed of in a landfill.

9.2.3.2   Plasma Arc Gasification   A recent development is the plasma-assisted gasification that volatilizes the waste at very high temperatures in an oxygen-deficient environment yielding syngas efficiently (Hrabovsky, 2011; Zhang et al., 2012). Plasma torches that produce arcs (>7000°F are reached in the plasma) are used as the heat source, and the temperature can usually be controlled regardless of variations in the feed stream. Because of the higher temperatures, even waste with high levels of moisture can be used as feedstock. It is an efficient and environmentally-benign gasification technology as only syngas and a vitreous slag residue (sometimes used as aggregate in concrete) are produced. Slag temperatures are around 3000°F and therefore no potentially toxic organic residues are present in the slag. Both air emissions (Zhang et al., 2013) as well as slag toxicity (Lapa et al., 2002) are minimal.

However, the capital cost for plasma-arc gasifiers is relatively high and explains why no commercial facilities are in operation in the United States at this time. The

TABLE 9.5   Environmental Features of Plastic Waste Management Options

Process	GHG emission	Waste	Toxic emissions	Product
Pyrolysis	Very low	~15–20% char, silica, ash	Very low	Oil/gas
Gasification	Very low	Char/tar	Very low	Syngas
Anaerobic digestion	Very low	Residue	Very low	Fuel gas
Incineration	High	Char/residue	Temperature dependent	Energy
Chemolysis	Very low	Residue	Low	Chemicals
Composting[a]	High	Residue	Moderate	Compost
Landfilling[b]	CH_4 gas[c]	Solid	Moderate	Fuel gas

[a]Only the biodegradable plastic can be composted and the features are common to biodegradation as well.
[b]A modern landfill or an anaerobic digestion facility where all or some of the methane is collected.
[c]Anaerobic digestion yields methane in place of CO_2, and the gas has more than 20 times the global warming potential compared to CO_2.

need to shred the waste feedstock and high operational costs are the other disadvantages of the process. Table 9.5 summarizes the main features of the thermal treatment methods applicable to plastics waste.

9.2.4   Feedstock Recycling

Feedstock recycling is the recovery of monomers (or other chemical raw material) through depolymerization or other chemical reaction. The recovery role played by feedstock recycling is illustrated in Figure 9.7. Controlled "cracking" (Paszun and Spychaj, 1997) of the waste breaks down the long-chain molecular structure of the plastic yielding shorter chains and smaller organic molecules. These might be used as feedstock in chemical processes including resin synthesis. This pathway of waste plastic to virgin resin is particularly attractive from a sustainability standpoint even though some losses are inevitable in the process. Both Nylon 6 and PET waste are recoverable efficiently using this approach. Depending on the chemistry used, feedstock recovery might be *via* hydrolysis, glycolysis, hydro-glycolysis, aminolysis, methanolysis, or acid cleavage. Examples of some of the potential types of reactions for feedstock recycling can be illustrated for PET as follows:

Hydrolysis: Hydrolysis of plastics (Lopez-Fonseca et al., 2009) with water or dilute alkali yields monomers or other compounds. Thermoset polyurethane foam for instance is hydrolyzed into polyols and amines by superheated steam at 230–315°C in about 15 min (Lonescu, 2005).

$$R' - NH - CO - O - R'' + H_2O \rightarrow R - NH_2 + HO - R + CO_2$$
$$R' - NH - CO - NH - R'' + H_2O \rightarrow 2R - NH_2 + CO_2$$

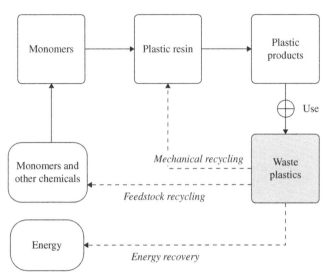

FIGURE 9.7   Basic recovery options available for plastics waste.

Similarly, PET can be hydrolyzed under acidic or basic conditions to yield terephthalic acid (Carvalho et al., 2006; Paszun and Spychaj, 1997; Sato et al., 2006).

Glycolysis: Reacting PET with excess ethylene glycol in the presence of sodium sulfate yielded BHET (see below) as the primary product (Shukla et al., 2009; Viana et al., 2011; Wang et al., 2009).

$$\text{HOCH}_2\text{CH}_2\text{O}-\overset{\overset{\text{O}}{\|}}{\text{C}}-\!\!\!\!\text{\Large\color{white}.}\!\!\!\!-\overset{\overset{\text{O}}{\|}}{\text{C}}-\text{OCH}_2\text{CH}_2\text{OH}$$

BHET

Aminolysis of PET with ethanolamine yields bis(2-hydroxy ethylene) terephthalamide (BHETA) (Shukla and Harad, 2006).

Alcoholysis of PET via methanolysis or ethanolysis yields dimethyl terephthalate and diethyl terephthalate respectively (Jie et al., 2006; Kurokawa et al., 2003). The reaction can also be carried out under supercritical conditions (Castro et al., 2006; Goto, 2009).

Selected chemolysis processes for PET are illustrated in Figure 9.8 and tabulated in Table 9.6. These reactions yield either the original monomers or products that can be converted to other monomers. Hydrolysis can be effectively used with PET and polyurethane waste plastics in feedstock recovery (Zia et al., 2007). Reaction conditions employed are varied and these selected references do not cover them exhaustively. Aromatic polyesters, PET and poly(butylene terephthalate), have been studied intensively for feedstock recovery. PET is extensively used in soda bottles and less than 30% of the product is mechanically recycled.

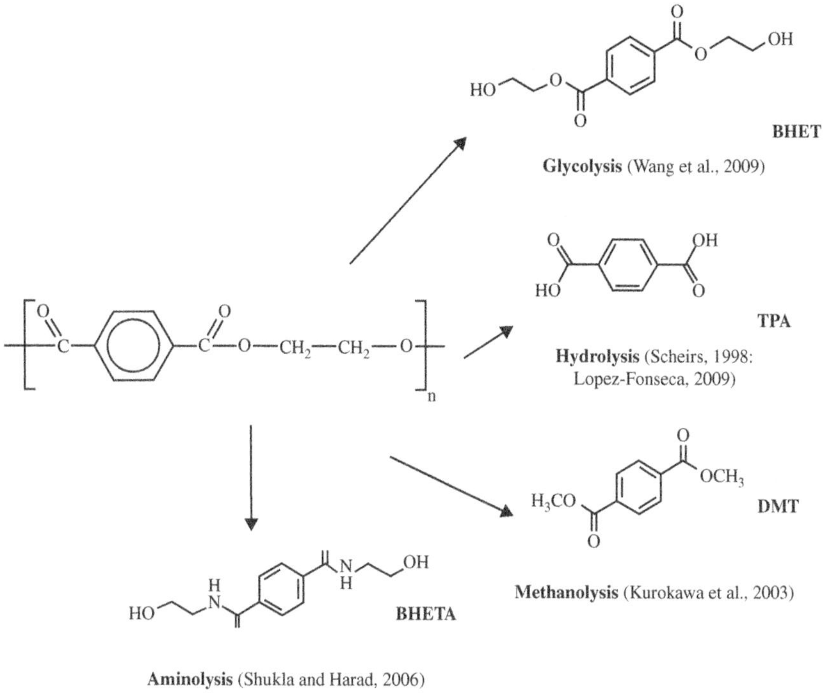

FIGURE 9.8   Chemolysis of poly(ethylene terephthalate) (PET) into chemical feedstock.

Plastics such as polystyrene undergo facile-catalyzed thermal depolymerization or pyrolysis yielding very high yields of styrene monomer. Base-catalyzed depolymerization of the polymer is as follows.

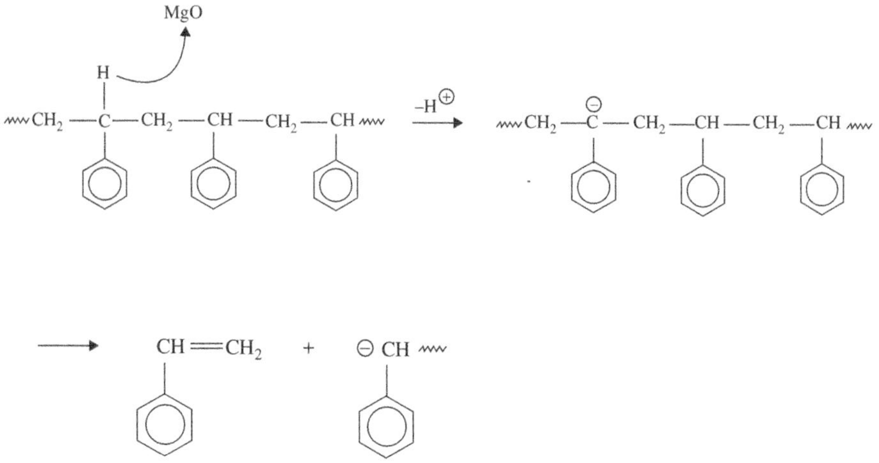

TABLE 9.6 Selected Examples of Thermolysis of Common Plastics Yielding Monomer and Mixed Fuel Gas/liquids

Plastic resin type		Approach	Product
Polystyrene (Woo et al., 2000)	PS	MgO-catalyzed thermal depolymerization: ~350°C	Styrene ~ 93%
Polystyrene (Lilac and Lee, 2001)	PS	Hydrolysis with supercritical H_2O: ~382°C and 28 MPa	Styrene ~ 90%
Polycarbonate (Huang et al., 2006)	PC	Supercritical ethanol solvolysis	Bisphenol A ~90, diethylcarbonate ~89%
Poly(methyl methacrylate)	PMMA	Pyrolysis (in the absence of air)	Methyl methacrylate ~90–98%
Mixed polyolefins (HDPE, LDPE, PP) (Nishino et al., 2005)		Pyrolysis catalyzed by Ga-ZSM-5 has been reported	~50% liquid fraction, predominantly benzene; toluene and xylene; gas fraction was mostly H_2
Polyethylene (Horvat and Ng, 2005)	HDPE	Pyrolysis under N_2 at 400–440°C	~80–90% straight-chain alkanes and 1-alkenes

9.2.5 Landfilling

Landfilling is the most-used MSW management option in the United States with around 3,100 active and over 10,000 old landfills across the nation.[3] It is the low cost of landfilling that makes it a popular management option.[4] But concerns about the increasing cost of hauling waste and the shortage of practical landfill space have been voiced (in the United States shortage of landfill space is less of an issue). Landfill is essentially a large in-ground excavation that is lined by a bottom liner (usually 100-mil HDPE or a clay lining), with a leachate collection system, and a cover. Clay liners are permeable to some extent while HDPE liners are usually not, but the latter can swell in the presence of solvents and leak leachate. Modern landfills use a double HDPE layer with a leachate detection system between layers, to address this limitation. A system of pipes collects most of the leachate. Leaking landfills contaminates the water table and can be a serious health hazard.

In the anaerobic environment within the fill-volume, organic biodegradable materials (but not plastics) slowly decompose anaerobically. Landfill gas (LFG) is approximately 50% CH_4 and 45% CO_2. In modern landfills, an effort is made to collect these gases but some invariably escape into the environment. Methane [CH_4] is 21 times as potent as CO_2 in its global warming potential (GWP). In modern landfills, 20–80% of the CH_4 is collected depending on its design. Global landfills in 2007 were estimated to release about 20% of the total (or 30–35 Tg) of CH_4 annually (Matthews and Themelis, 2007). In the United States, landfills contribute about 17.5% (84.1 MMT of CO_2 equivalents) of anthropogenic methane emissions.

Recovered LFG (primarily CH_4) can be burnt directly to generate liquid fuel (that can be used in waste collection fleet), or used in natural gas plants to produce electricity. At best, only 60–90% of the gas is typically recovered and burnt to produce electricity. Burning LFG produces CO_2 which is a less potent GHG compared to the CH_4. Landfill gas to energy (LFGE) projects were particularly popular in the 1990s. In co-generation projects, LFG is used to generate both electricity using turbines and heat. Using waste-derived energy not only conserves fossil fuels but also avoid the externalities associated with using an equivalent amount of the fossil fuel. According to the USEPA, presently there are about 620 operational LFG projects in the United States.

There is no justification for landfilling waste plastics; plastics do not break down anaerobically to any significant extent in landfills. Except for the minimal fraction of biodegradable plastics, they do not contribute to LFG generation. With no solar UV radiation available to initiate degradation, plastics survive for very long periods of time in the landfill essentially remaining in storage. But, the risk of additives in the plastics disposed of in landfills breaking down and polluting groundwater has been pointed out (Oehlmann et al., 2009; Teuten et al., 2009).

---

[3] The largest in the United States is the Puente Hills Landfill in City of Industry, CA, towering to a height of 500 ft (about that of the Washington monument) and covering a land area of over 700 acres.
[4] This may not be true where waste has to be transported long distances in search of a landfill, as in the case of New York City.

9.2.6 Plastics Waste Incineration

WTE plants burn MSW and convert some of the heat (in steam turbines) into useful energy.[5] In the United States, about 30 MMT of MSW is processed in about 80, mainly mass-burn, WTE facilities. The main advantage of incineration is that it does not require presorting the waste. Thus it is better suited for dealing with mixed waste streams such as automotive shredder residue. While some noncombustibles are recovered from the waste stream prior to incineration, the process generally results in fly ash as well as bottom ash. Metals, especially iron, is recovered from the bottom ash and recycled. Plants provide electricity as well as income in the way of tipping fees and via the sale of recycled metals. Operating on MSW (that include plastics), a WTE plant can be expected to yield at least 600 kWh/ton of MSW and can potentially provide cogenerated heat energy as well. The plastics fraction in MSW is readily incinerated and would contribute at least 10–12% of this energy. In the United States, MSW contains about ½% of chlorine, about half of it derived from PVC waste. The possibility of chlorine-containing waste being converted into toxic dioxins or dioxin-like chemicals has been discussed in the research literature (Shibamoto et al., 2007).

In late 1980s, the health hazards of dioxins and furans potentially emitted from burning mixed waste were recognized (Huang and Buekens, 1995). Polychlorinated dibenzo-p-dioxins (PCDDs), polychlorinated dibenzofurans (PCDFs), and polychlorinated biphenyls (PCBs) are classes of compounds potentially formed in during burning of wastes containing chlorine. Incinerating waste with organic or inorganic chlorides along with lignin (newsprint or wood) or plastics, especially those with aromatic structure, can yield these compounds (Yasuhara et al., 2002). About 90% of human exposure to these chemicals is via ingestion with food, especially in meats and dairy products. They are highly toxic[6] even at low levels, have very low solubility in water, bioaccumulative, and are persistent organic pollutants in the environment. While some human epidemiological studies are available, the weight of evidence on their toxicity is based on animal studies. TCDD, one of the congeners (see Fig. 9.9), is one of the most toxic substances known with a median lethal dose (LD$_{50}$) of 1 µg/kg body weight in guinea pigs. In general, dermal toxicity, hepatotoxicity, immunotoxicity, reproductive toxicity, teratogenicity, endocrine disruption, and carcinogenicity effects have been associated with dioxins.

MSW incineration was estimated to be the leading source of PCDDs in air (EuroChlor, 2003) in Europe as well as in the United States. There are 75 congeners[7] of PCDD and 135 congeners of PCDF. Most of the dioxins generated end up in the

---

[5] Waste with higher calorific values yield higher energy in incineration. The calorific value of waste varies with the country: In China, Korea, or Brazil, the values are ~4,000–6,000 kJ/kg, and in the United States (or Western Europe) it is 10,000–15,000 kJ/kg.

[6] The most toxic member of the dioxin family is 2,3,7,8-TCDD. Agent Orange is an approximately 1:1 mixture of two herbicides—2,4-dichlorophenoxyacetic acid and 2,4,5-trichlorophenoxyacetic acid. Dioxin TCDD was present as a contaminant in Agent Orange used as a defoliant in the Vietnam War.

[7] Up to eight positions on the ring structures of both PCDDs and PCDFs can be chlorinated yielding different isomers with the same numbers of C, H, and Cl atoms in the molecule. The toxicity of these congeners depend on the position of the Cl atoms in the molecule. 2,3,7,8-Tetrachlorodibenzodioxin (TCDD) is considered the most toxic of the congeners.

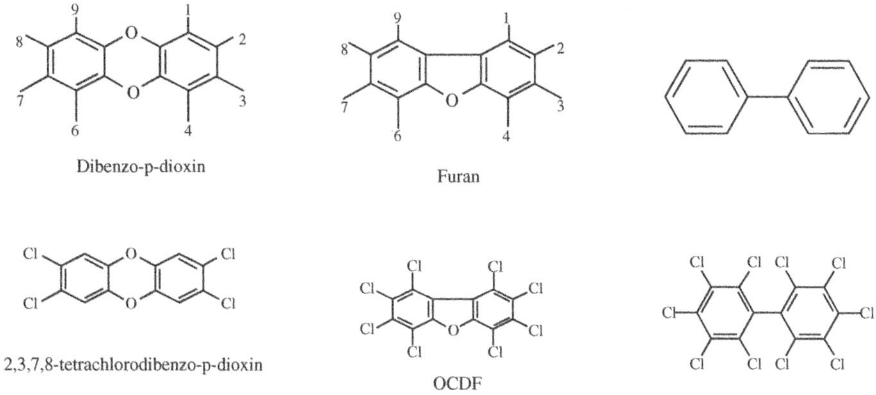

Dibenzo-p-dioxin

Furan

2,3,7,8-tetrachlorodibenzo-p-dioxin

OCDF

FIGURE 9.9 The general structure for PSDD, PCDF, and PCB are shown in the first row. An example of a congener derived from each of these is shown in the second row.

ash residue (at ppb and ppt levels) (Abad et al., 2000; EEA, 2000), and their disposal requires caution. Ash also carries a host of metal residues that can readily leach out, especially under low pH conditions.

Thanks to stringent regulation in mid-1990s, however, pollution control technologies used in WTE plants in the United States meet the strictest emission standards in the world. In modern municipal incinerators, the flue gas is subjected to a temperature above 850°C (1560°F) for few seconds to ensure the thermal breakdown of dioxins. WTE was claimed to be the process for generating electricity with the least environmental impact (Millrath et al., 2004)! The emissions from US waste incineration plants (see Table 9.7) have come down drastically from the 1990s values. Presently the dioxin release attributed to MSW combustion is minimal (USEPA, 2005), amounting annually to only 0.5 oz of dioxin toxic equivalents from combustion. Recent data suggest that modern WTE facilities are more environmentally acceptable than a coal-fired power plant contributing to dioxin emissions. But they emit about 33% more CO_2 per unit of electricity generated compared to the power plant. There is no universal agreement on the value of incinerators for waste treatment (Muller et al., 2011). Also, emissions such as dioxins and mercury (Van Velzen et al., 2002) depend on how well the plant is operated and can vary widely.

A comparison between landfilling and incineration shows the latter to be the more sustainable option. Burning organic material inevitably produces CO_2 contributing to global warming and this would suggest WTE to be inferior to landfilling in terms of its environmental footprint. However, most landfills do not capture the CH_4 emissions and even the ones that do capture only a fraction of it. Calculations taking this into account show that even the best-managed landfill will still have a GWP that can be as much as 50 times higher than that for a WTE plant (USEPA, 2005). Each ton of MSW diverted from landfill into a WTE plant would reduce the emission of GHG by 0.5–1.0 ton of CO_2 equivalents depending on the capture rate of CH_4 at the fill site. The foregoing assures that the WTE plants are state-of-the-art and well managed. The CO_2 emissions from combustion of plastics

TABLE 9.7   Average Emissions from 87 WTE Plants in the United States

Pollutants	Unit	Average emission	USEPA standard
Dioxin/furan	ng/dscm	0.05	0.26
Particulates	mg/dscm	4	24
SO_2	ppmv	6	30
NOx	ppmv	170	180
HCl	ppmv	10	25
Hg	mg/dscm	0.01	0.08
Cd	mg/dscm	0.001	0.020
Pb	mg/dscm	0.02	0.20
CO	ppmv	33	100

Source: Psomopoulos et al. (2009).

TABLE 9.8   Calculated GHG Emissions from incineration of Different Plastic Resins

Plastic material	CO_2 from incineration (MT CO_2/short ton of Plastic)	Avoided utility CO_2[a] (MT CO_2/ton of plastic)	Energy content of plastic (MBTU/ton)
HDPE	2.82	1.42	40.0
LDPE/LLDPE	2.82	1.41	39.8
PET	2.06	0.75	21.2
PP	2.82	1.42	39.9
PS	3.04	1.28	36
PVC	1.28	0.56	15.8

Source: WARM (Waste Reduction Model) Version 12 USEPA (2012).
[a] Net emissions after adjusting for avoided utility emissions from burning conventional fuel for utilities.

and the net emissions as well as the avoided utility emissions have been calculated for common plastics (Table 9.8).

9.2.7   Biological Recovery Technologies

Composting and anaerobic breakdown of waste with gas recovery are practical biological recovery options. Composting is the accelerated aerobic breakdown of material facilitated by thermophilic species of microbes. Commercial composting is carried out at temperatures of 40–62°C where no pathogens can survive. In anaerobic digestion, biomass is converted to biogas (CH_4 and CO_2) that is used as a replacement for natural gas or as a fuel to cogenerate heat and electricity. Unlike in composting, anaerobic digesters can operate under thermophilic, mesophilic (35–40°C), or psychrophilic (15–25°C) conditions. Even mesophilic temperatures are reported to reduce (but not eliminate) (Wright et al., 2001) any pathogen content in the waste; however, the digesters are enclosed and only the gas is removed for use. However, as already pointed out, the plastic content in the MSW is predominantly, if not exclusively, nonbiodegradable. They will survive either of the biological treatments, though with composting, some oxidation and even fragmentation can occur. As such, both these options are not strictly applicable to plastics fraction in MSW.

Biodegradable plastics are not a significant fraction of the MSW stream as yet. If they were indeed present in the waste and were anaerobically degradable, these too will be converted into biogas. Some bio-based polymers such as PLA, PHA, and others are promoted for packaging use as "compostable" plastics. Composting is an accelerated biodegradation process that converts waste into carbon dioxide and compost substrate. Complete mineralization of polymers in the waste, however, do not leave a recalcitrant humic residue and therefore do not contribute to the compost product, and are not suitable substrates for composting. There are useful applications for compostable plastics as in the case of compostable plastic yard waste bags that can be filled and composted conveniently without the need to remove the plastic bag from the waste stream. With some packaged foods, where plastic packaging is intimately associated with contents and is disposed along with food waste for composting, there is some justification for using compostable plastics. The quality of compost produced in these cases will be superior with no plastics residue in the product. This advantage, however, must be balanced with the risk of compostable plastics inadvertently contaminating a recycling stream.

9.3 SUSTAINABLE WASTE MANAGEMENT CHOICES

Post-use products often have a considerable service life, resource value, and extractible energy associated with them. Significant nonrenewable resources were used in their production by processes with adverse environmental externalities. Majority of post-use plastics in the US waste stream unfortunately end up in landfills. Disposal in landfills ranks low from a sustainability standpoint because of gas emissions, leachate, and waste of the valuable plastics resources. Waste plastic do not generate gas nor leachate in landfills. Only uncontrolled backyard burning (burn barrels) is worse than landfilling, as it emits particulates, air pollutants, and particulates creating a health hazard. Where any PVC is in the burnt waste, dioxins,[8] potent human toxins, as well as polyaromatic hydrocarbons (PAHs), a known carcinogen, are released into the air. Incineration of waste is closely controlled and modern plants do not release significant levels of dioxins.

Desirable waste management options must minimize three footprints that define sustainability (see Chapter 2); minimize energy use, conserve materials, and minimize pollution.

Material recycling (or reuse) does not preclude subsequent use of other waste management options. The recycled resin in a secondary product cannot be further recycled *ad infinitum* and has to be invariably disposed of at some stage in its lifecycle using one of these options. This option is discussed in the following section.

Feedstock recovery via gasification/pyrolysis is an attractive option as it is not carbon emitting and yields valuable feedstock chemicals and/or fuels. The carbon in

---

[8]Dioxins are a class of human carcinogens, ED agents, and cause severe developmental problems. The USEPA estimates that a single burn barrel can produce more dioxins than a full-scale municipal waste combustor burning 200 tons a day.

TABLE 9.9 Indicators of Principal Environmental Impact Categories, as Evaluated for Five Plastic Waste Management Approaches

Impact category (per kg PET waste)	Landfill	Combustion	Mechanical recycling	MR + low temp. pyrolysis	MR + hydrocracking
Energy consumption (MJ)	51.59	6.45	−5.41	12.14	**−11.40**
Crude oil consumption (g)	1462	995	−74.76	−145.86	**−210.65**
Water consumption (kg)	47.11	45.92	**3.48**	14.06	25.26
CO_2-equivalent (kg)	5.3	7.3	**1.4**	1.7	2.02
Air emissions of organic compounds (g)	26.8	14.3	**−0.05**	1.42	4.78
Waste production (kg)	2.49	0.19	**0.09**	0.20	0.26

Source: Reprinted with slight modifications with permission from Perguini et al. (2005).
Values in boldface type indicate the best environmental performance. MR = materials recycling.

the waste is not emitted as CO_2 but turned into fuel that can be used. As it is carried out in a closed system there are no polluting emissions except for inadvertent losses. The drawback here is that it requires source-separation of plastics if high-value products are to be expected. Waste to energy conversion via incineration when carried out under controlled conditions is preferable to landfilling in terms of environmental compatibility. It also has the advantage of not needing any source separation of plastic waste.

As environmental impacts and energy demands associated with different plastic waste management approaches are based on different LCA studies, they are not strictly comparable. Also, the specific technologies used in competing recovery methods influence these estimates. However, Perguini et al. (2005) found material recycling to be a particularly attractive option. Their findings summarized in Table 9.9 compares several different options. Low-temperature pyrolysis by BP polymer cracking process (Arena and Mastellone, 2005) and hydrocracking by the Veba Combi-Cracking process (Dijkema and Stougie, 1994) were considered. While these numbers are by no means applicable to all waste management scenarios, performance of recycling and combinations of recycling with resource recovery are particularly significant. It has been proposed that landfilling plastics can be more environmentally friendly compared to incineration (Finnveden et al., 2005), where plastics are assumed to biodegrade in the long term yielding GHGs or transport of waste over long distances is needed.

Figure 9.10 illustrates the hierarchy of waste management approaches based on this discussion as they apply to plastics waste. The order of the different options is consistent with the Lansink's Ladder[9] proposed for solid waste management in the Netherlands.

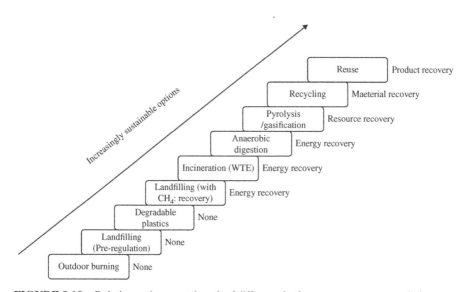

FIGURE 9.10   Relative environmental merit of different plastic waste management techniques.

[9]Named after the Member of Parliament in the Netherlands who proposed it. It is now specified in the Environmental Management Act in that country.

9.4   MECHANICAL RECYCLING OF PLASTICS

Mechanical recycling of plastics delivers energy efficiency, conserves fossil-fuel raw materials, and provides a convenient (albeit temporary) waste management approach (Scheirs, 1998). While it is not a waste disposal strategy *per se*, it in essence extends the useful lifetime of plastics at a highly discounted additional energy/environmental cost; the only better option (with near zero additional cost) is re-using the product with 100% materials recovery (in recycling, some material losses are inevitable). Furthermore, reuse and recycling do not preclude the eventual recovery of chemical feedstock and/or energy from post-use recycled plastic. Comparison of the energy and resources saved by using recycled resin (compared to using virgin resin) to fabricate products makes the argument for recycling an obvious one. This advantage in avoided expenditure of energy and carbon release is illustrated for PET resin in Figure 9.11. The values indicated are dependent on the LCA used, but the general argument holds for any plastic (or other material).

The recycling process can be either closed loop or open loop: in closed loop, recycling the plastic waste is recycled into the same product it was derived from, such as in bottle-to-bottle recycling. Generally, a fraction of the recycled resin is mixed in with virgin resin in the fabrication of products. In the United States, some PET bottles presently have up to 50% of recycled resin content.[10] In the more common open-loop recycling (sometimes called down cycling), the recycled plastic is used to replace at least a part of the virgin resin used to manufacture a different product; PET bottles are predominantly (72% in 2007) (Li Shen and Patel, 2010) recycled into fiber and only 10% in bottle-to-bottle conversions.

Based on a meta-analysis of LCA studies (Waste & Resources Action Program, 2006), the environmental impact of recycling was found to be much lower (by about 50%) compared to incineration or landfilling of plastics as long as the recycled resin can be substituted for at least approximately 33% of the virgin material. This ratio represented a break-even point where recycling and incineration have about the same overall impact. But, the meta-study also found incineration to be preferred over recycling where substantial washing and cleaning costs were involved prior to recycling. The specific impact categories used in the assessment were energy use, resource consumption, GHG emissions, waste generation, toxicity, and other energy-related impacts such as acidification or eutrophication.

The energy costs, GHG emissions, and other emissions are generally lower for manufacturing products using recycled resin as opposed to virgin resin of the same type. This "avoided energy, resource consumption, and emissions" (see Fig. 9.11) represent the main benefit of recycling. Savings in energy and GHG emissions vary with the type of resin (and the specific LCA used to calculate these). Recycling resin primarily saves the feedstock extraction and resin manufacturing energy, but adjusted down by a material loss factor inherent to the process. As the feed-stock related energy costs are avoided, a recent study found producing recycled PET pellets to takes only about 15% of the energy demand

[10] Nestlé Waters North America Inc. half-liter "ReBorn" bottles.

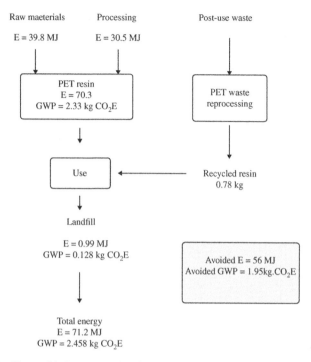

FIGURE 9.11 The avoided energy and carbon emissions per kilogram of PET mechanically recycled. GWP (Global warming potential in CO_2 equivalents). The numbers from other LCA studies can vary slightly. Source: Data from source Kuczenski and Geyer (2009).

associated with producing virgin resin pellets (Franklin Associates, 2011); the comparable estimate for HDPE is approximately 12%. The estimated avoided carbon values for producing PET and HDPE pellets compared to virgin resin pellet production was estimated to be 71% and 75% respectively. These are large differences and the energy, materials, and emission advantages in using recycled materials cannot be overstated.

Theoretically, all thermoplastics can be recycled. However, it is the availability of recycling infrastructure, rather than physicochemical feasibility, that determines the extent to which a given product will be recycled in practice. The cost of recycling, from collecting the post-consumer products at curbside to transporting the recycled plastic pellets to fabricators, invariably determines the market price of the recycled resin. Where this cost is well below that of virgin plastic, as with HDPE grades used in milk jug fabrication or PET grades used in blow-molded soda bottles, recycling makes good business sense. In the United States, these two postconsumer plastics products therefore enjoy significant levels of recycling. Figure 9.12 illustrates the general process.

In 1988, the Society of Plastics Industry developed the now-widely used set of seven identification codes for six classes of plastics (the seventh is a catch-all category). These symbols (see Fig. 9.13) molded into the products are intended to

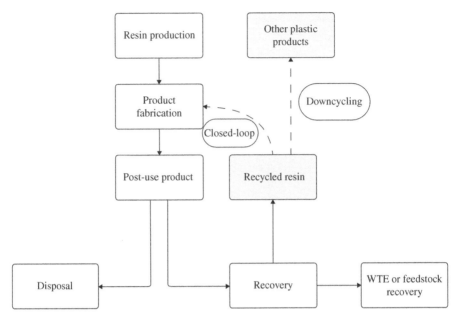

FIGURE 9.12 A general scheme for recycling of plastics recovered from MSW, illustrating closed- and open-loop pathways.

PETE   HDPE   V   LDPE   PP   PS   Other

FIGURE 9.13 Recycling symbols. PETE is polyester (PET), V is vinyl plastics, and the "other" category covers all other resins.

help consumer determine recyclability and encourage separation of plastics at source. But these can in fact confuse the consumer into assuming that "recyclable" means that the stamped product will in fact be recycled; all postconsumer plastic collected is not recycled.[11] Resin codes #3 to #7, for instance, are not even accepted for recycling in most communities. Food trays, blister packs, yogurt cups, and bottles (other than PET and HDPE) are not recycled because of insufficient demand for their recyclate product.

Over 94% of the US population have access to PET and HDPE recycling facilities. Therefore, in the United States, PET beverage bottles and HDPE milk jugs enjoy 32.6 and 22.2% rate of recycling, respectively, compared to the single-digit rates for other classes of plastics. Only a smaller group (57%) have access to facilities that can

[11] To assume that the presence of recycling "chasing arrow" logo or the printed word "recyclable" to mean that it will be recycled after use, is a common consumer misconception.

recover other plastics for recycling (Mouw and Penner, 2013). By comparison, plastic films, especially post-consumer bags,[12] the most visible of film-based products in MSW, are not recycled to any significant extent. Despite recently reported encouraging increases in US recycling rates for the general category of plastic films (that includes industrial shrink wrap that is recovered at a high rate), only a few percent of shopping bags are recycled (CalRecycle, 2009).

9.4.1 Recycling: A Sustainable Choice

The energy costs of using virgin resin to fabricate a product is $\{E_P + E_F + E_D\}$, where E_P, is the energy cost of virgin resin production, E_F, that of product fabrication, and E_D, that of postuse disposal. The cost of using recycled resin (that is subsequently disposed of) is $(E_C + E_R + E_F + E_D)$, where E_C, is the energy cost of collection of waste resin for recycling and E_R, the cost of recycling. Magnitude of E_C depends on the infrastructure available for pickup (e.g., curbside collection or centers), the distance the waste is transported. The value of E_R includes energy used in sorting, cleaning, grinding, and pelletizing the recycled resin.

The energy difference between using all-virgin resin and substituting $x\%$ of recycled resin (for virgin resin) in fabricating a product can be approximated as follows, assuming E_F and E_D to be the same in both cases. The transport energy costs are not included for simplicity (and because it is variable) for either case.

$$\Delta E = E_P - \left\{ x(E_C + E_R) + (1-x)E_P \right\}$$

Usually as $E_P \gg E_R + E_C$, recycling plastics almost always results in energy saving, that is, ΔE is positive and large. An exception (see Chapter 10) is small plastics debris in marine litter where E_C can be so large that the above does not hold. An analogous equation can be written for the pollution load (including carbon emission), ΔP associated with the two processes, again assuming the impact of either resin to be the same in the fabrication and disposal phases. Nonrenewable material conserved by using the recycled resin is x percent.

From the consumer's vantage point, recycling increases the product available per unit input of energy and materials. If recycling rate was 100%, at least in theory, one input load of polymer should continue to yield products repeatedly, at a highly discounted processing energy cost! In practice, recycling degrades the plastic, and some loss of material is incurred at each cycle and minimal externalities are associated with recycling as well.

The Waste and Resources Action Program (WRAP) study (2006) examined 10 LCAs covering 60 scenarios of plastics waste management. It found recycling to be the generally preferred waste management method with maximum benefits being incurred when the virgin resin is substituted for at near 100% levels. In a 2011 study, the energy costs of collecting, sorting, and reprocessing of plastics waste into recycled resin pellets was calculated for PET and HDPE. All virgin material production

---

[12]Except for PP which was recycled at a rate of 21% in 2011.

TABLE 9.10   A Comparison of Energy Used, GHG Emissions, and Solid Waste Generation to Produce Virgin and Recycled Resins

Resin	Energy (MBTU/10^3 lbs)		GHG emissions Equiv. (lbs of CO_2)		Solid waste (lbs/ 10^3 lbs)	
	Recycled	Virgin	Recycled	Virgin	Recycled	Virgin
PET	7.25	31.9	1169	2746	402	142
HDPE	3.87	35.8	628	1822	220	74.6

Source: Data from Franklin Associates (2011).

burdens were assigned to the first use of material and only collection, sorting, reprocessing, and transportation were assigned to the recycled stream. This is reasonable assuming that in the absence of the lower-cost recycled resin, more virgin resin would have to be used. The environmental advantage associated with recycling is considerable at 100% recycled resin use as reflected in Table 9.10.

9.5   RECYCLING BOTTLES: BEVERAGE BOTTLES AND JUGS

Bottles are the most-used package for beverages and the easiest postconsumer waste item to recover. The fact that 96% of the plastic bottles in the United States are made up of either PET or HDPE, helps the sorting process at MRFs. About a half of the rest is made up of PP. PET enjoys the highest rate of recycling in the United States primarily because of readily-available recycling infrastructure and a stable market demand for the product. In Europe, the recovery rate of PET bottles for recycling is higher (33–95% depending on country). It is an ideal candidate product for recycling, being easily separated from the recovery stream with no direct printing on plastic, with caps that can be separated by density, and no migratory additives. About 10–20% of bottles are recovered in high-value closed-loop recycling streams and the rest is down-cycled, mostly to manufacture polyester fiber.

Several LCA studies (Arena et al., 2003; Craighill and Powell, 1996; Eriksson et al., 2005; Finnveden et al., 2005) have concluded that recycling postuse PET is the preferred waste management option.

9.5.1   Bottle-to-Bottle Recycling

Closed-loop or bottle-to-bottle recycling has far reaching implications from a sustainability standpoint. Since post-consumer PET was approved by the FDA for food-contact uses in 1991, the technology has grown dramatically. In the United States, about 17% of the PET recyclate finds end markets in bottle applications. Bottles used for non-food items (such as bleach, detergent, pesticide, solvents) also end up in the recycling stream and pose a contamination risk (FDA 2006). The value of cleaning technologies is assessed on their efficacy in removing contaminants acquired by the bottles exposed to a cocktail of chemicals. Super-clean technologies

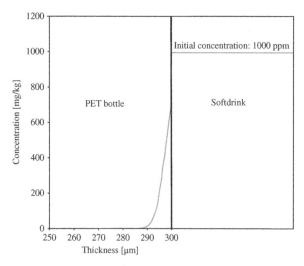

FIGURE 9.14 Calculated concentration profile of the flavor compound limonene in PET derived from a postconsumer bottle of wall thickness 300 µm, containing a beverage with 1000 ppm of limonene, after 365 days of exposure at 23°C. Source: Reproduced with permission from Welle (2011).

available today are able to deliver "food-grade" PET for bottle applications even where the PET waste streams are highly contaminated with non-food polyester containers (Blanchard et al., 2003).

Trace contaminants derived from the first use of the resin survive in the lower parts per million range in conventional recycling of bottles to flakes (Bayer, 2002; Franz et al., 2004). Bottle-to-bottle applications demand more stringent, super-clean recycling processes to reduce the level of contaminations to nearly the same levels as that in virgin resin. This is accomplished by a combination of thermal treatment (180–230°C), inert gas stripping or vacuum degassing, and re-extrusion (280–290°C) into pellets (Welle, 2011). The contaminants, such as flavor ingredients are generally localized in a thin layer of approximately 10–15 µm of the food-contact surface (see Fig. 9.14). Some super-clean techniques therefore rely on base-catalyzed hydrolytic depolymerization of the layer at approximately 150°C followed by washing, degassing, and reextrusion (Welle, 2008). The reaction produces ethylene glycol and terephthalic acid monomers that have to be removed by vacuum degassing. Alternatively waste PET might be glycolyzed (with ethylene glycol) to achieve partial depolymerization into oligomers that can supplement the prepolymer (prior to the melt condensation phase) in the PET manufacturing process (Welle, 2008). As the waste PET is not reduced to monomers, this is still considered a material recovery operation.

Bottle-to-bottle recycling is unlikely to present a cost advantage given the high-cost super-cleaning necessary to create a quality recyclate. But from a sustainability standpoint, it is a particularly valuable technology as it conserves material, energy use, and the associated pollution footprint (including the carbon emissions).

9.5.2 Open-Loop Recycling

In open-loop recycling, two "lives" of the plastic are very different as shown in Figure 9.15.

In the right-hand part of the figure, the mechanics of recycling are summarized in the second box as "Wash-Float-Dry." Generally, the bottles are prewashed, sorted, granulated, floated, and dried. Sorting is often done manually but automatic sorters based on reflective near-infrared (NIR) sensors operating at a speed of over 10 bottles per second are available. The rejected items are automatically ejected off the stream to be discarded. Modern sorting technologies are capable of providing recycled plastics that are nearly 100% of one a single polymer type (Franz and Welle, 2003).The sorted stream is flaked in water and floated to separate out contaminant plastics based on differences in density.

In the recycling sequence, some material is invariably wasted: each kilogram of post-use PET bottles yields only about 0.78 kg of recycled resin. A part of this loss is due to the removal of PVC, PE, or other contaminants during sorting and float separation. Shen et al. (2010) recently compared the environmental impact of mechanical recycling of PET with that for chemical recycling via depolymerization (glycolysis and methanolysis). Mechanical recycling yielded the largest energy savings and had the least impact in most of the other environmental categories. They reported savings

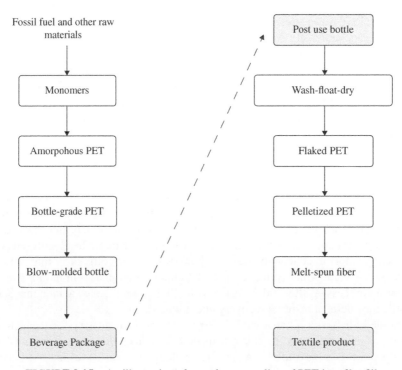

FIGURE 9.15   An illustration of open-loop recycling of PET into fiberfill.

of up to 85% nonrenewable energy used and 25–75% GWP, depending on the recycling technology used, compared to virgin PET.

Bottle grade PET can be recycled into fiber products in semi-closed loop recycling (100% of recycled resin used in the fiber) depending on the application. In any event, the fiber itself cannot be mechanically recycled a second time though chemical recovery of monomer is still possible. Repeated recycling of plastics cannot be continued indefinitely (even if 100% material recapture was possible) as the plastic resins degrade on repeated processing as in multiple extrusion cycles. This has been reported for HDPE (Canevarolo, 2000; Choudhury et al., 2005), PP (da Costa et al., 2005; González-González et al., 1998; Jansson et al., 2003), PVC (Yarahmadi et al., 2001), and PET (Assadi et al., 2004; Nait-Ali et al., 2011a, 2011b). The Mw of PET resin reduced from 60,000 (g/mol) to about 25,000 (g/mol) after three repeated extrusion cycles (La Mantia and Vinci, 1994). Others (Frounchi, 1999; Romao et al., 2009; Spinacé and De Paoli, 2001) have reported similar findings for PET resins of different molecular weights. With rigid PVC, five repeated extrusions reduced the lifetime of the product to 30% of the original extruded profile (Yarahmadi et al., 2001). As good mechanical properties of thermoplastics depend on their high degree of polymerization (DP) and therefore on high Mw (g/mol) values, these changes reflect substantially reduced mechanical integrity on multiple extrusion.

9.5.3   Recycling of HDPE

A majority of HDPE waste is downcycled into other useful products such as crates and storage bins. However, FDA-approved technology is available (Schut, 2009) in the US (as well as in Europe) to convert post-consumer HDPE bottles into a grade of recycled resin approved for food-contact applications. Technically, this is a significant achievement. Compared to PET waste, cleaning of used HDPE products free of contaminants, is more complicated given its lower processing temperature. A complicating factor is the contamination of HDPE intended for recycling by other bottles (such as detergent bottles) made of HDPE or copolymers of PE. These cannot be included in the closed-loop recycling stream for HDPE.

The cleaning in this case consists of multiple steps: washing, float/sink separation, and/or hydrocycloning stages (APC, 1999). Float/sink separation allows density-based removal of most non-HDPE debris from the granulated and the dried granules are pelletized. Design considerations that help recycling include not using PP components (caps) on HDPE bottles as they cannot be easily separated. Blends of (> ~5% PP) results in poor quality recyclate. The same is true of PVC labels.

9.6   DESIGNING FOR RECYCLABILITY

Designing products, especially plastic bottles, for easy recyclability is a key sustainable development goal. For instance, avoiding any PVC components in the bottle (such as labels or caps) is helpful as it simplifies cleaning especially in PET waste streams; the two resins have overlapping densities and PVC decomposes at a temperature lower than the melt temperature for PET, releasing HCl gas.

The same applies to other resins such as PLA (compostable plastic), vinyl alcohol copolymers (barrier layer and cap sealer), and poly(ethylene naphthenate) (used for additional barrier properties). These do not disperse well in the PET or HDPE base resins and can degrade the quality of the recycled resin (Bolin and Smith, 2011). Dark colors, any metal layers, adhesives not removed by water, non-plastic labels, and thermoset caps make it more difficult for PET to be recycled into a high-grade product.

Recycling mixed plastics yields a low-grade "plastic lumber" that can be used in a limited number of applications. Most classes of common plastics are not miscible with each other and the commingled resin therefore has poor mechanical properties (Chanda and Roy, 2007). This is avoided by using a compatibilizer (Ubonnut et al., 2007) at 2–5% level but it adds to the cost of the product; compounds such as maleic anhydride–grafted PE or PP (Lei and Wu, 2012; Leu et al., 2012) are often used (Nornberg et al., 2014). Alternatively, the wood fraction itself can be chemically functionalized (e.g., acetylated) (Hung et al., 2012) for better compatibility. Some of the recycled resin is mixed with 30–50% by weight of a natural filler such as wood flour and made into wood-plastic composite (WPC) (Najafi, 2006; Clemons, 2002). A higher-grade WPC product results when a single type of plastic such as virgin PP is used as the matrix polymer (Sobczak et al., 2012). However, the environmental advantage of the WPC as a material is not always clear: LCA studies on WPC material yield mixed results (Alves et al., 2010; Luz et al., 2010), depending on the particular blend.

With an annual global demand of 2.3 MMT (2010) and projected growth rate of 13.8% (BCC Research Market Forecasting, 2011), WPC is a growing composite product. Having a high fraction of hydrophilic wood (lignocellulose) in the polymer matrix restricts moisture absorption by wood avoiding any fungal growth and therefore its biodegradation.

REFERENCES

4R Sustainability, Inc. Conversion technology: a complement to plastic recycling. Portland: Apr 2011. Funded by the American Chemistry Council; 2011.

Abad E, Adrados MA, Caixach J, Fabrellas B, Rivera J. Dioxin mass balance in a municipal waste incinerator. Chemosphere 2000;40:1143–1147.

Al-Salem SM, Lettieri P, Baeyens J. Recycling and recovery routes of plastic solid waste (PSW): a review. Waste Manag 2009, Oct;29(10):2625–2643.

Al-Salem SM, Lettieri P, Baeyens J. The valorization of plastic solid waste (PSW) by primary to quaternary routes: from re-use to energy and chemicals. Prog Energy Combust Sci 2010, Feb;36(1): 103–129

Alves C, Ferrao PMC, Silva AJ, Reis LG, Freitas M, Rodrigues LB. Ecodesign of automotive components making use of natural jute fiber composites. J Clean Prod 2010; 18:313–327.

American Plastics Council (APC). Development of hydrocyclones for use in plastic recycling. Technical paper. Washington DC. 1999. Downloaded from http://infohouse.p2ric.org/ref/47/46174.pdf on October 21 2014.

Arena Umberto. Process and technological aspects of municipal solid waste gasification. A review. Waste Manag 2012, Apr;32(4):625–639.

Arena U and Mastellone ML. Fluidized bed pyrolysis of plastic wastes. In: Scheirs J, Kamimnsky W, editors. *Feedstock Recycling and Pyrolysis of Plastic Wastes*. Hoboken: Wiley; 2005.

Arena U, Mastellone ML, Perguini F. Life cycle assessment of a plastic packaging recycling system. Int J Life Cycle Assess 2003;8 (2):92–98.

Arena U, Zaccariello L, Mastellone ML. Fluidized bed gasification of waste-derived fuels. Waste Manag 2010;30:1212–1219.

Assadi R, Colin X, Verdu J. Irreversible structural changes during PET recycling by extrusion. Polymer 2004;45 (13):4403–4412.

Aznar MP, Caballero MA, Sancho JA, Francs E. Plastic waste elimination by co-gasification with coal and biomass in fluidized bed with air in pilot plant. Fuel Process Technol 2006;87 (5):409–420.

Bayer F. Polyethylene terephthalate recycling for food-contact applications: testing, safety and technologies: a global perspective. Food Addit Contam 2002;19 (supplement): 111–134.

BCC Research Market Forecasting. Global market for wood-plastic composites to pass 4.6 million metric tons by 2016. BCC, Wellesley, MA. 2011. Available at http://bccresearch. blogspot.com/2011/11/global-market-for-wood-plastic.html. Accessed Oct 6, 2014.

Blanchard F, Christel A, Gorski G, Welle F. Drinks from the detergent bottle. Plast Europe 2003;93 (9):42–45.

Bolin CA, Smith S. Life cycle assessment of ACQ-treated lumber with comparison to wood plastic composite decking. J Clean Prod 2011;19 (6–7):620–629.

CalRecycle. California department of resources recycling and recovery 2009 statewide recycling rate for plastic carryout bags. 2009. Available at http://www.calrecycle.ca.gov/ Plastics/AtStore/AnnualRate/2009Rate.htm#Summary. Accessed Oct 6, 2014.

Canevarolo SV. Chain scission distribution function for polypropylene degradation during multiple extrusions. Polym Degrad Stab 2000;70 (1):71–76.

Carvalho GM, Muniz EC, Rubira AF. Hydrolysis of post-consume poly(ethylene terephthalate) with sulfuric acid and product characterization by WAXD, ^{13}C NMR and DSC. Polym Degrad Stab 2006;91:1326–1332.

Castro RNE, Vidotti GJ, Rubira AF, Muniz EC. Depolymerization of poly (ethylene terephthalate) wastes using ethanol and ethanol/water in supercritical conditions. J Appl Polym Sci 2006;101:2009–2016.

Chanda M, Roy SK. *Plastics Technology*. Boca Raton: CRC Press; 2007.

Choudhury A, Mukherjee M, Adhikari B. Thermal stability and degradation of the post-use reclaim milk pouches during multiple extrusion cycles. Thermochim Acta 2005;430 (1–2):87–94.

Clemons C. Wood-plastic composites in the United States: the interfacing of two industries. For Prod J 2002;52 (6):10–18.

Conesa JA, Font R, Marcilla A, Caballero JA. Kinetic model for the continuous pyrolysis of two types of polyethylene in a fluidized bed reactor. J Anal Appl Pyrolysis 1997; 40–41:419–431.

da Costa HM, Ramos VD, Rocha MCG. Rheological properties of polypropylene during multiple extrusion. Polym Test 2005;24 (1):86–93.

Craighill AL, Powell JC. Lifecycle assessment and economic evaluation of recycling: a case study. Resour Conserv Recycl 1996;17:75–96.

Dijkema GPJ, Stougie L. *Environmental and Economic Analysis of Veba Oel Option for Processing of Mixed Plastic Waste*. Delft: Delft University Clean Technology Institute; 1994.

Ding WB, Liang J, Anderson LL. Thermal and catalytic degradation of high density polyethylene and commingled post-consumer plastic waste. Fuel Process Technol 1997;51 (1–2):47–62.

EEA. Dangerous substances in waste. Prepared by: Schmid J, Elser A, Strobel R, ABAG-itm, Crowe M, EPA, Ireland. European Environment Agency, Copenhagen: 2000.

Eriksson O, Carlsson RM, Frostell B, Bjorklund A, Assefa G, Sundqvist J-O. Municipal solid waste management from a systems perspective. J Clean Prod 2005;13:241–252.

EuroChlor. Dioxins and furans in the environment. (2003). Available at http://www.eurochlor. org/media/14933/sd3-dioxins-final.pdf. Accessed Oct 6, 2014.

FDA. Guidance for industry: use of recycled plastics in food packaging: chemistry considerations. HFS-275, Washington, DC: US Food and Drug Administration, US Department of Health and Human Services, Center for Food Safety and Applied Nutrition, 200 Independence Avenue, S.W. Washington, DC. 2006.

Finnveden G, Johannsson J, Lind P, Moberg A. Lifecycle energy from solid waste. J Clean Prod 2005;13:213–229.

Franklin Associates. Life cycle inventory of 100% postconsumer HDPE and PET recycled resin from postconsumer containers and packaging. Kansas: Franklin Associates, a division of ERG Prairie Village; 2011.

Franz R, Welle F. Recycling packaging materials. In: Ahvenainen R, editor. *Novel Food Packaging Techniques*. Cambridge: Woodhead Publishing; 2003. p 497–518.

Franz R, Mauer A, Welle F. European survey on post-consumer poly (ethylene terephthalate) materials to determine contamination levels and maximum consumer exposure from food packages made from recycled PET. Food Addit Contam 2004;21 (3):265–286.

Frounchi M. Studies on degradation of PET mechanical recycling. Macromol Symp 1999;144:465–469.

Gonçalves CK, Tenório ASJ, Levendis YA, Carlson JB. Emissions from premixed combustion of gasified polystyrene. Energy Fuels 2008;22:354–362.

González-González VA, Neira-Velázquez G, Angulo-Sánchez JL. Polypropylene chain scissions and molecular weight changes in multiple extrusion. Polym Degrad Stab 1998;60 (1):33–42.

Goto M. Chemical recycling of plastics using sub- and supercritical fluids. J Supercrit Fluids 2009;47:500–507.

He M, Xiao B, Hu Z. Syngas production from catalytic gasification of waste polyethylene: influence of temperature on gas yield and composition. Int J Hydrogen Energy 2009;34: 1342–1348.

Horvat N, Ng FTT. Tertiary polymer recycling: study of polyethylene thermolysis as a first step to synthetic diesel fuel. Fuel 2005;78 (4):459–470.

Hrabovsky M. Plasma aided gasification of biomass, organic waste and plastics. In: Proceedings of 30th ICPIG conference 2011 in Belfast; Northern Ireland; 2011. GEN-406. Available at http://mpserver.pst.qub.ac.uk/sites/icpig2011/406_GEN_Hrabovsky.pdf 2012. Accessed Oct 6, 2014.

Huang H, Buekens A. On the mechanisms of dioxin formation in combustion processes. Chemosphere 1995;31 (9):4099–4117.

Huang J, Huang K, Zhou Q, Chen L, Wu YQ, Zhu ZB. Study of depolymerization of polycarbonate in supercritical ethanol. Polym Degrad Stab 2006;91 (10):2307–2314.

Hung KC, Chen YL, Wu JH. Natural weathering properties of acetylated bamboo plastic composites. Polym Degrad Stab 2012;97:1680–1685.

Jansson A, Möller K, Gevert T. Degradation of post-consumer polypropylene materials exposed to simulated recycling—mechanical properties. Polym Degrad Stab 2003;82 (1):37–46.

Jie H, Ke H, Wenjie Q, Zibin Z. Process analysis of depolymerisation polybutylene terephthalate in supercritical methanol. Polym Degrad Stab 2006;91 (10):2527–2531.

Joo HS, Guin J, A. Hydrocracking of a plastics pyrolysis gas oil to naphtha. Auburn University, Chemical Engineering Department, Auburn, Alabama 36849. Energy Fuels 1997;11 (3):586–592.

Kaminsky W, Schlesselmann B, Simon C. Olefins from polyolefins and mixed plastics by pyrolysis. Anal Appl Pyrolysis 1995;32:19–27.

van Kasteren JMN. Co-gasification of wood and polyethylene with the aim of CO and H_2 production. J Mater Cycles Waste Manage 2006;8:95–98.

Kim JW, Mun TY, Kim JO, Kim JS. Air gasification of mixed plastic wastes using a two-stage gasifier for the production of producer gas with low tar and a high caloric value. Fuel 2011;90:2266–2272.

Kuczenski B, Geyer R. LCA and recycling policy—a case study in plastic. Life Cycle Assessment IX. University of California at Santa Barbara. Downloaded from http://www.lcacenter.org/LCA9/presentations/208.pdf on October 19, 2010.

Kurokawa H, Ohshima M, Sugiyama K, Miura H. Methanolysis of polyethylene terephthalate (PET) in the presence of aluminium triisopropoxide catalyst to form dimethyl terephthalate and ethylene glycol. Polym Degrad Stab 2003;79 (3):529–533.

La Mantia FP, Vinci M. Recycling poly (ethylene terephthalate). Polym Degrad Stab 1994;45:121–125.

Lapa N, Santos Oliveira JF, Camacho SL, Circeo LJ. An ecotoxic risk assessment of residue materials produced by the plasma pyrolysis/vitrification (PP/V) process. Waste Manag 2002;22 (3):335–342.

Lee K-H. Effects of the types of zeolites on catalytic upgrading of pyrolysis wax oil. J Anal Appl Pyrolysis 2012, Mar;94:209–214.

Lei Y, Wu Q. Wood plastic composites based on microfibrillar blends of high density polyethylene/poly (ethylene terephthalate). Bioresour Technol 2012;101 (10):3665–3671.

Leu SY, Yang TH, Lo SF, Yang TH. Optimized material composition to improve the physical and mechanical properties of extruded wood–plastic composites (WPCs). Construct Build Mater 2012;29:120–127.

Li Shen EW, Patel MK. Open-loop recycling: a LCA case study of PET bottle-to-fibre recycling. Resour Conserv Recycl 2010;55 (1):34–52.

Lilac W. Douglas, Lee Sunggyu. Kinetics and mechanisms of styrene monomer recovery from waste polystyrene by supercritical water partial oxidation Adv Environ Res 2001, Dec; 6, 1:9–16

Lonescu M. Chemistry and Technology of Polyols for Polyurethanes. Shropshire: Rapra Publishers; 2005.

López-Fonseca R, González-Velasco JR, Gutiérrez-Ortiz JI. A shrinking core model for the alkaline hydrolysis of PET assisted by tributyl hexadecyl phosphonium bromide. Chem Eng J 2009;146 (2):287–294.

Lopez-Urionabarrenechea A, de Marco I, Caballero BM, Laresgoiti MF, Adrados A. Catalytic stepwise pyrolysis of packaging plastic waste. J Anal Appl Pyrolysis 2012, July; 96:54–62

Luz SM, Caldeira-Pires A, Ferrão PMC. Environmental benefits of substituting talc by sugarcane bagasse fibers as reinforcement in polypropylene composites: ecodesign and LCA as strategy for automotive components. Resour Conserv Recycl 2010;54: 1135–1144.

Marcilla A, Gomez A, Reyes-Labarta JA, Giner A. Catalytic pyrolysis of polypropylene using MCM-41: kinetic model. Polym Degrad Stab 2003;80:233–240.

de Marco I, Caballero BM, Lopez A, Laresgoiti MF, Torres A, Chomon MJ. Pyrolysis of the rejects of a waste packaging separation and classification plant. J Anal Appl Pyrolysis 2009;85:384–391.

Mastellone ML, Arena U. Olivine as a tar removal catalyst during fluidized bed gasification of plastic waste. AIChE J 2008;54:1656–1667.

Matthews E, Themelis NJ. Potential for reducing global methane emissions from landfills. In: Proceedings Sardinia 2007, 11th international waste management and landfill symposium. 2007. Oct 1–5, 2007, Cagliari, pp. 2000–2030.

Millrath K, Roethel FJ, Kargbo DV. Waste-to-energy residues—the search for beneficial uses. In: 12th North American Waste to Energy Conference (NAWTEC 12); May 17–19; Savannah, GA: 2004. pp. 1–812.

Mouw Scott, Penner Rick. Moore Recycling Associates Inc. Plastic recycling collection national reach study: 2012 update Moore Recycling Associates Inc. Sonoma, CA: 2013.

Muller NZ, Mendelsohn R, Nordhaus W. Environmental accounting for pollution in the United States economy. Am Econ Rev 2011;101 (5):1649–1675.

Nait-Ali LK, Colin X, Bergeret A. Kinetic analysis and modelling of PET macromolecular changes during its mechanical recycling by extrusion. Polym Degrad Stab 2011a;96 (2):11.

Nait-Ali LK, Colin X, Bergeret A, Ferry L, Ienny P. Rheological modelisation of PET degradation during its recycling by extrusion. In: Proceedings 23rd annual meeting polymer processing society (cd-rom); 2011b. p 8.

Najafi SK, Hamidinia E, Tajvidi M. Mechanical properties of composites from sawdust and recycled plastics. J Appl Polym Sci 2006;100:3641–3645.

Namioka T, Saito A, Inoue Y, Park Y, Min T-j, Roh S-a, Yoshikawa K. Hydrogen-rich gas production from waste plastics by pyrolysis and low-temperature steam reforming over a ruthenium catalyst. Appl Energy 2011;88:2019–2026.

Nishino J, Itoh M, Fujiyoshi Y, Matsumoto Y, Takahashi, R, Uemichi Y. Development of a feedstock recycling process for converting waste plastics to petrochemicals. In: Muller-Hagedorn M, Buckhorn H, editors. Proceedings of the third international symposium on feedstock recycle of plastics & other innovative plastics recycling technique. Universitatsverlag Karlsruhe (25–29th; Karlsruhe; 2005. pp. 325–332

Nörnberg B, Borchardt E, Luinstra GA, Fromm J. Wood plastic composites from poly(propylene carbonate) and poplar wood flour—mechanical, thermal and morphological properties. Eur Polym J 2014;51:167–176.

Oehlmann J, Schulte-Oehlmann U, Kloas W, Jagnytsch O, Lutz I, Kusk KO, Wollenberger L, Santos EM, Paull GC, Van Look KJ, Tyler CR. A critical analysis of the biological impacts of plasticizers on wildlife. Phil Trans R Soc B 2009;364:2047–2062.

Panda Achyut K, Singh RK, Mishra DK. Thermolysis of waste plastics to liquid fuel: a suitable method for plastic waste management and manufacture of value added products—A world prospective. Renew Sustain Energy Rev 2010, Jan;141: 233–248.

Park DW, Hwang EY, Kim JR, Choi JK, Kim YA, Woo HC. Catalytic degradation of polyethylene over solid acid catalysts. Polym Degrad Stab 1999;65 (2):193–198.

Paszun D, Spychaj T. Chemical recycling of poly (ethylene terephthalate). Ind Eng Chem Res 1997;36:1373–1380.

Perguini F, Mastellone MA, Arena U. Recycling options for management of plastic packaging wastes. Environ Prog 2005;24 (2):137–154.

Pinto F, Franco C, Andre RN. Co-gasification study of biomass mixed with plastic wastes. Fuel 2002;81:291–297.

Plastics Europe, European Plastic Convertors (EuPC), European Plastic Recyclers (EuPR), Epro Plastic Recycling (EPRO). The Compelling facts about plastics—an analysis of European plastics production, demand and recovery for 2008. 2009.

Pohorely M, Vosecky M, Hejdova P. Gasification of coal and PET in fluidized bed reactor. Fuel 2006;85 (2006):2458–2468.

Psomopoulos CS, Bourka A, Themelis NJ. Waste-to-energy: A review of the status and benefits in USA. Waste Manag 2009;29:1718–1724.

Romao W, Franco MF, Corilo YE, Eberlin MN, Spinacé MAS, De Paoli MA. Poly (ethylene terephthalate) thermo-mechanical and thermo-oxidative degradation mechanisms. Polym Degrad Stab 2009;94 (10):1849–1859.

Sato O, Arai K, Shirai M. Hydrolysis of poly (ethylene terephthalate) and poly(ethylene 2,6-naphthalene dicarboxylate) using water at high temperature: Effect of proton on low ethylene glycol yield. Catal Today 2006;111:297–301.

Scheirs J. *Polymer Recycling: Science, Technology and Application*. Chichester: Wiley-Blackwell; 1998.

Schut JH. (2009) New processes for food-grade recycled HDPE. Plastics Technology, March Issue. Gardner Business Media, Cincinnati, OH. Available at http://www.ptonline.com/articles/new-processes-for-food-grade-recycled-hdpe. Accessed October 14, 2014.

Shen L, Worrell E, Patel MK. Open-loop recycling: A LCA case study of PET bottle-to-fibre recycling. Resour Conserv Recycl 2010;55 (1):34–52.

Shibamoto T, Yasuhara A, Katami T. Dioxin formation from waste incineration. Rev Environ Contam Toxicol 2007;190:1–41.

Shukla SR, Harad AM. Aminolysis of polyethylene terephthalate waste. Polym Degrad Stab 2006;91 (8):1850–1854.

Shukla SR, Harad AM, Jawale LS. Chemical recycling of PET waste into hydrophobic textile dyestuffs. Polym Degrad Stab 2009;94 (4):604–609.

Smolders K, Baeyens J. Thermal degradation of PMMA in fluidised beds. Waste Manag 2004;24 (8):849–857.

Sobczak L, Brüggemann O, Putz RF. Polyolefin composites with natural fibers and wood-modification of the fiber/filler–matrix interaction. J Appl Polym Sci 2012;127 (1):1–17.

Songip AR, Masuda T, Kuwahara H, Hashimoto K. Test to screen catalysts for reforming heavy oil from waste plastics. Appl Catal B 1993;2:153–164.

de Souza-Santos ML. *Solid Fuels Combustion and Gasification: Modeling, Simulation, and Equipment Operations*. Second ed. Boca Raton, FL: CRC Press; 2010. p 508.

Spinacé MAS, De Paoli MA. Characterization of poly (ethylene terephthalate) after multiple processing cycles. J Appl Polym Sci 2001;80:20–25.

Teuten EL, Saquing JM, Knappe DR, Barlaz MA, Jonsson S, Björn A, Rowland SJ, Thompson RC, Galloway TS, Yamashita R, Ochi D, Watanuki Y, Moore C, Viet PH, Tana TS, Prudente M, Boonyatumanond R, Zakaria MP, Akkhavong K, Ogata Y, Hirai H, Iwasa S, Mizukawa K, Hagino Y. Transport and release of chemicals from plastics to the environment and to wildlife. Philos Trans R Soc B 2009;364:2027–2045.

Themelis NJ, Castaldi MJ, Bhatti J, Arsova L. Energy and economic value of nonrecycled plastics (NRP) and municipal solid wastes (MSW) that are currently landfilled in the Fifty States. August 16, 2011; Columbia University. Earth Engineering Center, New York; 2011.

Tonjes David J, Greene Krista L. A review of national municipal solid waste generation assessments in the USA. Waste Manag Res 2012, Aug;308:758–771.

Tukker A, de Groot H, Simons L, Wiegersma S. Chemical recycling of plastic waste: PVC and other resins. European Commission, Final Report, STB–99–55 Final. TNO Institute of Strategy, Technology and Policy, Delft: 1999.

U.S. Environmental Protection Agency (USEPA). 2005. The inventory of sources and environmental releases of dioxin-like compounds in the United States: the year 2000 update External Review Draft, EPA/600/p-03/002A. . USEPA, Washington DC. Downloaded from http://www.epa.gov/ttn/chief/net/2005inventory.html.

Ubonnut L, Thongyai S, Praserthdam P. Interfacial adhesion enhancement of polyethylene-polypropylene mixtures by adding synthesized diisocyanate compatibilizer. J Appl Polym Sci 2007;104:3766–3773.

USEPA. 2011. Municipal solid waste in the United States tables and figures for 2010. United States Environmental Protection Agency Office of Solid Waste (5306P) EPA530-R-13-001; http://www.epa.gov/osw/nonhaz/municipal/pubs/MSWcharacterization_fnl_060713_2_rpt.pdf on Oct 15, 2014; 2011.

USEPA. 2012. WARM (Waste Reduction Model) Version 12. USEPA, Washington, DC. Available at http://epa.gov/epawaste/conserve/tools/warm/index.html. Accessed Oct 20, 2014.

Van Velzen D, Langenkamp H, Herb G. Review: mercury in waste incineration. Waste Manage. Res. 2002;20:556–568.

Viana ME, Riul A, Carvalho GM, Rubira AF, Muniz EC. Chemical recycling of PET by catalyzed glycolysis: kinetics of the heterogeneous reaction. Chem Eng J 2011;173:210–219.

Villanueva, A., Delgado, L., Zheng, L., Edser, P., Catarino, A.S., Litten, D. Study on the selection of waste streams for end-of-waste assessment. Final Report. European Commission, Joint Research Centre. Luxembourg: Publications Office of the European Union; (2010), pp. 367.

Wang H, Liu Y, Li Z, Zhang X, Zhang S, Zhang Y. Glycolysis of poly(ethylene terephthalate) catalyzed by ionic liquids. Eur Polym J 2009;45 (5):1535–1544.

Waste & Resources Action Program (WRAP). (2006). *Environmental Benefits of Recycling: An International Review of Life Cycle Comparisons for Key Materials in the UK Recycling Sector*. SAP097. BioIntelligence Service and Copenhagen Resource Institute, Banbury.

Welle F. Investigation into the decontamination efficiency of a new PET recycling concept. Food Addit Contam 2008;25 (1):123–131.

Welle F. Twenty years of PET bottle to bottle recycling—an overview. Resour Conserv Recycl 2011;55 (11):865–875.

Williams PT, Williams EA. Fluidised bed pyrolysis of low density polyethylene to produce petrochemical feedstock. J Anal Appl Pyrolysis 1999;51:107–126.

Wong HW, Broadbelt LJ. Tertiary resource recovery from waste polymers via pyrolysis: neat and binary mixture reactions of polypropylene and polystyrene. Ind Eng Chem Res 2001;40 (22):4716–4723.

Woo OS, Ayala N, Broadbelt LJ. Mechanistic interpretation of base-catalyzed de-polymerization of polystyrene. Catal Today 2000;55 (1–2):161–171.

Wright PE, Inglis SF, Stehman SM, Bonhotal J. Reduction of selected pathogens in anaerobic digestion. Paper presented at the Fifth Annual NYSERDA Innovations in Agriculture Conference. 2001. pp. 1–11

Xiao R, Jin B, Zhou H. Air gasification of polypropylene plastic waste in fluidized bed gasifier. Energy Convers Manag 2007;48:778–786.

Yang X, Sun L, Xiang J, Hu S, Su S. Pyrolysis and dehalogenation of plastics from waste electrical and electronic equipment (WEEE): a review. Waste Manag 2013;33 (2):462–473.

Yarahmadi N, Jakubowicz I, Gevert T. Effects of repeated extrusion on the properties and durability of rigid PVC scrap. Polym Degrad Stab 2001;73 (1):93–99.

Yasuhara A, Katami T, Okuda T, Shibamoto T. Role of inorganic chlorides in formation of PCDDs, PCDFs, and coplanar PCBs from combustion of plastics, newspaper, and pulp in an incinerator. Environ Sci Technol 2002;36 (18):3924–3927.

Zhang Q, Dor L, Fenigshtein D, Yang W, Blasiak W. Gasification of municipal solid waste in the Plasma Gasification Melting process. Appl Energy 2012;90 (1):106–112.

Zhang Q, Wu Y, Dor L, Yang W, Blasiak W. A thermodynamic analysis of solid waste gasification in the plasma gasification melting process. Appl Energy 2013;112:405–413.

Zia KM, Bhatti HN, Bhatti IA. Methods for polyurethane and polyurethane composites, recycling and recovery: a review. React Funct Polym 2007;67 (8):675–692.

10

PLASTICS IN THE OCEANS

Over 70% of the Earth's surface is covered with water, mostly bodies of salt water that together constitute the global ocean system. The few percent of fresh water on Earth is trapped within the arctic ice mass, in groundwater, or in rivers and lakes. The oceans support a variety of marine life ranging in size from the tiny nano-planktons to the giant whales. This intricate ocean ecosystem also supports life on land. Life on terra is only possible because of the numerous ecosystem services (benefits humans receive from ecosystems) provided by the oceans (Beaumont et al., 2008). These include carbon regulation (half the primary photosynthetic production harnessing solar energy, occurs in oceans), acting as a sink for acidic and carbonic gases produced on land, nutrient cycling, climate regulation, and provision of seafood. The ocean plankton produces over half the oxygen in the atmosphere and nearly 15% of the world's protein is from seafood. The global catch of fisheries is approximately 90 million tons per year (Roney, 2012) and the fishery serves as a perennial source of renewable food. Overfishing and general mismanagement of the resource, however, have resulted in more than half the global fish stocks being already fully exploited. Having a healthy, vibrant, functional ocean system is critical to sustainability of life on land.

But the health of world's oceans has already been seriously compromised and is rapidly deteriorating. Gradual acidification of oceans from burning of fossil fuel on land, climate change–driven increase in the average water temperatures, eutrophication in areas where nutritive pollutants contaminate the waters, and drastic changes

Plastics and Environmental Sustainability, First Edition. Anthony L. Andrady.
© 2015 John Wiley & Sons, Inc. Published 2015 by John Wiley & Sons, Inc.

in native ocean communities due to exotic invasive species, are all growing stresses on the ocean ecology. The pollution load on the oceans from chemicals, oils, and fertilizers has increased dramatically over the recent decades (Islam and Tanaka, 2004). A recent addition to this burgeoning list of ocean stressors is the influx of plastic waste into the oceans (Copello and Quintana, 2003; Derraik, 2002; Moore, 2008; Moore et al., 2001; Moret-Ferguson et al., 2010). The waste ranges from large plastic objects (for instance building debris from the Japanese tsunami) to the invisible plastic microparticles dispersed in seawater and in beach sand, worldwide (Saido, 2014).

The great majority of this plastic litter is derived either from post-use consumer plastics on land or from plastic fishing gear. With the world's fishing fleet almost exclusively relying on plastic gear, the latter is a significant contributor to marine plastics debris, and ghost fishing by derelict gear is a growing worrisome concern (Matsuoka et al., 2005). Crab pots are a particularly visible variety of derelict gear (Anderson and Alford, 2014). A survey of an area of 2 sq. mi in the Bering Sea (Stevens et al., 2002) found 200 derelict crab pots[1]! If extrapolated to the highly-fished areas of the Bering Sea, that would amount to 1.7 million derelict crabpots. In Chesapeake Bay area, 32,000 lost traps (over four seasons) were recently recovered and an estimated 900,000 marine animals were affected by their presence in the coastal ocean bottom (Bilkovic et al., 2012, 2014; Galgani et al., 2010).

Packaging plastics used in beach environments litter the coastal areas worldwide. Beach cleanups generally find the same set of common plastics used in packaging to be the major components of litter (Gregory, 1999; McDermid and McMullen, 2004; Ng and Obbard, 2006; Topçu et al., 2013). The size ranges of debris reported in such surveys are incomplete as they exclusively focus on beaches and on visible items overlooking those items smaller than a few millimeters in size. Enumerating plastics floating on surface water is even more difficult. Some of these plastics (polyethylene (PE), polypropylene (PP), and expanded polystyrene foam (EPS)) float in seawater, and others (such as poly(vinyl chloride) (PVC), poly(ethylene terephthalate) (PET), or poly(styrene (PS)) being relatively denser resins, sink and reach mid-water column or the benthic sediment. But as already discussed in Chapter 3, this observation on density pertains only to virgin resin prils; fillers and additives added to plastic materials used to make products, can change its density. Each year a small fraction of the 280 MMT of plastic resins produced in the world invariably enter the oceans as accidental spills, gear losses, or as postconsumer waste (Thompson et al., 2005). This fraction is not reliably known, though recent estimates (EC, 2013) suggest 10–20 MMT [4.8-12.7 MMT is a recent second estimate] of influx annually. Plastics as already discussed in Chapter 6 biodegrade extremely slowly, especially when they are in seawater or sediment. This suggests that all plastics that entered the oceans since the beginning of the plastics age (in 1950s) still very likely remain unmineralized, accumulated intact in the marine sediment.

---

[1] On the average, the survey found 1.5 crabs per pot, but some pots contained up to 125 crabs. The biodegradable panels and biodegradable hinges, either mandated or voluntarily used, do not appear to be a complete solution to this problem.

Plastic debris has been observed in the marine benthos (Barnes et al., 2010). The ecological impacts of plastics collecting in the benthos on the organisms living there and on the marine ecosystem, in general, remain unclear and unresearched.

The problem of plastic debris in the oceans will continue to grow in the future with more plastics being added each year into the ecosystem. Also, demographic patterns show preferential migration of human population to coastal regions. Already about half of the world's population live within 200 km of a coastline and that number will double by 2025 (Creel, 2003). The recreational use of beaches and the ocean (especially boating) is on the increase making it even more likely that plastics and other wastes will end up in the oceans. Once in water, plastics are dispersed in oceans worldwide; plastic debris has even been detected in waters in Antarctica (Barnes et al., 2010).

10.1   ORIGINS OF PLASTICS IN THE OCEAN

Plastics debris present in the oceans can be traced back to several sources:

1. Plastics derived from commercial fishing: With the entire global fleet now using plastic gear (Butler et al., 2013; Veenstra and Churnside, 2012; Watson et al., 2006), inadvertent losses and deliberate disposal of post-use gear at sea is an obvious source of debris. With an estimated 4.4 million fishing vessels (most in Asia) in operation, this can be a significant source of input of plastics into the ocean. In addition to gear, polystyrene foam (EPS) products are also disposed from vessels. EPS (as well as some polyurethane) is used in vessel insulation, floats on gear, baitboxes, and food service items.

2. Packaging waste from MSW and beach use: About a third of the plastics resin produced globally ends up as plastic packaging of which less than 9% of waste in the MSW is recovered for recycling. Plastic litter on land can be transported to the oceans via storm drain runoff. Recreational use of beaches also results in beach litter that can be picked up by wind and tidal movements and transported into the water (Corcoran et al., 2009; Rosevelt et al., 2013; Ryan et al., 2009). Beach cleanup operations help reduce the litter but there is virtually no mechanism to collect such debris once they enter the water.

3. Waste from naval[2] and commercial vessels: The US Naval ships no longer discard the plastic waste at sea but bring them back to the shore for recycling.

Common types of plastics found as ocean debris are listed along with their densities in Table 10.1. Those denser than seawater ($\rho \sim 1.025\,g/cm^3$) will sink and invariably end up in the marine sediment. The long-term fate of even the floating plastics, however, will be the same. A surface biofilm of bacteria and diatoms develops

---

[2] The United States has the world's largest naval fleet of 341 warships at sea. Larger US naval vessels are equipped with onboard plastic waste processors that melt the plastic and carry them back to port for recycling.

TABLE 10.1 Plastics Commonly Found in Ocean Debris

Plastic	Density (g/cm^3)	Typical products	Production (%)[a]
LDPE/LLDPE	0.90–0.94	Packaging film, six-pack rings, containers, gear	21
HDPE	0.94–0.97	Juice and milk containers, fishing gear	17
PP	0.90	Rope, packing bands, bottle caps, fishing gear	24
PVC	1.4	Plastic films and other packaging material	19
PS	1.04[a]	Food service utensils, packaging and EPS	6
PA (Nylon)	1.02–1.05	Fishing gear, monofilament line and film	<3
PET	0.96–0.97	Soda bottles and packaging bands	7
CA	1.28	Cigarette filters	—

Source: Data based on Andrady (2011).
[a] Expanded polystyrene foam (EPS) is much less denser than seawater.

on the samples almost immediately once in water (Lobelle and Cunliffe, 2011). This biofilm is soon colonized by a mix of hydroids, ectocarpales, bryozoans, and barnacles (Andrady and Song, 1991). How rich the foulant layer will grow to be and the course of succession of species therein, varies with the location of exposure. Under the weight of the fouling organisms, especially the encrusting epibionts, floating plastics will eventually sink in seawater (Andrady and Song, 1991; Costerton and Cheng, 1987; Railkin, 2003). However, once submerged, the rich surface growth on these is foraged by fish (or the algae dies off due to lack of light) and the plastic now rendered less dense reemerge may to the surface. A slow cyclic "bobbing" motion of floating plastic debris attributed to this cyclic change in density on submersion below a certain depth of water, first proposed by Andrady and Song (1991), was later confirmed by others (Stevens, 1992; Stevens and Gregory, 1996). Invariably, these plastics also end up in the mid-water column or the sediment. The ocean sediment already shows signs of significant plastic pollution (Backhurst and Cole, 2000; Katsanevakis et al., 2007; Stefatos and Charalampakis, 1999). For instance in the NE Mediterranean, 135 bottom trawls samples yielded on the average 0.72 kg of debris per hour of trawling and 73% of it was plastic (Eryaşar et al., 2014). Packaging material was the most abundant (53%) while fishing gear amounted to only 7%.

Most plastics debris encountered in oceans tends to be the commodity thermoplastics, primarily PE and PP, in the surface water samples. The expected high levels of nylons (PA) from fishing gear are not reflected in floating plastic debris because of their high density. Products such as cigarette filters or empty bottles, though made of denser varieties of plastics, often float in seawater for awhile because of entrained air and are at times counted in floating debris.

TABLE 10.2 Summary of Impacts on Marine Animals

	Species worldwide	Species with reports of entanglement (%)	Species with reports of ingestion (%)
Sea turtles	7	86	86
Sea birds	312	51	36
Penguins	16	38	6
Albatross, petrels, shearwater	99	10	63
Shorebirds	122	18	33
Marine mammals	115	28	23
Baleen whales	10	60	20
Toothed whales	65	8	32
Fur seals and Sea lions	14	79	7
True seals	19	42	5

Source: Data compiled from Mudgal et al. (2011).

Large floating plastic debris items such as net fragments (fishing gear), rope, six-pack rings, and plastic bands are well documented to entangle marine animals including large marine mammals (Hofmeyr et al., 2006; Waluda and Staniland, 2013). Often, the entanglement restricts the freedom of the distressed animal to graze and grow. An early review (Laist, 1997) reported at least 135 marine species to suffer from entanglement in plastics debris. Smaller fragments of plastics, especially those covered with algae and biota, are mistaken for food and ingested by a wide range of marine animals (Ryan et al., 2009; van Franeker et al., 2011), sea turtles (Bugoni et al., 2001), and fish (Jantz et al., 2013; Lusher et al., 2012; Possatto et al., 2011). Most extensively documented cases of ingestion are for marine bird species (Tanaka et al., 2013). About a third of the fish sampled (across 10 species) both in the English Channel (Lusher et al., 2012) as well as the North Pacific Central Gyre (Boerger et al., 2010) region showed 1.9–6 pieces of plastics per animal. While plastics are not toxic in the conventional sense, they not only cause physical obstruction of the digestive tract but also their presence in the stomach elicits a sense of satiation, discouraging feeding (Wright et al., 2013). This can interfere with growth and reproduction of these animals. Recent data from studies on fish (*Medka*) suggest, however, that ingestion of even virgin plastics can also result in histological changes associated with stress (Rochman et al., 2013b). A summary of the different marine species known to ingest plastics is given in Table 10.2.

10.2 WEATHERING OF PLASTICS IN THE OCEAN ENVIRONMENT

Of the different mechanisms of degradation discussed in Chapter 6, photo-initiated oxidation (solar-UV facilitated oxidation) plays a dominant role in degrading plastics in the marine environment (Andrady, 2011). Weathering of products such as netting, bands, or six-pack rings reduces damage to marine animals entangled in them as the degradation and weakening reduces the likelihood of distress in entanglement. It is also an important

TABLE 10.3   Degradation Agencies Available in Different Zones in the Marine
Environment

Zone	Description	Agencies	
Supralittoral	Beach above water line	Photodegradation (accelerated)[a]	~20% oxygen
Intertidal	Beach between tidal marks	Photodegradation	~20% oxygen
Surface water	Water surface	Photodegradation	~5 ml/l of seawater
Deep water and sediment	Ocean bottom	Very slow biodegradation	—

[a] The degradation is accelerated because of high sample temperatures.

phenomenon because weathering likely generates microplastics from larger fragments of plastic waste and the presence of these in oceans affects a wide range of smaller marine invertebrates. The rates of weathering degradation of plastics obtained in different zones within the marine environment are very different due to differences in temperatures and solar irradiation. Table 10.3 shows the different zones and their characteristics with respect to weathering of plastics.

10.2.1   Beach (Supralittoral) Zone

Plastic debris lying on beaches, exposed to direct sunlight, undergo photooxidative degradation facilitated by the UVR in sunlight. Plastics also absorb solar infrared radiation causing their surface temperature to rise well above the ambient (the process being called "heat build-up"), especially in darker colored materials. Depending on the season, plastic debris lying on sandy beach surfaces can reach relatively high temperatures that promote faster oxidative degradation of plastics (François-Heude et al., 2014; Ho et al., 1999). The activation energy for degradation of polyolefin being $\Delta E \sim 50\,kcal/mol$, a small change in temperature, can make a significant difference in the rate of degradation.

The combination of UV-B radiation and elevated temperature results in rapid breakdown of plastic debris on beaches. Numerous studies have been carried out on weathering of common plastics in beach environments. Andrady and Pegram (1991) and Andrady et al. (1993a) have reported degradation rates for plastics commonly found in ocean debris, exposed in beach environments. The degradation mechanisms and kinetics for beach exposure are not any different from exposures of plastics samples on land. The photodegradation of fishing gear (Al-Oufi et al., 2004; Meenakumari and Radhalakshmi, 1995, 1988), polyethylene netting and twine (Meenakumari and Ravindran, 1985a, b), and nylon monofilament (Meenakumari and Radhalakshmi, 1988; Thomas and Hridayanathana, 2006) in beach environments have been reported. Mechanistically and kinetically, the weathering of plastics exposed in the supralittoral zone is very similar to that on land. These mechanisms and factors affecting degradation on land are well known (Andrady, 1996; Andrady et al., 2009; Hamid, 2000).

10.2.2 Surface Water Zone

Weathering rates for plastic waste floating in surface waters of the ocean are markedly lower compared to that exposed on beaches or on land (Andrady and Pegram, 1989b, 1990b, 1993). There are several reasons responsible for this retardation in degradation in water.

1. The plastics floating in seawater precludes heat build-up and consequent increase in sample temperatures. The plastic remains at the lower seawater temperatures slowing down degradation relative to debris exposed on beaches.
2. The surface of the plastic in seawater is readily fouled by a variety of marine species. This thick opaque foulant layer shields the plastic from solar UV-B radiation and retards oxidation.
3. Oxidation rates depend on the availability of oxygen in the environment. Unlike in air ($\sim$20% O_2), seawater has a much lower concentration of oxygen (6.6 mg/l at 25°C).

10.2.3 Deep Water and Sediment Zones

Beyond the photic zone, photodegradtion is much slower; the cold, anoxic deep-water environment is not conducive to oxidative degradation of plastics. The same is true of the marine sediment where plastics are believed to ultimately accumulate. While, the occurrence of plastics debris in the sediment has been demonstrated, interaction of such debris with bottom-dwelling organisms is not well understood at the present time (Watters et al., 2010).

10.2.3.1 Comparison of the Weathering Rates in Different Zones The difference in the rates of weathering is illustrated when the rate of change in tensile extensibility of polypropylene sheets, exposed floating in seawater in Biscayne Bay (FL) is compared to that on land at the same location (Fig. 10.1). The figure also shows the marine exposure setup used, consisting of a frame with floats, the test samples were tethered to. The same retardation of degradation is also seen with a range of different types of plastics products commonly found in beach litter (Andrady, 2003; Andrady and Pegram, 1989a, 1990a) as well as with enhanced photodegradable plastics (such as the degradable six-pack rings). Consistently, the rate of loss in tensile elongation (%) of samples was reported to be slower in samples-exposed floating in seawater compared to those exposed in air. For instance, the decrease in the average tensile extensibility of plastic laminates exposed on the beach and floating in sea water for a period of 12 months in Beaufort, NC, was as follows: low density polyethylene (LDPE) films lost 95% in air and approximately 2% in seawater; PP tape lost 99% in air but only 25% in the seawater exposure. Not only is the average tensile elongation a particularly sensitive indicator of weathering or degradation but it is also directly relevant to entanglement hazard posed by plastic debris. Animals entangled in weathered and therefore weakened plastic debris (e.g., netting) can free themselves with minimal harm.

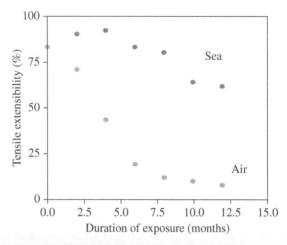

FIGURE 10.1 *Upper:* Change in percent original tensile extensibility of polypropylene laminate exposed in air and floating in seawater at a beach location Biscayne Bay, FL. *Lower:* A floating rig used to expose plastics to surface water environment (in Miami Beach, FL). Reproduced with permission from Andrady and Pegram (1989b).

Generally, the tensile extensibility of plastics decreases exponentially with the time of exposure allowing the process to be modeled as follows.

$$\log\left\{\text{extensibility }(\%)\right\} = A + 10^{\left\{-B \cdot t (\text{days})\right\}}$$

where t is the duration of exposure.

The value of B (days^{-1}) listed in Table 10.4 for films of LDPE, ethylene-carbon monoxide (1%) copolymer (ECO), and photodegradable LDPE (containing a metal catalyst pro-oxidant) quantify their rates of degradation. The ECO copolymer is the same as that used in photodegradable six-pack rings (and supplied by the manufacturer

TABLE 10.4 Summary of Results for Degradation of LDPE Control Samples, ECO Copolymer, and Metal-catalyzed Polyethylene Exposed in Air and Floating in Water at Different Locations

Location	Air		Floating	
	B (days^{-1})	r^2	B (days^{-1})	r^2
Biscayne bay				
LDPE	14	0.91	—	—
ECO copolymer	69	0.92	45	0.86
LDPE with metal catalyst	72	0.97	27	0.97
Puget sound (WA)				
LDPE	4	0.77	—	—
ECO copolymer	40	0.88	11	0.65
LDPE with metal catalyst	54	0.96	18	0.87
Freshwater lake (VA)				
LDPE	14	0.91	1	0.66
ECO copolymer	89	0.83		
LDPE with metal catalyst	8	0.78	8	0.83

Source: Based on data from Andrady et al. (1993a, b).

ITW HiCone Company) and is a product intended to breakdown at a faster rate outdoors. The metal-catalyzed LDPE film contained low levels of an iron catalyst and was supplied by the manufacturer (Plastigone Inc., Miami, FL). Again, the rates of degradation of these enhanced photodegradable plastics (as well as control LDPE laminates) are much slower for samples floating in water compared to those exposed in air (on land), at the same location. The table also shows that enhanced photodegradable plastic technologies do perform with plastics floating in water, though the rates of weakening are relatively slower compared to exposures in air. The values of coefficients A and B can be estimated by fitting the tensile data with the above equation.

Latex rubber balloons, though not made of plastics, still contribute significantly to marine debris (O'Shea et al., 2014). Toy helium balloons and the larger weather balloons are released into the atmosphere every day (Whiting, 1998). As a balloon rises in air it expands as the air pressure gradually decreases, and at a height of approximately 4–5 miles, these bursts with the rubber falling back to Earth. Weather balloons rely on bursting to recover the radiosonde equipment with atmospheric data recorded during the ascent of the balloon. Where the balloons will fall is difficult to assess; with greater than 70% of Earth's surface covered with oceans, a large fraction of these invariably ends up in oceans. Latex rubber balloons do biodegrade slowly in the ocean environment (Lambert et al., 2013), but the rate of loss in mechanical integrity due to such degradation or their rate of mineralization are not known. In an outdoor exposure study, natural rubber latex balloons were found to lose approximately 94% of their extensibility in air but only approximately 38% when exposed floating in seawater for a 6 month period in Beaufort, NC (Andrady and Pegram, 1989a) (see Fig. 10.2).

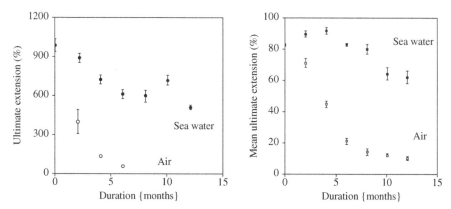

FIGURE 10.2 A comparison of the rate of loss in extensibility of latex rubber balloons in Beaufort, NC (left) polypropylene tape in Biscayne Bay, FL (right) exposed outdoors in air and in sea water. Source: Reproduced with permission from Pegram and Andrady (1992).

10.3 MICROPLASTIC DEBRIS

A relatively recent concern in the marine environment is the widespread occurrence of microscale particulate plastics or microplastics (Cole et al., 2011; Thompson et al., 2004). While the size range that is considered "micro-" is not well defined in the literature, there is growing agreement on "microplastics" being defined as debris in the size range of $1\,\mu m$ to $1\,mm$. Other sizes of debris would then fall in different categories as follows: nanoplastics $= <1\,\mu m$; mesoplastics $= 1\,mm$ to $2.5\,cm$ and macroplastics $2.5\,cm$ to $1\,m$. Virgin resin pellets, an important component of beach debris, will according to this scheme be classified as mesoplastics.[3] Defining these size ranges is important because their interactions with marine plankton, especially via ingestion, will be size-dependent.

Microplastics have been isolated from surface water, mid-water, marine sediment, and beach surface in all the world's oceans (Moore, 2003). Majority of the studies have used neuston plankton nets ($333\,\mu m$ mesh size) on surface waters to collect and enumerate samples. Because of the mesh size of nets used, most of the fragments collected are greater than $300\,\mu m$ in size. Depending on the location, anywhere from less than 10 particles/m^3 to 10^3 particles/m^3 of seawater has been reported (highest ever count is $\sim 10^7\,m^2$ surface). The collection and quantification of microplastics in the ocean has been recently reviewed (Hidalgo-Ruz et al., 2012).

Isolation of micro- and mesoplastics from debris samples that also include biomass is particularly difficult. Floatation (or density-based separation) alone does not completely isolate the plastic fragments from marine debris. Digestion of sample to solubilize and remove the biomass has been suggested: acid digestion (Andrady,

---

[3] What is generally referred to as "microplastics" in the research literature is mostly mesoplastics with a smaller fraction of microplastics.

2011; Claessens et al., 2013), alkaline hydrolysis (Jin et al., 2009), and peroxide digestion (Nuelle et al., 2014) have been used. Recently, proteinase-K digestion was demonstrated to solubilize greater than 97% (wt) of the biomass with no apparent effect on the plastic fraction (Cole et al., 2013). The technique might also be used to isolate ingested microplastics from zooplankton samples as well. Chemical digestion does not affect most of the common thermoplastics except for some condensation polymers such as polyamides and polyesters.

10.3.1 Primary and Secondary Microplastics

Meso- and microplastics can be of either primary or secondary in origin. The primary microplastics enter the ocean already as microscale particulate material. The same is true for some of the mesoplastics; for instance, virgin plastic prils or pellets enter the oceans as accidental spills during transport and also with runoff from plastic processing operations (Doyle et al., 2011; Ogata et al., 2009). Industry is making an effort at minimizing the loss of virgin pellets to control their influx into the sea. Small plastic beads that are used in sandblasting (beads used in place of sand) are also washed into the oceans. These microplastics may even carry metal residues picked up from their use. Some cosmetic formulations include microbeads as well used as an exfoliant in their formulations (Fendall and Sewell, 2009). Another significant source of micro-particles and microfibers is that from laundering synthetic fiber clothing (Browne et al., 2011). A single garment laundered in a washing machine can shed nearly 2000 microfibers. In all these instances, pre-formed microscale plastic enters the ocean environment.

Secondary meso- and microplastics on the other hand are generated in the marine environment itself, derived from the weathering degradation of larger plastic debris (Andrady, 2011; Gregory and Andrady, 2003). Breakdown of plastics in the environment due to extensive photo-induced oxidation facilitated by sunlight is well known. Oxidative processes in plastics are likely localized to surface layers exposed to solar UV radiation. Plastics products and virgin prils typically have smooth surfaces. Plastic debris on beaches as well as in water, however, are well known to show uneven surface textures including pits, cracks, and flaking (Cooper and Corcoran, 2010) due to weathering degradation. The microcrack pattern is generally indicative of weathering damage as opposed to mechanical fragmentation (Zbyszewski et al., 2014). Similar surface cracks often develop in plastics exposed to UV radiation in the laboratory (Akay et al., 1980) with LDPE (Küpper et al., 2004; Tavares et al., 2003), polycarbonate (Blaga and Yamasaki, 1976) and polypropylene (Qayyum and White, 1993; Yakimets et al., 2004). Micrographs illustrating the surface textures are shown in Figure 10.3. It is reasonable to assume that missing fragments from these surface textures contribute to marine micro- and nanoplastics. With virgin plastic pellets that have no UV stabilizers in them, the weathering process is relatively fast compared to that in plastic products. A weakened, embrittled, microcracked surface can readily fracture, shedding off microplastics due to water movement, abrasion against sand, or due to encounters with marine animals (Fendall and Sewell, 2009). The smallest dimension of

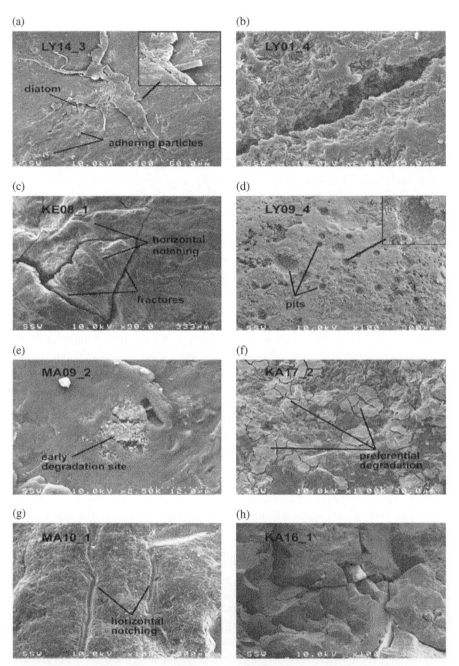

FIGURE 10.3 SEM images of different surface textures on plastic beach debris samples. (a) Flaking of surface, (b) vermiculite texture, (c) microfracture of surface, (d) surface pitting, (e) signs of initial degradation, (f) regions of preferential degradation, (g) horizontal notching from cracks, (h) deep cracks and fractures. Source: Reproduced with permission from Cooper and Corcoran (2010).

microplastic particles so far observed in the marine environment is 1.6 µm (Galgani et al., 2010), but smaller particles including nanoscale microparticles should also be present in seawater. The smaller the particle size, the more difficult it is to isolate and image the particle.

Extensive photothermal oxidation occurs at higher temperatures in plastics on beaches or at the waterline. Plastics disposed directly into water, however, are unlikely to be extensively degraded as discussed previously. The identification of the beaches as the primary potential site for generation of microplastics underlines the importance of beach cleanup efforts as a mitigation strategy. Removing larger debris items from the beaches likely reduces the chance of the debris generating microplastics via weathering degradation.

10.3.2 Persistent Organic Pollutant in Microplastics

The most serious concern relating to micro- and mesoplastics is their potential ingestion by marine organisms. Ingestion of plastic fragments can sometimes cause distress to marine animals from physical obstruction of their gut passage. Much more serious is the hazard posed by their potential delivery of the toxic pollutant species concentrated from seawater to the ingesting organism to the marine food web and perhaps even to human consumers.

Numerous organic compounds from industrial activity and agricultural runoff are present in seawater at very low concentrations. These have been detected in all oceans as well as in bodies of fresh water. Most persistent organic pollutants (POPs) of interest being hydrophobic and less denser than seawater tend to concentrate in the surface layers of the ocean. The common micro plastics (PE, PP, and EPS) debris also reside in the same stratum of water because of their relatively density, allowing facile partitioning of POPs into the hydrophobic plastics. At equilibrium, concentration of POPs are several orders of magnitude higher in the plastic particles compared to that in seawater (Endo et al., 2005; Koelmans et al., 2013; Teuten et al., 2007). Therefore, meso- and microplastics are a credible means of transport of POPs in the oceans (Zarfl and Matthies, 2010). The crucial question, however, is the bioavailability of the sorbed POPs to organisms ingesting microparticles and how serious the consequent adverse impacts might be at species and population levels.

The term *partition coefficient* refers to the equilibrium distribution of a solute between two phases separated by a boundary. In this case, the two phases are the solid plastic and seawater. Any compound (such as a POP) in seawater will readily diffuse into the hydrophobic plastic as well resulting in an equilibrium concentration of the compound in the plastic will be very much higher than that in seawater (Ogata et al., 2009). The partition coefficient $K_{P/W}$ is the ratio of the concentrations, and in the case of plastics in seawater can be expressed as follows:

$$K_{P/W} = \frac{[q_e]_P}{[C_e]_W}$$

TABLE 10.5   Estimated Values of Log $K_{PE/sw}$, Log $K_{PP/sw}$, and Log $K_{PS/sw}$ for Selected Model POPs

Chemicals	log $K_{ow}{}^a$	log $K_{PEs/w}$	log $K_{PP/sw}$	log $K_{PS/sw}$
Phenanthrene	4.52	4.44	4.00	5.39
Fluoranthene	5.20	5.52	4.79	5.91
Anthracene	4.50	4.77	4.29	5.61
Pyrene	5.00	5.57	4.80	5.84
Chrysene	5.86	6.39	5.51	6.63
Benzopyrene	6.35	7.17	6.10	6.92
Dibenzanthracene	6.75	7.87	7.00	7.52
Benzoperylene	6.90	7.61	6.69	7.15

Source: Based on Table 1 in Lee et al. (2014).

where q_e and C_e are the equilibrium concentrations of the POP compound in the plastic and in water.[4] The higher the $K_{P/W}$ value, the more efficient is the plastic in concentrating the POP from seawater. Typical values of $K_{P/W}$ for some common organic pollutants are given in Table 10.5. As seen from the table, the values are large and plastic-specific. Rochman et al. found PS to sorb higher equilibrium concentrations of polycyclic aromatic hydrocarbons (PAH) compared to PP, PET, or PVC and about the same as for high density polyethylene and LDPE (Rochman et al., 2013c). The relatively higher available free volume and aromaticity of PS were probably responsible for the enhanced solubility.

A value of log $K \sim 5$ for instance shows that the equilibrium concentration of the compound in plastic to be 100,000 higher than that in seawater. Essentially, the presence of plastics at sea tends to clean the water of the POPs! Despite their low concentration in water, large K values suggest very high equilibrium concentrations of POPs to be reached in micro- and mesoplastics (Rochman et al., 2013a). At least in theory, plastics debris can sorb POPs to the saturation point at a highly polluted locale, and on drifting to a second cleaner-water environment subsequently releases the contaminant into water (Endo et al., 2013). Contribution to the local concentrations of POPs in seawater due to this mechanism, however, is likely not significant because of the minimal mass fraction of carrier microplastics in sea water.

Rios and Moore (2007) found mesoplastic samples on four Hawaiian, one Mexican, and five Californian beaches to have high levels of POPs; the ranges of values reported were Σ PAH = 39–1200 ng/g, ΣPCB PCB = 27–980 ng/g, and ΣDDT 27–7100 ng/g (cumulative values for all congeners.) A US beach study (mainly in Californian beaches) (Ogata et al., 2009) found somewhat lower values: ΣPAH

---

[4] Alternatively, the following equation based on the Freundlich sorption isotherms can also be used

$$\log[q_e]_P = \log K_f + \left(\frac{1}{n}\right)\log C_e.$$

PAH = 32–605 ng/g; ΣPCB = 2–106 ng/g, ΣHCH HCH = 0–0.94 ng/g. Had the samples been collected from beaches in industrial areas, these concentrations might have been much higher.

10.3.3   Ingestion of Microplastics by Marine Species

The potential bioavailability of POPs to organisms, especially zooplanktons, that ingest microplastics is an important issue as it presents the potential for contaminating the marine food web. The POPs dissolved in the plastics ingested, might be released in the gut and be bioavailable to the organism, depending on the residence time of the plastic in the gut. A wide range of species from the zooplanktons to baleen whales[5] ingest the plastic particles. Some species such as birds are believed to mistake plastics fragments for food items (Robards et al., 1995) and has been observed to feed plastics to their young ones. In the case of marine birds a robust positive relation between the ingestion of meso- and macroplastic and tissue concentrations of PCBs been demonstrated (Teuten et al., 2009; Yamashita et al., 2011). Data recently available for lugworms (*Arenicola marina*) and (Besseling et al., 2013) fish species medaka (*Oryzias latipes*) (Rochman et al., 2013b) also indicates the same.

Zooplanktons are well known to ingest plastic fragments (Cole et al., 2011; Frias et al., 2014; Murray and Cowie, 2011); plastic beads have been used in grazing studies in the literature (Powell and Berry, 1990). Pacific krill when presented with a mix of its stable phytoplankton and microbeads of LDPE were observed to ingest both (with little or no grazing bias (Andrady, Unpublished data)). Some typical examples of species known to ingest microplastics are summarized in Table 10.6. However, these are generally laboratory studies and observing ingestion of microplastics in the field is a challenging undertaking.

Ingestion of POPs-laden microplastics by zooplankton- and phytoplankton-feeding fish is a concern because the high loading of POPs in them may transfer to body tissue and potentially move up the trophic levels (Farrell and Nelson, 2013), possibly eventually reaching the human consumer. The concern is based on the assumption that POPs in microbeads will be released in the gut of ingesting species. Studies suggest that gut surfactants tend to enhance (Bakir et al., 2014; Sakai et al., 2000; Wright et al., 2013) the release of the POPs increasing the likelihood of their bioavailability. Bioavailability also depends primarily on the residence time of the microplastics and the hydrophobicity of the gut contents of the ingesting species. Species-specific information on gut-environment and residence times are not available to reliably quantify the bioavailability of the POPs and to estimate possible bio-magnification of their concentration on transfer across trophic levels.

However, the contribution to the body-burden of POPs in human consumers via this mechanism is likely to be low. Other types of media (air, water, dust) in combination with ingestion with food will generally be more significant routes of exposure.

---

[5] Using phthalate levels (derived from plastics) in the tissue of larger filter feeding organisms such as whales as indicators of microplastics pollution has been suggested (Fossi et al., 2012). However, plasticized PVC that incorporate phthalates is only a small fraction of the plastics debris sampled at sea.

TABLE 10.6   A Summary of Selected Studies on the ingestion of Microparticles
by Marine invertebrates

Organism(s)	Microplastics (μm)	Identification techniques	Publications
Copepods (*Acartia tonsa*)	7–70	Microscopy	Wilson (1973)
Echinoderm larvae	10–20	Video observation	Hart (1991)
Trochophore larvae (*Galeolaria caespitosa*)	3–10	Microscopy	Bolton and Havenhand (1998)
Scallop (*Placopecten magellanicus*)	16–18	Detection of ^{51}Cr labeled particles	Brillant and MacDonald (2002)
Amphipods (*Orchestia gammarellus*), lugworm (*Arenicola marina*), and barnacle (*Semibalanus balanoides*)	20–2000	Dissection and wormcast examination	Thompson et al. (2004)
Mussels (*Mytilus edulis*)	2–16	Dissection and fluorescence microscopy	Browne et al. (2008)
Sea cucumbers	Various	Excrement analysis	Graham and Thompson (2009)
Mysid shrimps, copepods, cladocerans, rotifers, polychaete larvae and ciliates	10	Dissection and fluorescence microscopy	Setälä et al. (2014)

Source: Reprinted with permission from Cole et al. (2011).

10.4   OCEAN LITTER AND SUSTAINABILITY

The presence of plastics debris in general and microplastics in particular in the ocean environment is a variant of the urban litter problem. Unlike urban litter, however, both beach litter and microplastics are transported by currents, eventually ending up in coastal communities or on pristine beaches far away from their point of origin. They particularly concentrate in several gyres (the North Pacific Gyre "garbage patch" is widely quoted in the press[6]). In the gyre regions, the incidence of microplastics in water is statistically higher than in other areas. It is also a far more difficult problem to deal with than urban litter as collection of these is both expensive and impractical. A majority of plastics eventually end up in the benthic sediment, and their impact on the bottom-dwelling fauna is important, but has not been extensively studied as yet.

The potential transfer rates of POPs within the marine food web and the adverse impact it can have on marine ecosystem services provided by the oceans is a serious

[6]There is no "patch" or island of floating plastic debris that is visually discernible. The edges of the "patch" are diffused and cannot be easily established to reliably assess the area of the "patches."

concern. Such impacts might manifest at two different levels. With some species (for instance in fish bivalves in seafood (Van Cauwenberghe, 2014)), POPs might be present at a level where they might be of concern to the human consumers. More importantly, the major impact might be on zooplankton populations that constitute the base of the marine food web supporting the ocean fishery and perhaps even perturb the ability of the oceans to fix carbon. If such trophic level transfer does occur at all is not known and the magnitude of the adverse impacts have not been reliably estimated. There is not enough information to assess the likelihood or the magnitude of potential adverse impacts on the oceans.

Sustainable use of plastics requires that oceans are not polluted by litter of any size range, but especially by microoplastics (and those of a smaller dimension) that can interact with the organisms at lower levels of the food web. It also demands research efforts into understand the extent of pollution in different ocean regions at the present time, potential impacts of plastics already in the oceans, and developing viable mitigating strategies to manage this environmental problem. Provided the functionality of the gear is not compromised by their use, this might be a niche application to use biodegradable plastics in synthetic fiber fishing gear and other plastics used on beaches. The value of enhanced photodegradation of plastics on beaches, however, is not entirely clear; while it avoids entanglement of animals, the material also fragments at an accelerated rate yielding microplastics.

REFERENCES

Akay G, Tinçer T, Ergöz HE. A study of degradation of low density polyethylene under natural weathering conditions. Eur Polym J 1980;16 (7):601–605.

Al-Oufi H, McLean E, Kumar AS, Claereboudt M, Al-Habsi M. The effects of solar radiation upon breaking strength and elongation on of fishing nets. Fish Res 2004;66 (1):115–119.

Anderson JA, Alford AB. Ghost fishing activity in derelict blue crab traps in Louisiana. Mar Pollut Bull 2014;79 (1–2):261–267.

Andrady AL. Chapter 40: Effects of UV Radiation on Polymers. In: *Handbook of Polymer Properties*. American Institute of Physics Press; 1996. p 547–557.

Andrady AL. Microplastics in the marine environment. Mar Pollut Bull 2011;62 (8):1596–1605.

Andrady AL, Pegram JE. Outdoor weathering of selected polymeric materials under marine exposure conditions. Polym Degrad Stab 1989a;26:333.

Andrady AL, Pegram JE. Outdoor weathering of selected polymeric materials under marine exposure conditions. Polym Degrad Stab 1989b;26:333.

Andrady AL, Pegram JE. Weathering of polyethylene (LDPE) and enhanced photodegradable polyethylene in the marine environment. J Appl Polym Sci 1990a;39:363–370.

Andrady AL, Pegram JE. Weathering of polyethylene (LDPE) and enhanced photodegradable polyethylene in the marine environment. J Appl Polym Sci 1990b;39:363–370.

Andrady AL, Pegram JE. Weathering of polystyrene foam on exposure in air and in sea water. J Appl Polym Sci 1991a;42 (6):1589.

Andrady AL, Song Y. Fouling of floating plastic debris under Biscayne Bay exposure conditions. Mar Pollut Bull 1991b;22 (12):608–613.

Andrady AL, Pegram JE, Nakatsuka S. Studies on controlled lifetime plastics: 1. The geographic variability in outdoor lifetimes of enhanced photodegradable polyethylene. J Environ Polym Degrad 1993a;1 (1):31.

Andrady AL, Pegram JE, Song Y. Studies on enhanced degradable plastics: II. Weathering of enhanced photodegradable polyethylenes under marine and freshwater floating exposure. J Environ Polym Degrad 1993b;1 (2):117–126.

Andrady AL, Pegram JE. Studies on enhanced degradable plastics: II. Weathering of enhanced photodegradable polyethylenes under marine and freshwater floating exposure. J Environ Degrad 1993c;1 (2):117–126.

AL Andrady, P. J. Aucamp, A. Austin, A. F. Bais, C. L. Ballaré, P. W. Barnes, G. H. Bernhard, J. F. Bornman, M. M. Caldwell, F. R. de Gruijl, D. J. Erickson III, S. D. Flint, K. Gao, P. Gies, D.-P. Häder, M. Ilyas, J. Longstreth, R. Lucas, S. Madronich, R. L. McKenzie, R. Neale, M. Norval, K. K. Pandy, N. D. Paul, M. Rautio, H. H Redhwi, S.A. Robinson, K. Rose, M. Shao, R. P. Sinha, K. R. Solomon, B. Sulzberger, Y. Takizawa, X. Tang, A. Torikai, K. Tourpali, J. C. van der Leun, S-Å. Wängberg, C.E. Williamson, S. R. Wilson, R. C. Worrest, A. R. Young, R. G. Zepp. Environmental effects of ozone depletion and its interactions with climate change: progress report, 2008. Photochem Photobiol Sci 20098(1), 13.

Backhurst MK, Cole RG. Subtidal benthic marine litter at Kawau Island, North-eastern New Zealand. J Environ Manage 2000;6:227–237.

Bakir A, Rowland SJ, Thompson RC. Enhanced desorption of persistent organic pollutants from microplastics under simulated physiological conditions. Environ Pollut 2014;185:16–23.

Barnes DKA, Walters A, Goncalves L. Macroplastics at sea around Antarctica. Mar Environ Res 2010;70:250–252.

Beaumont NJ, Austen MC, Mangi SC, Townsend M. Economic valuation for the conservation of marine biodiversity. Mar Pollut Bull 2008;56:386–396.

Besseling E, Wegner A, Foekema EM, van den Heuvel-Greve MJ, Koelmans AK. Effects of microplastic on fitness and PCB bioaccumulation by the Lugworm *Arenicola marina* (L.). Environ Sci Technol 2013;47 (1):593–600.

Bilkovic DM, Havens KJ, Stanhope DM, Angstadt KT. Use of fully biodegradable panels to reduce derelict pot threats to marine fauna. Conserv Biol 2012;26:957–966.

Bilkovic DM, Havens K, Stanhope D, Angstadt K. Derelict fishing gear in Chesapeake Bay, Virginia: spatial patterns and implications for marine fauna. Mar Pollut Bull 2014;80 (1–2):114–123.

Blaga A, Yamasaki RS. Surface microcracking induced by weathering of polycarbonate sheet. J Mater Sci 1976;11:1513–1520.

Boerger CM, Lattin GL, Moore SL, Moore CJ. Plastic ingestion by planktivorous fishes in the North Pacific Central Gyre. Mar Pollut Bull 2010;60:2275–2278.

Bolton TF, Havenhand JN. Physiological versus viscosity-induced effects of an acute reduction in water temperature on microsphere ingestion by trochophore larvae of the serpulid polychaete *Galeolaria caespitosa*. J Plankton Res 1998;20:2153–2164.

Brillant M, MacDonald B. Postingestive selection in the sea scallop (*Placopecten magellanicus*) on the basis of chemical properties of particles. Mar Biol 2002;141:457–465.

Browne MA, Dissanayake A, Galloway TS, Lowe DM. Ingested microscopic plastic translocates to the circulatory system of the mussel, *Mytilus edulis*. Environ Sci Technol 2008;42 (13):5026–5031.

Browne MA, Crump P, Niven SJ, Teuten E, Tonkin A, Galloway T, Thompson R. Accumulation of microplastic on shorelines worldwide: sources and sinks. Environ Sci Technol 2011;45 (21):9175–9179.

Bugoni L, Krause L, Petry MV. Marine debris and human impacts on sea turtles in southern Brazil. Mar Pollut Bull 2001;42:1330–1334.

Butler JRA, Gunn R, Berry HL, Wagey GA, Hardesty BD, Wilcox C. A value chain analysis of ghost nets in the Arafura Sea: identifying trans-boundary stakeholders, intervention points and livelihood trade-offs. J Environ Manag 2013;123:14–25.

Claessens M, Van Cauwenberghe L, Vandegehuchte MB, Janssen CR. New techniques for the detection of microplastics in sediments and field collected organisms. Mar Pollut Bull 2013;70:227–233.

Cole M, Lindeque P, Halsband C, Galloway TS. Microplastics as contaminants in the marine environment: a review. Mar Pollut Bull 2011;62 (12):2588–2597.

Cole M, Webb H, Lindeque PK, Fileman ES, Halsband C, Galloway, TS. Isolation of microplastics in biota-rich seawater samples and marine organisms. Scientific reports 4, Article number: 4528. Nature Publishing Group; 2013.

Cooper DA, Corcoran PL. Effects of mechanical and chemical processes on the degradation of plastic beach debris on the island of Kauai, Hawaii. Mar Pollut Bull 2010;60 (5):650–654.

Copello S, Quintana F. Marine debris ingestion by Southern Giant Petrels and its potential relationships with fisheries in the Southern Atlantic Ocean. Mar Pollut Bull 2003;46:1513–1515.

Corcoran PL, Biesinger MC, Meriem G. Plastics and beaches: a degrading relationship. Mar Pollut Bull 2009;58 (1):80–84.

Costerton JW, Cheng KJ. Bacterial biofilms in nature and disease. Annu Rev Microbiol 1987;41:35–464.

Creel L. 2003. Ripple effects: population and coastal regions. Population Reference Bureau. Population, Health, and Environment Program. Downloaded from http://www.prb.org/pdf/rippleeffects_eng.pdf. Accessed on October 2, 2014.

Derraik JGB. The pollution of the marine environment by plastic debris: a review. Mar Pollut Bull 2002;44:842–852.

Doyle MJ, Watson W, Bowlin NM, Sheavly SB. Plastic particles in coastal pelagic ecosystems of the Northeast Pacific Ocean. Mar Environ Res 2011;71 (1):41–52.

EC. European Commission. On a European Strategy on plastic waste in the environment. Green Paper. [Report]. Brussels: European Commission; 2013.

Endo S, Takizawa R, Okuda K, Takada H, Chiba K, Kanehiro H, Ogi H, Yamashita R, Date T. Concentration of polychlorinated biphenyls (PCBs) in beached resin pellets: variability among individual particles and regional differences. Mar Pollut Bull 2005;50 (10):1103–1114.

Endo S, Yuyama M, Takada H. Desorption kinetics of hydrophobic organic contaminants from marine plastic pellets. Mar Pollut Bull 2013;74 (1):125–131.

Eryaşar AR, Özbilgin H, Gücü AC, Sakınan S. Marine debris in bottom trawl catches and their effects on the selectivity grids in the north eastern Mediterranean. Mar Pollut Bull 2014;81 (1):80–84.

Farrell P, Nelson K. Trophic level transfer of microplastic: *Mytilus edulis* (L.) to *Carcinus maenas* (L.). Environ Pollut 2013;177:1–3.

Fendall LS, Sewell MA. Contributing to marine pollution by washing your face. Microplastics in facial cleansers. Mar Pollut Bull 2009;58 (8):1225–1228.

Fossi MC, Panti C, Guerranti C, Coppola D, Giannetti M, Marsili L, Minutoli R. Are baleen whales exposed to the threat of microplastics? A case study of the Mediterranean fin whale (Balaenoptera physalus). Mar Pollut Bull. 2012;64 (11):2374–2379.

François-Heude A, Richaud E, Desnoux E, Colin X. Influence of temperature, UV-light wavelength and intensity on polypropylene photothermal oxidation. Poly Degrad Stab 2014;100:10–20.

Frias PGL, Otero V, Sobral P. Evidence of microplastics in samples of zooplankton from Portuguese coastal waters. Mar Environ Res 2014;95:89–95.

Galgani F, Fleet D, Franeker JV, Katsanevakis S, Maes T, Mouat J, Oosterbaan L, Poitou I, Hanke G, Thompson R, Amato E, Birkun A, Janssen C. Task group 10 report: marine litter. In: Zampoukas N, editor. Marine Strategy Framework Directive. Ispra: European Commission Joint Research Center; 2010.

Graham ER, Thompson JT. Deposit- and suspension-feeding sea cucumbers (Echinodermata) ingest plastic fragments. J Exp Mar Biol Ecol 2009;368:22–29.

Gregory MR. Plastics and South Pacific island shores: environmental implications. Ocean Coast Manag 1999;42 (6–7):603–661.

Gregory MR, Andrady AL. Plastics in the marine environment. In: Andrady AL, editor. Plastics and the Environment. Hoboken: John Wiley & Sons; 2003.

Hamid SH. Handbook of Polymer Degradation, Revised and Expanded 2nd ed, Editor. New York: Marcel Dekker; (2000).

Hart MW. Particle captures and the method of suspension feeding by echinoderm larvae. Biol Bull 1991;180:12–27.

Hidalgo-Ruz V, Gutow L, Thompson RC, Thiel M. Microplastics in the marine environment: a review of the methods used for identification and quantification. Environ Sci Technol 2012;46:3060–3075.

Ho KLG, Pometto AL, Hinz PN. Effects of temperature and relative humidity on polylactic acid plastic degradation. J Environ Polym Degrad 1999;7:83–92.

Hofmeyr GJG, Bester MN, Kirkman SP, Lydersen C, Kovacs KM. Entanglement of Antarctic fur seals at Bouvetøya, Southern Ocean. Mar Pollut Bull 2006;52:1077–1080.

Islam MS, Tanaka M. Impacts of pollution on coastal and marine ecosystems including coastal and marine fisheries and approach for management: a review and synthesis. Mar Pollut Bull 2004;48 (7–8):624–649.

Jantz LA, Morishige CL, Bruland GL, Lepczyk CA. Ingestion of plastic marine debris by longnose lancetfish (Alepisaurus ferox) in the North Pacific Ocean. Mar Pollut Bull 2013; 69:97–104.

Jin Y, Li H, Mahar RB, Wang Z, Nie Y. Combined alkaline and ultrasonic pretreatment of sludge before aerobic digestion. J Environ Sci 2009;21:279–284.

Katsanevakis S, Verriopoulos G, Nicolaidou A, Thessalou-Legaki M. Effect of marine litter on the benthic mega fauna of coastal soft bottoms: a manipulative experiment. Mar Pollut Bull 2007;54:771–778.

Koelmans AA, Besseling E, Wegner A, Foekema EM. Plastic as a carrier of POPs to aquatic organisms: a model analysis. Environ Sci Technol 2013;47:7812–7820.

Küpper L, Gulmine JV, Janissek PR, Heise HM. Attenuated total reflection infrared spectroscopy for micro-domain analysis of polyethylene samples after accelerated ageing within weathering chambers. Vib Spectrosc 2004;34 (1):63–72.

Laist DW. Impacts of marine debris: entanglement of marine life in marine debris including a comprehensive list of species with entanglement and ingestion records. In: Coe JM, Rogers

DB, editors. Marine Debris: Sources, Impacts, and Solutions. New York: Springer Verlag; 1997. p 99–139.

Lambert S, Sinclair CJ, Bradley EL, Boxall AB. Environmental fate of processed natural rubber latex. Environ Sci Process Impacts 2013;15:1359–1368.

Lee H, Shim WJ, Kwon J-H. Sorption capacity of plastic debris for hydrophobic organic chemicals. Sci Total Environ 2014;470–471:1545–1552.

Lobelle D, Cunliffe M. Early microbial biofilm formation on marine plastic debris. Mar Pollut Bull 2011;62:197–200.

Lusher AL, McHugh M, Thompson RC. Occurrence of microplastics in the gastrointestinal tract of pelagic and demersal fish from the English Channel. Mar Pollut Bull 2012;67:94–99.

Matsuoka T, Nakashima T, Nagasawa N. A review of ghost fishing: scientific approaches to evaluation and solutions. Fish Sci 2005;71:691–702.

McDermid KJ, McMullen TL. Quantitative analysis of small plastic debris on beaches in the Hawaiian archipelago. Mar Pollut Bull 2004;48:790–794.

Meenakumari B, Radhalakshmi K. Induced photo oxidative degradation of nylon 6 fishing net twines. Ind J Text Res 1988;13:84–86.

Meenakumari B, Radhalakshmi K. Weathering of PA netting yarns. Fish Technol 1995; 32:85–88.

Meenakumari B, Ravindran K. Tensile strength properties of polyethylene netting twines under exposure to out-door and artificial UV radiation. Fish Technol 1985a;22:82–86.

Meenakumari B, Ravindran K. Effect of sunlight and UV radiation on mechanical strength properties of PA netting twines. Ind J Text Res 1985b;10:15–19.

Moore CJ. Trashed: across the Pacific Ocean, plastics, plastics everywhere. Nat Hist 2003;112 (9):46–51.

Moore CJ. Synthetic polymers in the marine environment: a rapidly increasing, long-term threat. Environ Res 2008;108 (2):131–139.

Moore CJ, Moore SL, Leecaster MK, Weisberg SB. A comparison of plastic and Plankton in the North Pacific Central Gyre. Mar Pollut Bull 2001;42:1297–1300.

Moret-Ferguson S, Law KL, Proskurowski G, Murphy EK, Peacock EE, Reddy CM. The size, mass, and composition of plastic debris in the Western North Atlantic Ocean. Mar Pollut Bull 2010;60:1873–1878.

Mudgal S, Lyons L, Bain J, Dias D, Faninger T, Johansson L, Dolley P, DShields L, Bowyer C. Plastic waste in the environment—revised final report for European Commission DG Environment. Paris: Bio Intelligence Service; 2011.

Murray F, Cowie PR. Plastic contamination in the decapod crustacean *Nephrops norvegicus* (Linnaeus, 1758). Mar Pollut Bull 2011;62 (6):1207–1217.

Ng KL, Obbard JP. Prevalence of microplastics in Singapore's coastal marine environment. Mar Pollut Bull 2006;52:761–767.

Nuelle M-T, Dekiff JH, Remy D, Fries E. A new analytical approach for monitoring microplastics in marine sediments. Environ Pollut 2014;184:161–169.

O'Shea OR, Hamann M, Smith W, Taylor H. Predictable pollution: an assessment of weather balloons and associated impacts on the marine environment—an example for the Great Barrier Reef, Australia. Mar Pollut Bull 2014;79:61–68.

Ogata Y, Takada H, Mizukawa K, Hirai H, Iwasa S, Endo S, Mato Y, Saha M, Okuda K, Nakashima A, Murakami M, Zurcher N, Booyatumanondo R, Zakaria MP, Dung L,

Gordon M, Miguez C, Suzuki S, Moore C, Karapanagioti HK. International pellet watch: global monitoring of persistent organic pollutants (POPs) in coastal waters. Initial phase data on PCBs, DDTs, and HCHs1. Mar Pollut Bull 2009;58 (10):1437–1446.

Pegram JE, Andrady AL. Outdoor weathering of polystyrene foam. In: Fornes R, Gilbert R, Mark H, editors. *Polymers and Fiber Science—Recent Advances*. New York: VCH Publishers; 1992. p 287–298.

Powell MD, Berry AJ. Ingestion and regurgitation of living and inert materials by the estuarine copepod Eurytemora affinis (Poppe) and the influence of salinity. Estuarine Coast Shelf Sci 1990;31:763–773.

Possatto FE, Barletta M, Costa MF, do Sul JA, Dantas DV. Plastic debris ingestion by marine catfish: an unexpected fisheries impact. Mar Pollut Bull 2011;62:1098–1102.

Qayyum MM, White JR. Effect of stabilizers on failure mechanisms in weathered polypropylene. Polym Degrad Stab 1993;41:163–172.

Railkin AI. *Marine Biofouling: Colonization Processes and Defenses*. Boca Raton: CRC Press; 2003.

Rios LM, Moore C. Persistent organic pollutants carried by synthetic polymers in the ocean environment. Mar Pollut Bull 2007;54 (8):1230–1237.

Robards MD, Piatt JF, Wohls KD. Increasing frequency of plastic particles ingested by seabirds in the subarctic North Pacific. Mar Pollut Bull 1995;30 (2):151–157.

Rochman CM, Hoh E, Hentschel BT, Kaye S. Long-term field measurement of sorption of organic contaminants to five types of plastic pellets: implications for plastic marine debris. Environ Sci Technol 2013a;47 (3):1646–1654.

Rochman CM, Hoh E, Kurobe T, The SJ. Ingested plastic transfers hazardous chemicals to fish and induces hepatic stress. Sci Rep 2013b;3:3263.

Rochman CM, Manzano C, Hentschel BT, Simonich SLM, Hoh E. Polystyrene plastic: a source and sink for polycyclic aromatic hydrocarbons in the marine environment. Environ Sci Technol 2013c;47 (24):13976–13984.

Roney JM. Taking Stock: World Fish Catch Falls to 90 Million Tons in 2012. Earth Policy Institute; 2012. Downloaded from http://www.earth-policy.org/indicators/C55/fish_catch_2012. Accessed October 30, 2014.

Rosevelt C, Los Huertos M, Garza C, Nevins HM. Marine debris in central California: Quantifying type and abundance of beach litter in Monterey Bay, CA. Mar Pollut Bull 2013;7:299–306.

Ryan PG, Moore CJ, van Franeker JA, Moloney CL. Monitoring the abundance of plastic debris in the marine environment. Philos Trans R Soc Lond B Biol Sci 2009;364: 1999–2012.

Saido K. 1.6—Ocean contamination generated from plastics. In: Satinder A, editor. *Comprehensive Water Quality and Purification*. Waltham: Elsevier; 2014. p 86–97.

Setälä O, Fleming-Lehtinen V, Lehtiniemi M. Ingestion and transfer of microplastics in the planktonic food web. Environ Pollut 2014;185:77–83.

Sakai S, Urano S, Takatsuki H. Leaching behavior of PCBs and PCDDs/DFs from some waste materials. Waste Manag 2000;20 (2–3):241–247.

Stefatos A, Charalampakis M. Marine debris on the seafloor of the Mediterranean sea: examples from two enclosed Gulfs in Western Greece. Mar Pollut Bull 1999;38 (9):389–393.

Stevens LM. Marine plastic debris: fouling and degradation [Unpublished MSc Thesis]. University of Auckland; 1992. p 110.

Stevens LM, Gregory MR. Fouling bryozoa on pelagic and moored plastics from Northern New Zealand. Bryozoans in space and time. In: Proceedings of the 10th International Bryozoology Conference, University of Wellington, Wellington, Victoria, New Zealand, January 1, 1996.

Stevens BG, RA MacIntosh, Haaga JA. Report to industry on the 2002 eastern Bering Sea crab survey. Alaska Fisheries Science Center, Processed report 2002–05. NMFS, NOAA, 7600 Sand Point Way NE, Seattle, WA 99115; 2002.

Tanaka K, Takada H, Yamashita R, Mizukawa K, Fukuwaka M, Watanuki Y. Accumulation of plastic-derived chemicals in tissues of seabirds ingesting marine plastics. Mar Pollut Bull 2013;69 (1–2):219–222.

Tavares AC, Gulmine JV, Lepienski CM, Lepienski A. The effect of accelerated aging on the surface mechanical properties of polyethylene. Polym Degrad Stab 2003;81 (2):367–373.

Teuten EL, Rowland SJ, Galloway TS, Thompson RC. Potential for plastics to transport hydrophobic contaminants. Environ Sci Technol 2007;41:7759–7764.

Teuten EL, Saquing JM, Knappe DRU, Barlaz MA, Jonsson S, Björn A, Rowland SJ, Thompson RC, Galloway TS, Yamashita R, Ochi D, Watanuki Y, Moore C, Viet PH, Tana TS, Prudente M, Boonyatumanond R, Zakaria MP, Akkhavong K, Ogata Y, Hirai H, Iwasa S, Mizukawa K, Hagino Y, Imamura A, Saha M, Takada H. Transport and release of chemicals from plastics to the environment and to wildlife. Philos Trans R Soc Lond B Biol Sci 2009;364 (1526):2027–2045.

Thomas SN, Hridayanathana C. The effect of natural sunlight on the strength of polyamide 6 multifilament and monofilament fishing net materials. Fish Res 2006;81 (2–3):326–330.

Thompson RC, Olsen Y, Mitchell RP, Davis A, Rowland SJ, John AWG, McGonigle D, Russel AE. Lost at sea: where is all the plastic? Science 2004;304:838.

Thompson R, Moore C, Andrady A, Gregory M, Takada H, Weisberg S. New directions in plastic debris. Science, 2005, Nov 18;310:1117.

Topçu EN, Tonay AM, Dede A, Öztürk AA, Öztürk B. Origin and abundance of marine litter along sandy beaches of the Turkish Western Black Sea Coast. Mar Environ Res 2013;85: 21–28.

Van Cauwenberghe L, Janssen CR. Microplastics in bivalves cultured for human consumption. Environ Pollut 2014;193:65–70.

Van Franeker JA, Blaize C, Danielsen J, Fairclough K, Gollan J, Guse N, Hansen P-L, Heubeck M, Jensen J-K, Le Guillou G, Olsen B, Olsen KR-O, Pedersen J, Stienen EWM, Turner DM. Monitoring plastic ingestion by the northern fulmar Fulmarusglacialis in the North Sea. Environ Pollut 2011;159:2609–2615.

Veenstra TS, Churnside JH. Airborne sensors for detecting large marine debris at sea. Mar Pollut Bull 2012;65:63–68.

Waluda CM, Staniland IJ. Entanglement of Antarctic fur seals at Bird Island, South Georgia. Mar Pollut Bull 2013;74 (1):244–252.

Watson R, Revenga C, Kura Y. Fishing gear associated with global marine catches I. Database development. Fish Res 2006;79 (1–2):97–102.

Watters DL, Yoklavich MM, Love MS, Schroeder DM. Assessing marine debris in deep sea-floor habitats off California. Mar Pollut Bull 2010;60:131–138.

Whiting SD. Types and sources of marine debris in Fog Bay, Northern Australia. Mar Pollut Bull 1998;36:904–910.

Wilson DS. Food size selection among copepods. Ecology 1973;54:909–914.

Wright SL, Thompson RC, Galloway TS. The physical impacts of microplastics on marine organisms: a review. Environ Pollut 2013;178:483–492.

Yakimets I, Lai D, Guigon M. Effect of photooxidation cracks on behaviour of thick polypropylene samples. Polym Degrad Stab 2004;86:59–67.

Yamashita R, Takada H, Fukuwaka M-A, Watanuki Y. Physical and chemical effects of ingested plastic debris on short-tailed shearwaters, Puff inustenuirostris, in the North Pacific Ocean. Mar Pollut Bull 2011;62 (12):2845–2849.

Zarfl C, Matthies M. Are marine plastic particles transport vectors for organic pollutants to the Arctic? Mar Pollut Bull 2010;60:1810–1814.

Zbyszewski M, Corcoran PL, Hockin A. Comparison of the distribution and degradation of plastic debris along shorelines of the Great Lakes, North America. J Great Lakes Res 2014;40 (2):288–299.

INDEX

Plastics and Environmental Sustainability, First Edition. Anthony L. Andrady.
© 2015 John Wiley & Sons, Inc. Published 2015 by John Wiley & Sons, Inc.

Printed and bound by CPI Group (UK) Ltd, Croydon, CR0 4YY